Russia's Cosmonauts

Inside the Yuri Gagarin Training Center

Rex D. Hall, David J. Shayler and Bert Vis

Russia's Cosmonauts

Inside the Yuri Gagarin Training Center

Rex D. Hall, MBE
Education Consultant
Chairman of the BIS
London
UK

David J. Shayler
Astronautical Historian
Astro Info Service
Halesowen
West Midlands
UK

Bert Vis
Firefighter, Dutch Fire Service
Den Haag
The Netherlands

SPRINGER–PRAXIS BOOKS IN SPACE EXPLORATION
SUBJECT *ADVISORY EDITOR*: John Mason B.Sc., M.Sc., Ph.D.

ISBN 0-387-21894-7 Springer Berlin Heidelberg New York

Springer is a part of Springer Science + Business Media (*springeronline.com*)

Library of Congress Control Number: 2005922814

Cover design: Jim Wilkie
Project Copy Editor: Mike Shayler
Typesetting: BookEns Ltd, Royston, Herts., UK

Printed in Germany on acid-free paper

This book is dedicated to the staff of the Cosmonaut Training Centre named for Yuri Gagarin who, over the last 40 years, have created a facility that has enabled science fiction to be turned into science fact.

They have used their skills and knowledge, gathered from a wide range of disciplines, to enable humans to not only fly into space, but to learn to live and work there for long periods of time.

Other books by the same authors in this series

David J. Shayler and Rex D. Hall
The Rocket Men (2001), ISBN 1-85233-391-X
Soyuz: A Universal Spacecraft (2003), ISBN 1-85233-657-9

David J. Shayler
Disasters and Accidents in Manned Spaceflight (2000), ISBN 1-85233-225-5
Skylab: America's Space Station (2001), ISBN 1-85233-407-X
Gemini: Steps to the Moon (2001), ISBN 1-85233-405-3
Apollo: The Lost and Forgotten Missions (2002), ISBN 1-85233-575-0
Walking in Space (2004), ISBN 1-85233-710-9

With Ian Moule
Women in Space: Following Valentina (2005), ISBN 1-85233-744-3

With Andrew Salmon and Michael D. Shayler
Marswalk One: First Steps on a New Planet (2005), ISBN 1-85233-792-3

Table of Contents

Foreword

I was always proud and I am proud at present of what we created 45 years ago in the woods, 25 kilometres to the North East of Moscow. Being selected into the cosmonaut team in March 1960, I did not realise that the next 30 years of my life would be dedicated to the exploration of outer space, including duties of the Deputy Chief of the Cosmonaut Training Centre named for my friend and colleague Yuri Gagarin and training the new generation of cosmonauts for future space missions. I

Cosmonaut Major-General Aleksey Leonov in his study in the Cosmonaut Training Centre. Note the model of the planned extensions for the training centre and Star City behind him

have lived in and worked close to the centre here, so I am the representative of the generation having the right to be called the pioneers of space.

We created a unique facility, meeting the demands of the training of space flyers that would spend years in orbit, capable of controlling and managing complicated space vehicles to conduct and carry out various scientific research and experiments. To solve all those tasks they created a highly skilled team who found ways of selecting and training people in all aspects of their life, capable of enduring the load both physical and psychological, and to neutralise negative influences of space flight factors.

I and my colleagues of the Gagarin selection were the first to test and obtain the unique training skills, sharing them with the following generations. New demands of space missions needed new approaches to training technology for space missions, many of them became the unique ones.

As well as the Yuri Gagarin Training Centre, we have created a community including the secondary school named after V. Komarov, the house of cosmonauts, shops, homes and recreation facilities for all ages.

Our cosmonaut training centre has bound all of us together forever in a way which we could not have expected 45 years ago, when the first manned mission was performed.

I am glad that this book has been written and it represents fundamental hard work in history, research development of the center and dedicated to the founders of the centre and the inhabitants of Zvyozdnyy Gorodok.

Major-General of Aviation Aleksey A Leonov
Twice Hero of the Soviet Union
Pilot cosmonaut of the Soviet Union 1965
Cosmonaut 1960 to 1975 flying 2 missions, Voshkod 2 and Soyuz-Apollo
First Deputy Director for Pilot and Space Training of the centre named for Yuri Gagarin 1975–1991

Предисловие

(Издание посвящено 40летию Звездного городка)
Британское Межпланетное Общество
Президент Рекс Холл

Я всегда гордился и горжусь тем, что сорок пять лет назад мы заложили и затем создали в живописном лесном массиве в 25 километрах на северо-востоке от Москвы.

Меня зачислили в отряд космонавтов в марте 1960 года, и я тогда в полной мере не осознавал, что предстоящие 30 лет посвящу открытию и исследованию космического пространства, что долгое время буду заместителем начальника Центра Подготовки Космонавтов имени моего друга и товарища Юрия Гагарина, и буду готовить к космическим полетам моих товарищей и новых космонавтов. В течение этого времени, я , работая в ЦПк, живу в Звездном городке и представляю поколение, которое по праву можно назвать поколением первопроходцев освоения космического пространства.

Нами была создана уникальная база для подготовки человека к полету в космос, которому предстояло научиться эксплуатировать сложнейшие космические аппараты, проводить многоплановые научные исследования и эксперименты. Для решения этих, во многом неизвестных задач в Центре сформирован коллектив единомышленников высококвалифицированных специалистов, разработавших методики отбора и подготовки человека для жизнедеятельности в космосе, способного переносить физические и психологические нагрузки негативных факторов космического полета.

Я и мои коллеги «Гагаринского набора» впервые на себе испытали и приобрели уникальный опыт подготовки, что позволило передать его последующим поколениям, научить их жизни и работе в космосе. Поэтому мы стремились дать оценку и рекомендации особенности всем предстоящим экспедициям. Новые требования к космическим полетам потребовали сформировать новые технические средства подготовки космонавтов. Многие из созданных тренажерно-моделирующих комплексов в дальнейшем стали поистине уникальными. Все это стало возможным сделать лишь опираясь на квалифицированный научный и инженерно-технический потенциал в нашем Центре.

Для развития ЦПК им. Ю.А. Гагарина, для полнокровной жизни, работы и отдыха его сотрудников нами была создана соответствующая инфраструктура. Это жилые дома, школа имени В. Комарова и детский сад. Это магазины, предприятия обслуживания и культурный центр для развлечения и отдыха жителей всех возрастов.

Наш ЦПК и жилой городок связал нас на всю жизнь, как мы и не предполагали 45 лет назад, когда был осуществлен первый полет человека в космос.

Рад. Что данное издание представляет собой фундаментальный труд исследования истории и развития Центра, в который вложены труд и знания и посвящено его основателям и жителям Звездного городка.

Генерал-майор авиации
Дважды герой Советского союза
Летчик-космонавт СССР 1965 г.
Первый заместитель начальника ЦПК им. Ю.А. Гагарина по летной и космической подготовке (1975 -1991 гг.)

А. Леонов

Authors' preface

This book is about the training and selection of Russia's cosmonauts. The story is mainly about the training centre created outside Moscow at Star City, but this was not where the team initially reported or started training. The dream of human space exploration started somewhere which is now neither marked or recognised, except by those of the first selection and those who trained them.

In the grounds of the Central Army Sports Club is a neglected, light yellow building, located on the Leningradskiy prospect near modern blocks of flats and a sports complex. It was here, on 14 March 1960, that the first Soviet cosmonaut team reported to start training for their world-changing missions. The building was a two-storey construction. On the ground floor were the headquarters and the classrooms where they studied space theory, mechanics and medicine. On the second floor was a lecture theatre in which they had lectures, but also party meetings, political conferences and entertainment.

It was here that they were introduced to the specialists who would mould the team into cosmonauts, and their first lessons showed that they had very little knowledge of space flight or what it would mean. They met Colonel Nikolay Nikitin, who would teach them parachute jumping, and heard lectures from some of the most qualified engineers within the space program. Academician Oleg Gazenko, the founder of the theory of space physiology, explained to them the effects of space on human organisms, but he also explained what it was like to live in Moscow, which for most was a new experience. The cosmonauts also used the sports facilities, such as the swimming pool, gymnasium, and a roofed shed where they played basketball and football. It is probably here that they considered that their first set of lectures had been dull and very difficult to grasp. When Sergey Korolyov heard this, he drafted in many of his young engineers to make the lectures relate to the work being undertaken within his bureau. They included prospective future cosmonauts such as Oleg Makarov and Vitaliy Sevastyanov.

A major part of their training was the ideological education. Yuri Gagarin was accepted as a member of the Communist Party (and German Titov as a candidate) at some of the early meetings. This reflected the very high priority given to the goal of putting a man in space.

During this time they did do a lot of training away from this site, but it was practical or medical such as parachute training, weightless flights in an aircraft, medical testing and similar activities. Eventually, in July 1960, the team moved closer to its permanent base at Star City.

While this book focuses on the Cosmonaut Training Centre (TsPK), it is worth reviewing the Russian approach to spaceflight training and how this has evolved throughout almost 50 years of operations. The training philosophy for Vostok missions was to select a small group from the twenty candidates and focus their training on specific missions, with the final prime and back-up pilots being assigned shortly before the mission. The back-up could replace the prime crewmember if called upon to do so, or would utilise this experience to rotate to a prime slot on a subsequent mission, taking advantage of the training he had already undergone. Of course, the only time this role could not be developed was for the first flight, when a rookie cosmonaut had to be selected to take that very first step into space. Others in this training group would support the flight and hope to secure a mission of their own. When not flying, the cosmonauts, like American astronauts, would fill a range of support roles in planning and developing new missions, flight hardware, systems and experiments. Unlike the Americans, these background roles often went unrecognised and, in many cases, never led to a flight assignment. The Vostok training group philosophy has continued throughout the history of Soviet and Russian manned spaceflight into the present operations on ISS.

While there have been fifteen selections of Air Force candidates for spaceflight training (mostly pilots) and a number of 'engineer' selections, there have been very few 'scientist' selections. In contrast, NASA has selected five pilot groups, two scientist groups (all in the 1960s) and twelve mixed groups of pilots and mission specialists for the Shuttle programme. Both the US and Russia have selected candidates for specific flights, experiments or objectives, who have then returned to their previous careers after the mission. The Russians have also selected a cadre of highly experience pilots for the cancelled Buran programme. From the 1960s to the 1980s, both the Soviets and the Americans also selected spaceflight candidates for classified military programmes with strategic objectives, but these never materialised to their full potential. It will be interesting to see how the pattern of cosmonaut selection continues in the future.

In the early years, training on Vostok was limited because the vehicle was essentially an automated one. Their training really focused on survival and emergency procedures, habitability and a few scientific experiments or observations. A new training element entered the programme for Voskhod 2; that of spacewalking, or EVA. By contrast, the American astronauts fought hard to be allowed to fly and control the spacecraft wherever possible, and the scientific experiments were not always met with enthusiasm. With the introduction of Soyuz, the Soviets had to train their crews in the art of rendezvous and docking and, with the creation of a manned lunar programme, training for landing on the Moon, deploying experiments and collecting lunar samples would have been included in any mission preparation. Since the manned lunar programme was terminated prior to any crews participating, however, it is not clear just how much 'lunar surface training' the cosmonauts received. Several references indicate that the cosmonauts were undergoing helicopter training in the 1960s, and this was linked to the fact that Apollo landing crews also took helicopter training as part of their preparations. But there is no evidence of geological field trips by cosmonauts comparable to those undertaken by the Americans from 1962/63 to 1972.

When the space station programme was formed, with the prospect of long flights and participation in scientific, engineering, technological and military experiments, a new training philosophy had to be incorporated into the programme. For Buran, experienced test pilots were initially assigned to support the development of the programme and its early orbital flights, but the programme was cancelled after only a single unmanned orbital test flight. It is not clear exactly how the Buran training and crewing philosophy would have worked, but it may have resembled the NASA Shuttle programme (where mission training takes about twelve months). This would have meant pilots 'flying' the vehicle, engineers assigned to 'missions specialist' roles such as satellite retrieval and deployment and EVA, and probably one-flight specialists to work with a specific 'payload' or experiment.

New cosmonauts complete a period of basic training lasting up to two years, which includes a programme of academic work in space sciences, rocketry and space navigation, plus life support, survival and wilderness training, environmental training in isolation, stress and weightless simulations (such as parabolic aircraft and water tanks) and a vast programme of medical and psychological testing and training.

As most crews since 1971 have been assigned to space station missions, this is by far the largest programme in Russian manned spaceflight, and the Russians have become masters at preparing crews. Psychological issues and crew compatibility are high on the training syllabus and are instrumental in the final selection and assignment of each crewmember.

Training for a flight to a space station includes medical and physical training to ensure the cosmonaut's body can withstand launch, long duration flight, landing and post-flight recovery. Technical training is conducted in a fleet of 'flying laboratories' – aircraft fitted out to simulate various degrees of gravity – and EVA training is conducted in the water tank and in 1-g facilities. They practice both nominal and non-standard situations in 1-g simulators and mock-ups of parts of their mission; prepare for contingency operations on the launch pad by evacuating the spacecraft or simulating a launch abort; undergo survival training and recovery by helicopter from remote wilderness or water landings; work on flight documentation and flight plans; practice using their pressure suits (Sokol and Orlan); run through unpacking, stowing and activating equipment, experiments and stores; become familiar with crew systems (waste management, personal clothing, food and hygiene facilities), life support and space station control systems; test emergency procedures, radio and communications gear and photographic equipment; study the night sky and the theory of spaceflight; and work with mission controllers to plan daily activities, physical exercise programmes, communications, and data gathering and reporting.

To achieve all of this successfully and safely, a wide range of facilities, procedures and equipment has been developed over the years to support each flight into space.

The success of the Soviet/Russian manned space station programme was built upon foundations laid at TsPK. Further insight into cosmonaut mission training can be gleaned from accounts of specific programmes and from some of the international cosmonauts, who have helped generate a broader understanding of a once-secret facility, one that still remains restricted to western visitors.

In 2004, the Americans changed the direction of their future space programme by announcing the retirement of the Shuttle, the inauguration of a new Crew Exploration Vehicle, completion of human spaceflight research on ISS, and a return to the Moon, all in support of a possible human mission to Mars by 2030. In Russia, this was greeted with disappointment and frustration. It revitalised calls for a new Russian domestic space station programme, the replacement of Soyuz with a more capable and versatile vehicle, and reigniting the dreams of Tsiolkovskiy, Korolyov and others of placing cosmonauts on the Moon and on Mars. It is clear that, despite numerous and significant difficulties with finance and maintenance, TsPK will feature prominently in any national programme as a training facility for Russian cosmonauts, while still retaining the international links to secure not only cooperation, but also much needed funding. It is clear that, despite over 40 years of use, the full potential of TsPK is still to be realised, and its potential for expansion remains a possibility with support and funding.

As pioneers of the Russian manned space programme, Konstantin Tsiolkovskiy, Sergey Korolyov, Yuri Gagarin and the events of the 1957-1965 period have become legends in spaceflight history. The success of what followed those pioneering years has to be credited to the staff and facilities of one of the world's leading space training facilities – the Cosmonaut Training Centre named for Yu. A. Gagarin.

Rex D. Hall
London
England

David J. Shayler
Halesowen
West Midlands
England

Bert Vis
Den Haag
The Netherlands

www.astroinfoservice.co.uk

April 2005

Acknowledgements

This book is a cooperative project by three authors with a long association with the study of the Soviet/Russian cosmonaut team, their training, flight assignments and career accomplishments. Central to this research is understanding the departments and operation of the Cosmonaut Training Centre (TsPK) near Moscow, and initial thanks should be given to those who work behind the scenes, without recognition, ensuring that each Russian crew is trained to their peak at the point they leave Earth.

At TsPK, we acknowledge the assistance of Colonel-General Pyotr I. Klimuk and Lieutenant-General Vasiliy V. Tsibliyev, whose help over many years has been much appreciated. Our special thanks also go to Major-General Valeriy G. Korzun and his staff, who arranged a tour of the facilities that is not usually available to non-space agency personnel; the Director and staff of the Museum of the Cosmonaut Training Centre, who arranged a number of interviews with the staff of the Centre; and the large number of former and current cosmonauts who have given us their time and knowledge freely over many years. We particularly acknowledge Ivan Kolotov, Lev Vorobyov, Gennadiy Kolesnikov and Sergey Gaydukov for their friendship, help and encouragement. We would also like to thank the staff at IMBP, under the direction of Dr Valeriy G. Polyakov, who showed us the facilities that have supported human flights for 40 years during a number of visits to the bureau.

Though this is a book on the training of Russian cosmonauts, considerable information has been obtained from the archives of NASA, primarily the JSC history collection (ASTP, Shuttle-Mir) at the University of Clear Lake, Houston, Texas, and the Public Affairs Office at JSC. In addition, the PAO staff of NASA JSC were helpful in giving access to their Contact Files during several research trips about Shuttle missions. Special thanks should go to John Charles, David Portree and to Media Services Corp. at NASA JSC. Thanks, also, to the staff of ESA public affairs, who assisted with information requests.

Thanks is due to the ongoing research of fellow Russian space watchers: Colin Burgess, Mike Cassutt, Phil Clark, Brian Harvey, Bart Hendrickx, Gordon Hooper, Neville Kidger, James Oberg, Andy Salmon, Lida Shkorkina and Natalya Talanov.

Appreciation is due to the Council and Staff of the British Interplanetary Society for access to their library and photo archive. Photos have also been loaned and used with permission by Tim Furniss, Daniel Tromeur and Cap Espace. We would also like to thank Mark Shuttleworth for the use of photographs from his website; Eduard Buinovskiy for access to his personal photo archive; former NASA astronauts who have provided insight into the American experiences of training at TsPK and flying joint missions over the years; and to many astronauts from various countries, who have assisted us with interviews and insights into their training in Russia and work with their Russian colleagues.

Thanks are due to Major-General Aleksey Leonov for his Foreword and to Elena Esina, curator of the museum in the House of Cosmonauts at Star City. We also acknowledge the staff and work of *Novosti Kosmonavtiki,* Russia's leading magazine on space exploration, who also provided a number of pictures used in this book.

Special thanks go to our project editor Mike Shayler for hours of extra effort in preparing the illustrations and text for publication, and to Clive Horwood and the staff of Praxis for their continued support and encouragement of a protracted project. We must thank Jim Wilkie for his cover design and Arthur and Tina Foulser at BookEns for typesetting.

This book is the companion to the earlier works in this series *The Rocket Men (2001)* and *Soyuz: A Universal Spacecraft (2003)* by Rex Hall and Dave Shayler, and to numerous articles by all three authors in the publications of the British Interplanetary Society and via Astro Info Service Publications. The research for this book forms part of a long and ongoing programme of research encompassing these other works.

List of Illustrations

Simulators

Other National and International Facilities

Survival Training

Image Pages (1)

The Cosmonaut Group of the RGNII TsPK

The Cosmonaut Group of RKK Energiya

Other Soviet and Russian Cosmonaut Selections

Cosmonaut Selected to Fly the Buran Shuttle

International Training

Image Pages (2)

Joint Programmes

Zvyozdnyy Gorodok – The Town

Appendix 3

Appendix 5

Front cover (Top left) Statue on the main road to Star City; (top right) Two cosmonauts in the Soyuz TM simulator at Star City.

Back cover (Top) Logo of the Training Centre, which is worn by staff on uniforms and training overalls; A memorial to the first spacewalk on the steps of the House of Cosmonauts; (centre) A cosmonaut wearing an Orlan training suit is lowered into the Hydrotank; (right) A cosmonaut egresses a training capsule while undertaking survival training

Prologue

> *'Yuri A. Gagarin showed, during training and practice for the flight, a high degree of precision in various experimental psychological tests. He displayed a high degree of immunity under both sudden and loud stimuli. His reactions to 'novelties' (weightlessness, long periods of isolation in the anechoic chamber, parachute jumps and others) were always active... he was able to control himself in various unexpected situations... in practice sessions on the training model [he would] work in a calm and assured manner'* – Our Gagarin, p 138.

These skills, along with his strong personality, allowed Yuri Gagarin to become the first person to put into practice months of space flight preparation, to be forever known as the first person to fly into space, on 12 April 1961. Since then, 98 Russian cosmonauts have followed in his footsteps on a variety of missions in Earth orbit, mostly aboard the Soyuz series of spacecraft. The actual flight into space is the very visible phase of a mission that starts with preparations in the gravity environment of Earth, using a variety of facilities and methods that have evolved with the programme over the past four decades. The direct application of this training to activities in space was first used by Russian cosmonauts aboard the Vostok series of spacecraft, at a time when training for a flight into space was in its infancy.

Russia and the United States have developed extensive programmes for training their citizens to fly into space over a period of almost five decades. Although many other countries have selected their citizens to fly on either Russian or American missions, and in some cases have developed their own basic and specialist training programmes, the final mission training phase has always been carried out in either the US or Russia, or more recently, both countries. In 1998 a third country, China, developed a space training programme, but it was based on the Russian programme. It supported a domestic manned space flight, but they are still in their infancy compared to the US and Russia and have not yet reached the scale of their training programmes. While the American astronaut training programme has been widely explained, the selection and training of Russian cosmonauts has largely been overlooked. The methodology of training Russians for space flight is the main topic

The logo of the Cosmonaut Training Centre named for Yuri A. Gagarin

of this book, but an overview of the overall direction of Russian human space flight and how this differs from that of the United States is worth recording here.

As the focal point for the preparation of Russia's cosmonauts, the Cosmonaut Training Centre has evolved to reflect the development of the programme. When the first cosmonauts were selected in 1959/1960 for the Vostok programme, there was no dedicated 'centre' for space flight training, but rather a number of locations across a selection of mainly aviation facilities, designed to provide rudimentary training devices until a more central location could be established. This was similar to the American selection of astronauts for the Mercury programme during the same period.

As the crew preparation schedules for Mercury and Vostok developed, plans for follow-on missions suggested that a more centralised and specialist crew preparation area would be required. In Russia, the cosmonaut team moved into the new Star City facility near Moscow, which continues to be the main location for cosmonaut training today. The original American 'manned spacecraft centre' for astronaut training was located at the Langley Research Center in Virginia from 1959, until the Houston facility opened in 1962. Since then, Houston has been the central location for astronaut training and remains the premier space flight preparation facility in the United States. Established in 1960, the Russian Star City facility became the world's first dedicated space flight training facility, and though both Star City and Houston were aimed at the training of space flight crews, each was operated completely differently, reflecting both the programmes they would support and the nature of the country they were located in. The Manned Spacecraft Center (MSC, later the Johnson Space Center, or JSC) in Houston had, up to the early 1990s, mostly resembled a university campus, although it was still closed to the general public, except for official and guided tours. A significant increase in security was seen at the time of the Shuttle-Mir programme, and was strengthened in the wake of the 9/11 terrorist incident in 2001. In contrast, Star City has always been part of a functional military base and has always had an effective security system preventing unofficial access. This 'closed-base' approach has restricted the flow of information about its history and the operating infrastructure of the facility, even to those who work there.

The main focus of NASA astronaut training during the 1960s was the development of techniques for sending Americans to the Moon by the end of the decade, under the Mercury, Gemini and Apollo programmes. There was also a separate core of military astronauts selected to train for classified programmes under the auspices of the USAF (X-20 Dyna Soar space plane; Blue Gemini; the Manned Orbiting Laboratory, or MOL). In Russia also, there was a dual purpose to the training, for Earth orbital and lunar missions (Vostok/Voskhod, Soyuz, Zond) and for military-orientated programmes (Soyuz, Zvezda, Almaz).

The training philosophy for both the Americans and Russians was also very different. At NASA, groups of experienced military (later civilian) test or jet pilots were selected, in clear groups, for specific phases of the programme: Mercury in 1959, Gemini and Apollo in 1962 and 1963 and Apollo Applications (post-initial landings) in 1966. Two groups of mainly civilian scientist astronauts were selected in 1965 and 1967, to broaden the research field experience in the astronaut office for the Apollo Applications Program (AAP). In 1969, following the cancellation of the military manned space programme, several former Department of Defense (DoD) astronauts from MOL were transferred to NASA to support the remaining AAP programme (later Skylab). Each of these groups completed a programme of basic academic, wilderness (survival) and systems training, as well as taking a variety of technical assignments to support other missions and programmes, having a hands-on approach to each mission.

Over in the Soviet Union, the selection and training of cosmonauts initially fell under the auspices of the Soviet Air Force, with candidates undergoing a series of medicals and evaluations by military, political and administrative committees, who selected a group of candidates for further consideration (though this did not always mean they would start training). This was followed by a period of basic and survival training, after which the candidates took their exams and qualified (or failed) to be awarded a cosmonaut certificate. Experience varied on these selections and their enrolment into the cosmonaut team was dependent upon attrition, programme scheduling and other factors. Some cosmonauts were selected, completed basic training and worked as cosmonauts, but were never assigned to a flight and remained in support roles for years. Unlike the Americans, who were identified publicly on selection, Russian cosmonauts were not normally identified until they were launched into space, making their official role anonymous even to their own family members, particularly for those who never achieved a space flight.

In addition to the number of Air Force selections, the Soviets also created a 'civilian' cosmonaut team by the mid 1960s, mainly comprising engineers of the primary spacecraft design bureau, OKB-1 (headed by Sergey Korolyov). There were also smaller teams created for specific missions or programmes, with varying degrees of success in achieving a space flight. These included scientists, doctors, journalists and representatives of other design bureaus and institutes. Many completed a basic cosmonaut training programme but, except for a handful, they were never called upon to perform more advanced training until selected to train for a flight. Of course, some never received the call for crew assignment and remained a 'cosmonaut' in name only.

During the 1970s, the American astronaut team chosen for the Mercury–Apollo programmes also supported the Skylab space station and the first 'international manned space flight', the joint US/USSR docking mission designated the Apollo-Soyuz Test Project (ASTP). This flight reflected a time of change in both programmes and a willingness to exchange data and experiences, more for national objectives than international relations. Prior to Soviet involvement in ASTP, the facilities at the training centre were already being expanded to accommodate the change of direction towards long duration space station missions. All indications of a Soviet manned lunar programme were hidden for another two decades. While the training cycle for the Apollo missions was revealed in some detail during the 1960s and early 1970s, those for the Russian lunar programme of the same period have never been clearly defined. There was obviously some training conducted for the flight to and from the Moon, possibly including the lunar orbital phases of the mission, and some simulations of surface activities were presumably performed, at least in the early stages of EVA development.

With the demise of the American Apollo programme in the early 1970s, the training simulators were mothballed and replaced with a new suite of simulators and training schedules to reflect the introduction of the Space Shuttle. The new intake of astronauts in 1978 was the first of a fairly regular intake over the next twenty years, for crew members to fly the Shuttle (Pilots) and those who would conduct space walks and operate equipment peculiar to each mission (Mission Specialists). The Russians were also developing a Space Shuttle programme, called Buran, but relied on experienced test pilots (civilian and military) to provide potential crew members. Unfortunately, although test flights in the atmosphere and a single unmanned orbital flight were completed, no manned orbital missions were mounted, even though several were planned and some Buran pilots were assigned to short Soyuz space flights to gain experience prior to flying a Buran in space.

The change to long duration space station missions created new training procedures for the cosmonaut team, both for long duration residencies and short visiting missions. A scientifically-orientated civilian space station programme (Salyuts 1, 4, 6 and 7) was operated between 1971 and 1986, and a classified military space station programme (Almaz – Salyuts 3 and 5) was mounted between 1972 and 1977. From 1986 until 2001, the highly successful Mir space complex represented the peak of Russian/Soviet manned space flight, creating a host of records in space flight endurance and regular space operations. Since the 1980s, the selection of new cosmonauts to crew these missions has slowed, reflecting a changing programme, longer-lasting space stations, reduced funding and political upheavals.

Despite this, and at a time of great change of priorities in Russia, the honour of holding the title 'Cosmonaut' remains high, though perhaps not as high as in the days of Yuri Gagarin.

The assignment of potential crew members to the American or Soviet programmes differed greatly right from the start. In the US, a system was devised for Mercury in which a back-up crew member supported a primary crew member, then rotated to fly a subsequent mission. This was refined for Gemini and Apollo (under the direction of the Director of Flight Crew Operations, and former astronaut, Deke Slayton), so that a flight crew occupied the primary position and their back-up crew skipped the next two missions but flew the third. A back-up crew or single crew member could replace a grounded prime crew member if called upon to do so. From Apollo, a third, support crew was assigned to represent the prime crew in meetings and visits. This system changed with the Shuttle programme, once the test flights had been completed. Here, a prime crew was selected based upon experience, the mission and availability, with relatively few back-up crew members assigned. A displaced prime crew member would merely be replaced from the pool of experienced astronauts. In contrast, right from the early 1960s, the Soviets formed a small group of cosmonauts from the larger training corps, creating a 'training group of immediate preparedness' from which the prime and back-up crews were formed, each training together in a two- or three-person unit. If called upon, the whole back-up crew could (and sometimes did) replace the whole prime crew. A third crew would also be assigned to provide additional support and they could rotate to a later back-up or prime position if required. This system continued through to the end of the Mir programme in 2001 and elements of it remain today.

In 1975, ASTP afforded both countries the opportunity to experience each other's training regimes at that time. Though space station missions were considered (Soyuz/Skylab or Apollo/Salyut docking), these were not pursued and no training sessions were completed. The success of ASTP and the prospect of the Americans flying international astronauts as one-flight specialists on the US Shuttle led the Soviets to offer flights to Salyut stations to members of the Eastern Bloc, then to other international 'partners'. As a result, an abbreviated cosmonaut training programme was devised to incorporate these 'non-career' cosmonauts, some of which included long duration and EVA training.

In the early 1990s, the introduction of the joint US/Russian Shuttle-Mir programme in preparation for the International Space Station allowed American astronauts to train for long duration assignments on main Mir crews. Following the 1992/1993 agreements between Russia and the International Partners who were developing the Freedom Space Station (USA, ESA, Canada, Japan and Brazil), the new partners developed what has become known as the International Space Station (ISS). As a result, the new ISS crew training cycle has necessitated a reassessment of Russian cosmonaut training to reflect visits to US, Canadian and European (and eventually Japanese) training facilities and mastering the English language. It has also led to commercial marketing of the spare seats on the Soyuz in order to secure much needed funding, and to try to retain some national pride now that there is no longer a Russian national space station programme.

With the announcement in 2004 that the US Shuttle was to be retired and replaced with what has been termed the Crew Exploration Vehicle (CEV, designed to return astronauts to the Moon as well as re-supplying the ISS), the American programme has also to undergo significant changes, and as part of this process, the selection of astronauts and their training is under review. Meanwhile in Russia, although there are plans to create a Soyuz replacement (Klipper), the basic cosmonaut training programme and crew assignment methodology continues to focus on this highly successful and reliable spacecraft and on ISS systems and procedures. Furthermore, the basic cosmonaut training programme for these missions continues to operate essentially unchanged, apart from its more international slant, since the early 1960s, basically because it has proven to be adaptable and highly successful.

Training philosophy in the US prior to ISS reflected the short mission durations (up to three weeks) flown between 1961 and 2001, with intensive training for a minutely detailed time line. The only long missions Americans had flown prior to 1995 were the three Skylab missions, where such a detailed time line proved to be unworkable and problematic. In Russia, where long duration missions have been pursued since 1970, repetitive training has allowed for more relaxed preparation to long missions, giving better flexibility in dealing with real-time situations. Central to this programme of space flight preparation remains the Cosmonaut Training Centre named for Yuri Gagarin (TsPK) that has also been the home of cosmonauts for over four decades. The story of how this specialised and largely unknown training centre evolved to become the leading facility for extended duration space flight and regular space station operations over the past four decades is also the story behind the achievements of all of Russia's cosmonauts since Yuri Gagarin began the journey.

The Cosmonaut Training Centre: birth and growth

Soon after the government decision of 11 January 1960 to create a centre for training cosmonauts, a commission that included the head of cosmonaut training, General Nikolay Kamanin, began searching for a suitable site for the new centre, which was planned to become not just the central location for the training of Soviet cosmonauts, but also the place they would live.

FINDING A LOCATION

The commission had set a number of criteria that the site had to meet in order to be selected. It had to be in reasonable proximity to Moscow, with an air base nearby, and both a railway and roads had to be available.[1] However, the site still had to be remote enough to keep curious outsiders away. Finally, it had been decided that it would have to be a pleasant place for the cosmonauts and trainers to live, so a location would be selected that was out in the woods, away from Moscow's city atmosphere.

Eventually, the ideal place was found, in a forest in the Shchelkovo region, some forty kilometres north-east of Moscow. It was only a few kilometres from the Chkalovskiy Air Base, the largest military airfield in the Soviet Union, and near the railroad between Moscow and Monino. It was also conveniently located only a short distance from Sergey Korolyov's OKB-1 design bureau in Kaliningrad and Vladimir Chelomey's OKB-52 in the town of Reutov. The fact that the Air Force academy was situated in nearby Monino was considered a bonus.

Another factor in this site's favour was that the land was already home to a military installation. In the late 1930s it had been the site of a radio range, and a few small buildings had been built at that time. One of these (built in 1938) has survived and is now being used as a garage for buses from the TsPK motor pool.

TsPK's first buildings

Construction of other buildings and facilities started in 1960 and by the following year, the first new building was completed. The Headquarters and Staff Building was a two-storey office structure that was big enough to house the centre's entire staff, including the cosmonaut detachment. In 1962, a second building was constructed, which now houses the TsPK technical library, and in 1964, a new wing was added to the staff building, containing a gymnasium and other facilities.

The main entrance to the Headquarters and Staff Building. Either side of the door are reliefs honouring Yuri Gagarin and Sergey Korolyov. The Order of Lenin and the Order of Friendship of Peoples can be seen next to the name of the centre

Another new complex built was a barracks for the soldiers serving in Unit 26266, as the Cosmonaut Training Centre was officially known. Right next to that was the first medical department, which in the 1960s and early 1970s also acted as the profilactorium (the facility in which crews re-adapted to gravity after space flights) and where they were debriefed. This function was later taken over by the Profilactorium that is still in use today. Once the switch had been made, the old facility became a policlinic and pharmacy.

The simulator building

It was also in 1964 that the first structure that was meant to house simulators was built, dubbed 'Korpus D'. Up to that point, mission training had taken place at Sergey Korolyov's design bureau OKB-1 in Kaliningrad. In fact, in the 1960s, there would be only a few simulators in TsPK and there was a constant struggle between Korolyov's design bureau and the Air Force over who should be in charge of crew training and where this training was to take place.

This was finally resolved in the early 1970s, but the conflict had an effect on the development of TsPK and on the Soviet space programme as a whole, which suffered delays and setbacks because of the rivalry between the Air Force and the design bureaus, all of whom wanted to have a decisive role in the planning and preparation of manned space flight. This went so far that the design bureaus were not even willing to

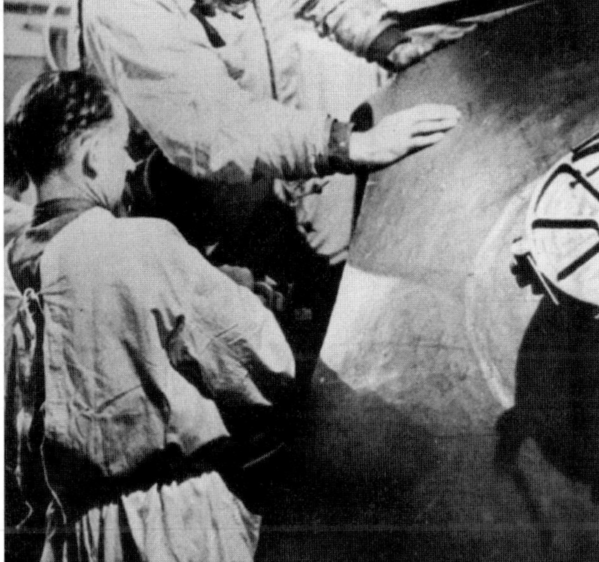

Vostok back-up cosmonaut Vladimir Komarov entering a Vostok simulator

supply TsPK with the necessary simulator technology to get crews trained for missions.

Another aspect that had a negative influence on the early years of the Soviet programme was the ad-hoc style in which missions were planned. It seemed that there was no long-term vision and this was reflected in the decisions that were made in the 1960s. When the government resolution was passed that described what the new Soyuz spacecraft would be used for, on 3 December 1963, there was no mention whatsoever about the simulators that would be needed to train the crews to fly the missions.[2]

In the very early years of TsPK's existence, activities were limited to physical training and general management of the new unit. Medical issues were dealt with in hospitals and medical institutes in Moscow, while the actual flight training took place at the Korolyov design bureau.

When simulators were finally delivered to TsPK, they were placed in Korpus D, until they were relocated to the new Engineering and Simulator Building when that was constructed in 1973.

One of the small shops near the House of Cosmonauts in the Star City town

Early EVA training

In 1965, one major new facility – a swimming pool – was added to the training centre. Although this pool was used to train for the early EVA missions such as Voskhod 2 and Soyuz 4/5, it is unclear if this was the reason for actually building it. Other EVA training was conducted during parabolic flights on a Tupolev Tu-104 aircraft that had been modified for cosmonaut training.

Gymnasium

Although the Headquarters and Staff Building had a gymnasium added to it in 1964, within a few years it was clear that that facility no longer met the needs of the ever growing cosmonaut detachment. Therefore, a new gymnasium was fitted out on the top floor of the building housing the pool. Here, the crewmembers and other cosmonauts could work on their physical fitness in peace and quiet.

This gym is still in use today and when western astronauts started training in Russia in the early 1990s, photos were released on a number of occasions showing them working out on modern training equipment in the facility.

Struggle for power

In the meantime, there was the ongoing struggle between TsPK and the Korolyov design bureau (notably Korolyov himself) about who should train cosmonauts and where that training should occur. In September 1965, Korolyov reacted furiously to an Air Force proposal to have one cosmonaut fly a mission lasting between 20 and 25 days in order to conduct military experiments. He accused Kamanin of having instigated the proposal (which had been signed by Air Force commander-in-chief Konstantin Vershinin) and threatened that he was perfectly capable of training cosmonauts himself and would do so, both flight-engineers and commanders.[3]

On the square in front of the Headquarters and Staff Building, this obelisk with a mosaic of Lenin is one of the few remaining reminders from the days of the USSR

At that time, TsPK was still under threat from the initiatives of the design bureaus. In August 1966, Kamanin obtained a document that showed that the former OKB-1, which was now called TsKBEM (for Central Construction Bureau of Experimental Machine Building), had developed its own programme to train cosmonauts, without discussing things with the military and without having TsPK participate. Kamanin indicated that it was clear to him that TsKBEM's new chief designer, Vasiliy Mishin, and the head of the Soviet Academy of Sciences, Mstislav Keldysh, were attempting to create their own, civilian, cosmonaut training centre to prepare cosmonauts for the L-1 and L-3 manned lunar programmes.[4]

Kamanin and other Air Force officials fought these attempts and were eventually successful. On 8 December 1966, Kamanin noted in his diary that Mishin and his deputy, Pavel Tsybin, had signed what he described as a 'peace treaty' with the Air Force. The document was the start of improved relations between TsKBEM and the Air Force and was the result of a successful defence against the attempts of the design bureaus, the Ministry of General Machine Building and the Academy of Sciences to create a new training centre for civilian cosmonauts. Kamanin stated that for at least the next three or four years, cosmonauts would only be trained at TsPK, without any significant influence from non-military organisations and ministries.

Kamanin used the new developments to try and improve the situation for TsPK. Since 1967 would be a special year for the Soviet Union because of the 50th anniversary of the October Revolution, the government and the Central Committee of the Communist Party of the Soviet Union (CPSU) were pressing for space spectaculars to mark the occasion. Although he blamed the industry for the setbacks and delays that had occurred in recent years, Kamanin believed that in the current situation, cosmonaut training might become another delaying factor to achieving the government's goals. He felt it necessary that TsPK be further expanded; that a large 16-metre centrifuge called TsF-16 be built; that a special air wing be formed; and that a second training building be constructed. Furthermore, he thought that the staff of the centre should be increased significantly, although he didn't give an indication of numbers. Kamanin also thought that the status of TsPK should be elevated to that of a Scientific and Testing Institute (NII).[5]

TsPK gets renamed

On 27 March 1968, Yuri Gagarin was killed when the MiG 15 he was flying with Vladimir Seryogin crashed near the village of Novoselovo, some seventy kilometres north-east of Moscow. The death of the Soviet Union's number one hero was a blow to the nation as a whole, and to those in the Cosmonaut Training Centre in particular. It had been expected that before long, Gagarin would have been a General and would have become the commander of the centre.

While the investigation was still ongoing, the Ministry of Defence decided that TsPK would be renamed after Gagarin on 1 April 1968. Furthermore, the 70th Air Force Fighter Training Regiment which, although an independent Air Force unit, was basically TsPK's air wing, would be named after Seryogin. General Kamanin used the same meeting to suggest a reorganisation of the Cosmonaut Training Centre into a 'Research and Test Centre for Flights with Manned Spacecraft'. Ordered to put his request in writing, the next day Kamanin handed it in to Marshal Ivan Yakubovskiy, who told Kamanin that he would go along with the proposal. Kamanin's request to have the staff of the centre increased to 500 was not fulfilled, however. This was something that would have to be discussed with the Party.

On 4 May, the TsPK staff was officially informed of the decision to rename the centre and the air wing.[6] A week later, Pavel Belyayev was put in charge of having a monument for Gagarin designed and put up either in the training centre itself, or in the living area of Star City.

The centre expands

One of the plans Kamanin had for the centre was to have a separate air base especially for the cosmonauts. The cosmonauts themselves were opposed to the plan, but Kamanin felt that training flights by cosmonauts were being hindered by the increasingly overcrowded air space around Moscow. The Chkalovskiy Air Base, which was the largest base in the country, was also getting much busier, and Kamanin foresaw the problems getting worse, especially as the number of

An unusual view of the training centre showing the Hydrolaboratory (left), the Planetarium (right) and the boiler facility with its smokestack in the background

cosmonauts expanded. Kamanin had a meeting with Marshal Sergey Rudenko, who in principle agreed with the idea of building a separate air field. He suggested either the Crimea or Central Asia as a location, but although he didn't rule out the possibility of having an air field there someday, Kamanin was thinking of a site between 100 and 200 kilometres from Moscow.[7] In the end however, these plans would not be carried out and to this day, the Seryogin Regiment is stationed at Chkalovskiy Air Base, only a few kilometres from TsPK.

Proof that things were not always easy for Kamanin was a paragraph in his diary entry for 21 June 1968. He had had to ask Vershinin to send a letter to the Chief of the General Staff in order to ask the State Planning Commission to allocate twelve Volga staff cars for the cosmonauts to use for business travel. Four years earlier, six cars had been obtained for that purpose, but they badly needed to be replaced. Kamanin had stressed that in other locations that were frequented by cosmonauts, notably Baykonur and the Crimea (mission control in Yevpatoriya and the water survival training area in Feodosiya), they had no cars at their disposal at all, which meant that they regularly had to try and get about by hitchhiking! Since several of those hitched rides had ended in road accidents, Kamanin stressed that stationing three cars in the Crimea and three at the cosmodrome was also in the interest of cosmonaut safety.[8]

Two plaques by sculptor Ivan Misko were mounted next to the entrance of the Headquarters and Staff Building. This one commemorates Sergey Korolyov

The twin plaque, mounted on the other side of the entrance, commemorates Yuri Gagarin

REORGANISATION

In late 1968, TsPK underwent a major restructuring. The Central Committee of the CPSU issued a resolution to convert the Cosmonaut Training Centre into the 'Test and Research Centre for Space Flights named after Yu. A. Gagarin', having the centre operate on the same level as test and research institutes of the first category.[9] Kamanin and the TsPK management then drafted a new structure for the centre. It would be divided into five departments:

- Orbital Spaceships
- Military Spaceships
- Lunar Spaceships
- Engineering Department, and
- Medical Department

Although it seemed that TsPK was past its most difficult years, not everything would go as smoothly as Kamanin had hoped. In early February 1969, Kamanin and Vershinin discussed plans for TsPK during a flight from the Baykonur Cosmodrome back to Moscow. The centre was to consist of 600 officers, including eight generals (at that time, there was only one), three directors (also only one), and six deputy positions (there were three). TsPK was to become the country's centre for both cosmonaut training and scientific research, but according to Kamanin's diary, Vershinin was not able to get anything going on these plans, despite promises to implement them by higher officers.[10]

In March 1969, Vershinin, who had always supported Kamanin, lost his post as Commander of the Soviet Air Force, and was replaced by General Pavel Kutakhov. According to the Kamanin diaries, Kutakhov was bad news for space flight in general, and for TsPK in particular. When looking back at 1969, Kamanin wrote that it hadn't been a bad year, but concluded that Kutakhov was totally indifferent to civilian space flight. He didn't support TsPK, and refused to help Kamanin strengthen the centre. In fact, when Kutakhov had been ordered to cut personnel in the general staff by ten per cent, he had done so almost exclusively by reducing space unit staff by 25–30 per cent. However, the ministry quickly intervened.

Although it is uncertain to what degree Kamanin's ideas of January 1967 influenced the events in the years to follow, it must be said that the development of TsPK did keep track with those ideas. After four years without any major facilities having been constructed, a building to house the medical department was completed in 1969. Designated Korpus 3, it was the first structure to be built outside the fenced-off TsPK perimeter and it led to the enlargement of that perimeter to the current size. Most of the new facilities that have been built since are situated outside the original fence.

Given the time it takes to build a four-storey structure like this, it seems likely that construction started in 1967, not long after Kamanin put his visions on paper. In the following years, more new facilities were opened. In 1970, a new wing was added on the west side of the medical department building, which would become the home of TsF-7, the centre's first centrifuge. Until that time, all centrifuge testing was done

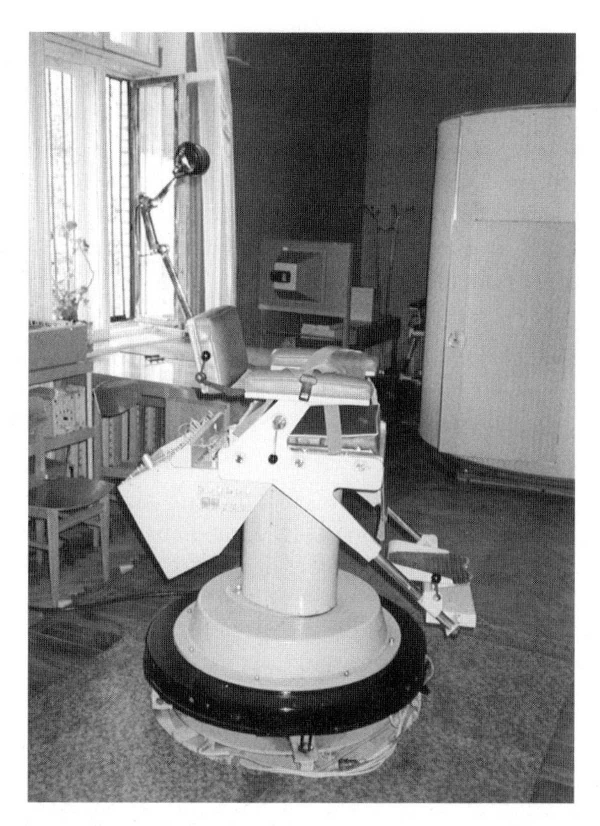

A rotating chair is used by the medical department to test and train cosmonauts for the effects of motion sickness

either at the Institute of Medical-Biological Problems, or in Air Force hospitals in Moscow. Since centrifuge rides were not exclusive to cosmonaut selection, but were also a regular part of mission training, it made sense to set one up in TsPK as well. TsF-7, which has a radius of seven metres, was manufactured in the Soviet Union. Commissioned in 1973, it can pull a maximum of 20-g.

In January 1971, Vasiliy Mishin made one last attempt to set up a separate training centre for civilian cosmonauts at the Moscow Aviation Institute. In fact, Mishin and the civilian cosmonauts even had the nerve to come to view the TsPK premises to get ideas, but once again nothing came of it. As far as is known, this was the last time an attempt was made to try and train cosmonauts outside of TsPK.

Later that year, Kamanin was informed that the Cosmonaut Training Centre had been awarded the Order of Lenin, the Soviet Union's highest award.

Boiler installations

Construction work in TsPK continued and a new feature was a boiler facility, consisting of two large boiler installations supplying hot water and central heating. It was built in two stages and after the second boiler had been added in 1973, the

TsF-7, the smaller of the two centrifuges in TsPK

The TsF-7 control room

facility provided heat both to Star City and to the nearby town of Bakhchivandzhiy, next to Chkalovskiy Air Base. Since that town didn't have enough heating facilities and TsPK had constructed a few apartment buildings there to house the staff of the Seryogin Regiment, they were obligated to supply heat.[11]

By now, there were major building operations going on in TsPK, as was seen first hand by a number of American aerospace reporters who were shown around by the Director of Cosmonaut Training, Vladimir Shatalov, during preparations for the Apollo-Soyuz Test Project (ASTP) in June 1973.[12]

Profilactorium

One of the new facilities that was being constructed at that time was the 'American Hotel', which was especially built as a hotel for the American delegations that were visiting TsPK for ASTP. It was constructed on the bank of an artificial lake that had been excavated at the order of Georgiy Beregovoy, then head of TsPK. It was conveniently located outside the fence of the actual training centre, undoubtedly in order to prevent the Americans from getting too nosy.

The NASA office in TsPK, which is housed in the Profilactorium

The lobby of the Profilactorium

After ASTP, the facility was renamed the Profilactorium and from then on would serve as the place where returning crews could readapt after their space flights. In the early 1990s, when NASA and ESA astronauts began participating in Mir missions, both space agencies set up liaison offices in Star City, which are located on the second and third floors of the Profilactorium.

Centrifuge TsF-18

From Kamanin's diaries, it was clear that there had been attempts to set up a large centrifuge in the centre since 1962.[13] Constructing this facility had no connection with ASTP as was suggested by the reporters that were shown around in 1973.[12] While ASTP was only agreed upon in May 1972, the decision to procure the centrifuge dated back to 1970. From the outset, the plan had called for a 16-metre centrifuge, and Kamanin had always referred to it as TsF-16 in his diaries. On 3 August 1970, a meeting was held to finally get the centrifuge into the state's new five-year plan. It was clear that it would have to be bought abroad, and the choice was between a French and a Swedish model. The Swedish model was the one the Soviets preferred, but the price the Swedes had indicated (12 million roubles) was more than the budget allowed. The following month, a Soviet delegation travelled to Sweden in order to negotiate with the prospective builders of a centrifuge, ASEA, an ABB subsidiary. For a while, the Soviet plans called for one with a 20-metre radius, TsF-20, but negotiations dragged on and in May 1971, despite three trips to France and

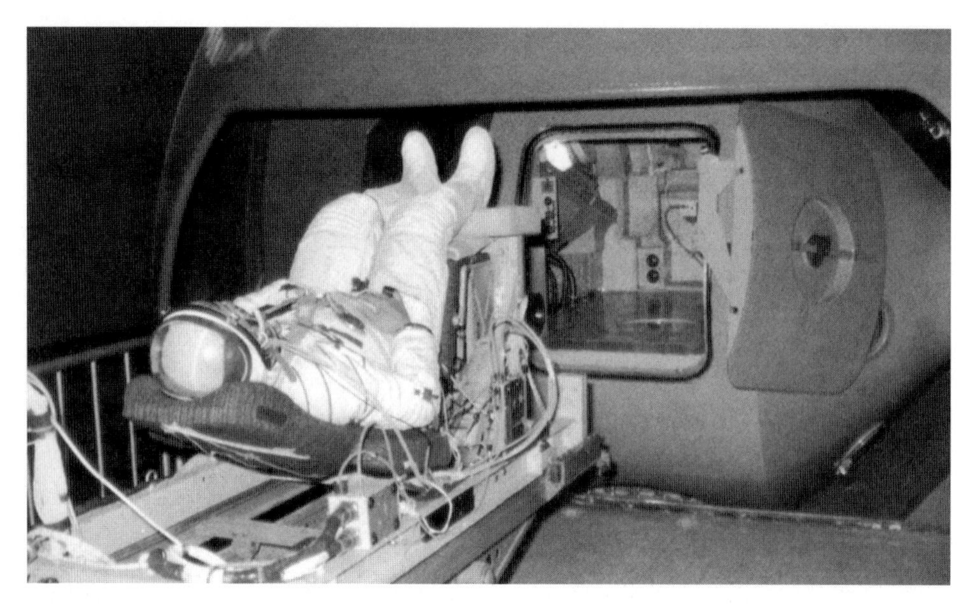

A test subject wearing a Sokol pressure suit is waiting to be placed in the TsF-18 centrifuge's cabin

Cosmonaut Anatoliy Filipchenko undergoing medical tests in the medical department

Korpus 3A, the home of the TsF-18 centrifuge, under construction in the early 1970s

The TsF-18 centrifuge, built by the Swedish company ASEA, was delivered by barge in 1974

Sweden, no order had been placed. In the end, the order went out to ASEA for a centrifuge with a radius of eighteen metres, which therefore would be known as TsF-18. The biggest differences between TsF-7 and TsF-18 were that the new facility could pull 30-g, ten more than its counterpart, and that it was capable of simulating Soyuz landing patterns.

Once the order had been placed and it was determined what the dimensions would be, ground could be broken for the building that would house it. The conspicuous circular facility was built as an extension to the medical department and was dubbed Korpus 3A. (Prior to this, once the wing that contained the TsF-7 had been built, that wing was known as 'BTsF', and the oldest part of the building as 'ATsF'.)

Construction started in 1971 and was in full swing at the time of the visit of the American journalists in 1973. The building was completed the following year and shortly after that the centrifuge was delivered by barge. Once it was installed, the first systems testing took place in 1976, but those early tests only consisted of starting it and stopping again immediately, making the centrifuge move only several dozen centimetres at a time. In December 1980, ASEA officially handed the centrifuge over to the Russians and in September 1981, the first cosmonaut took a ride.

Engineering and Simulator Building

As well as the new centrifuge facility, the five-year plan called for the construction of some 5,000 square metres of new laboratories and a neutral buoyancy facility for EVA training. Although it would be another ten years before the latter would be built, the construction of a new simulator building began in 1971 and was finished just in time for the visit by the American journalists. It consisted of two wings, which were named Korpus 1 (the east wing) and Korpus 1A (the west wing). A separate facility that was to provide the power for the simulators was built in between these two wings and finished in 1974.

Once it was finished, the simulators that were located in the old Korpus D were transferred to the new facility, after which Korpus D was renamed the Experimental Plant (Eksperimentalnyy Zavod) and turned into a workshop where engineers could build new training devices on site.

When the American journalists were shown around TsPK in June 1973, they also visited the eastern hall and were shown the Salyut and Volga simulators and the Soyuz 2 Descent Module that were placed there. According to *Aviation Week*'s Donald Winston, Shatalov had said: 'We have no secrets here. The green wooden fence you see is only to keep out the Press. We want privacy and tranquillity here so we can concentrate on training for the tasks we must perform.'[12] However, the Americans weren't shown the western hall of the complex, where the simulators that were connected to the still highly secret Almaz and lunar programmes had been relocated to, nor did Shatalov mention that they were there!

From this point on, all simulator training for Soyuz, Salyut and Mir was conducted in the Engineering and Simulator Building.

Administration Building

Another major structure that was built was an office building. Dubbed Korpus 2, it

The main entrance of the Engineering and Simulator Building

would become the home of the training department and included offices for the cosmonauts and a cafeteria. In 1981, a new wing was added to the Korpus 2 Training Department building. One of its features was a 12.5-metre planetarium dome and a highly sophisticated Zeiss planetarium that was used to train cosmonauts in navigating with the help of the stars. A crew cabin was put up on the centre of the dome's floor, containing the various devices that cosmonauts used for navigating while in orbit. Since the planetarium instrument was capable of depicting the skies from any point in orbit, it was no longer necessary to have cosmonauts travel to southern Africa in order to acquaint them with the southern skies. The training device was officially christened the 'Astronavigation Functional Modelling Stand'.

Hydrolaboratory

Once the centrifuge and simulator buildings, the administration offices, and the new Profilactorium had been finished, building activities died down again in the training

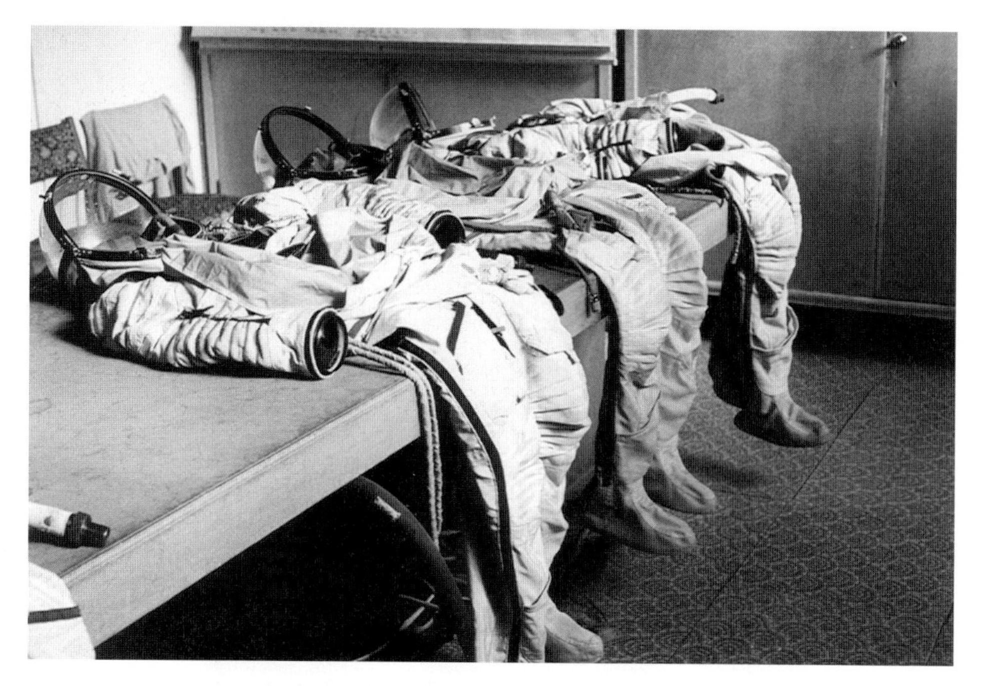

Three Sokol spacesuits, used for training purposes, are lying ready to be donned by cosmonauts for a training exercise

The Administration Building, which houses the cosmonaut offices. The wing on the left is the home of the Zeiss Planetarium that is used for training to navigate in space

The Hydrolaboratory

Inside the Hydrolaboratory. On the left is one of the cranes used to lower suited cosmonauts into the water. On the right are two mock-ups of space station modules

centre, although Star City itself saw several new large apartment complexes being built. Within the training centre, a small wing was added to the Headquarters building in 1975, which would become the home of the Communications Department. After this, no new facilities would be built until 1980, when a Neutral Buoyancy Laboratory, called the Hydrolaboratory, was finally constructed. It was situated at a strategic location near a small lake, which is known in Star City as 'Soldier's Lake', because it is a popular place for the conscripts to swim and have barbecue parties. The lake, which is situated just outside the fence surrounding TsPK, is used to dump the water when the Hydrolaboratory's basin is emptied once every year or so.

In 1999–2000, an annex was added to the Hydrolaboratory, which is the storage place for the various ISS modules that are not being used during specific training sessions. Until that time, it had been the custom to park the modules that were not needed outside the building, in the open air.

The Buran programme

Apart from the Hydrolaboratory, the 1980s also saw a lot of other building activities that were instigated by the Buran programme. A large facility, known as KTOK (Kompleksnyy Trenazher Orbitalnogo Korablya, or Simulator Complex for Orbital Ships), was completed in 1988. The north part of the complex consisted of a large hall, in which a fixed-base and a motion-base Buran simulator were placed. In addition, a Buran flight deck simulator was located in the hall and there was still room left for a full-scale mock-up of an orbiter. When it became clear that Buran

Old training modules for the Soyuz, Salyut and Mir programmes are stored outside the Hydrolaboratory and left to the elements

The KTOK building

would be cancelled, TsPK management decided to use the hall to house the simulators of the Spektr and Priroda modules for the Mir space station. Later, the Zarya and Zvezda modules for the International Space Station were also placed in the hall. In spite of the fact that the programme was cancelled in the early 1990s, the three Buran simulators were still present in 2004, although by then, the motion-base one had been partially dismantled.

The southern part of the complex, which mainly consists of offices, is currently the home of the photo department and of the survival training and recovery departments, currently led by former cosmonaut Yuri Gidzenko, who took over this position from Nikolay Grekov, a cosmonaut who had been medically disqualified before having the chance to fly in space. Grekov, in turn, had taken over from Yevgeniy Khludeyev, another unflown cosmonaut.

Construction of a second Buran training facility was started, but by the time the main structure had been completed, it had become clear that the future of the Buran programme was very much in doubt. Work on the building was put on hold indefinitely, and eventually it was abandoned. The building had been put up outside the TsPK perimeter and it was foreseen that the fenced-off area would be expanded. In fact, a new gate had already been built at the north side of TsPK, but after construction of the new Buran facility was halted, this new gate was situated way outside TsPK and was also abandoned.

The Buran building itself is still prominent, given the fact that, at seven storeys, it is the highest structure of the centre. However, regular visitors have noted that it is slowly deteriorating. It was planned to have a full-scale mock-up of the orbiter

The Buran training facility is the highest structure in TsPK. When the programme was cancelled, construction was halted and the building abandoned

placed into it, which could be placed in both a horizontal and a vertical position, and the facility was also meant for training cosmonauts in the use of the manipulator arm for Buran.

Another restructuring

In May 1995, in a new reorganisation, TsPK and the Seryogin Regiment were merged into one new unit; the Yuri A. Gagarin Russian State Scientific Research Centre for Cosmonaut Training.[14] Now, in 2005, the composition of the training centre seems to be fixed. For the immediate future, there appear to be no major building operations planned. The Hydrolaboratory annex has been the most recent addition to TsPK and no additional facilities seem to be needed, so no significant additions to the buildings are foreseen. The current TsPK staff numbers around 1100 people, of which some 700 are military officers.[15]

Monuments, memorials and statues in TsPK

As was common in the Soviet Union, a number of works of art were put up over the years that the centre was built. However, most of these are never seen by visitors to the centre. The road leading up to the Headquarters and Staff Building has a large number of Soviet-style displays and mosaics; among them a more than life-size one showing Gagarin in his spacesuit and a large obelisk with a portrait of Lenin, the last remaining obvious memento of the Soviet era.

Another monument on this access road is a wall that consists of 45 granite slabs. On eighteen of them, all Soviet manned space flights up to and including Soyuz 11 are commemorated.

A mosaic portrait of Yuri Gagarin was erected along the road from the entrance of the training centre to the Headquarters and Staff Building

After the deaths of Korolyov and Gagarin, bronze memorial plaques were made, which were put up on both sides of the Headquarters and Staff Building's official entrance. The sculptor of the plaques was Ivan Misko, a Belorussian who is regarded as one of the Soviet Union's most important sculptors of space-related statues and monuments and designer of commemorative medals.

In the lobby of the Engineering and Simulator Building stands a life-size bronze statue of Yuri Gagarin greeting visitors with his right hand. The lobby is also decorated with a number of leaded-glass mosaics.

Finally, on the road leading up to the Hydrolaboratory, in between the KTOK and Korpus 2, there are two statues. The female and male figures, named 'Science' and 'Technology' respectively, had initially been placed in the VDNKh (the All-Union Exhibition Centre) in Moscow where the Soviet Union showed off its economic accomplishments, but the sculptor donated them to TsPK and they were relocated to their present places.

A remarkable picture of the complete 1976 cosmonaut selection. This was only the second photo ever made of a complete selection and was clearly meant as a publicity photograph. It was not released until after the fall of the Soviet Union

Lieutenant-General Mikhail Odintsov, TsPK's second commander (1963)

Major-General Nikolay Kuznetsov, TsPK's third commander (1963-1972)

Table 1: Chiefs of the Cosmonaut Training Centre (TsPK)
Military Unit Number 26266

Director	
1960 Feb 24–1963 Jan 26	Colonel Yevgeniy Anatoliyevich Karpov
1963 Jan 26–1963 Nov 2	Colonel-General Mikhail Pyotrovich Odintsov
1963 Nov 2–1972 Feb 26	Major-General Nikolay Fedorovich Kuznetsov
1972 Feb 26–1987 Jan 3	Lieutenant-General Georgiy Timofeyevich Beregovoy
1987 Jan 3–1991 Sep 12	Lieutenant-General Vladimir Aleksandrovich Shatalov
1991 Sep 12–2003 Sep 25	Colonel-General Pyotr Ilyich Klimuk
2003 Sep 25–date	Lieutenant-General Vasiliy Vasileyevich Tsibliyev
The First Deputy Chief	
1960 Mar 30–1961 Mar 6	Colonel Vladimir Vasiliyevich Kovalev
1961 Mar 6–1969 Apr 9	Post not filled
1969 Apr 9–1972 Jun 26	Major-General Georgiy Timofeyevich Beregovoy
1972 Jul–1992 Sep	Major-General Andriyan Grigoriyevich Nikolayev
1992 Sep–2000 Apr	Major-General Yuri Nikolayevich Glazkov
2000 Apr–2003 Sep 25	Major-General Vasiliy Vasileyevich Tsibliyev
2003 Sep 25–date	Major-General Valeriy Grigorievich Korzun

The training centre was organised in 1969 into a number of Directorates. These were headed by cosmonauts until recently

1st Directorate was Soyuz, Salyut-Mir and now ISS operations and simulators. Responsible for all Module and Soyuz simulators.

1969 Mar–1970 Jan	Colonel Pavel Ivanovich Belyayev
1970 Feb–1972 Dec	Colonel Aleksey Arkhipovich Leonov
1972 Dec–1978 Jan	Colonel Pavel Romanovich Popovich
1978 Jan–1988 May	Major-General Anatoliy Vasilyevich Filipchenko
1988 Jul–1997 Aug	Major-General Vladimir Aleksandrovich Dzhanibekov
1997 Nov–date	Colonel Y.I. Zhuk

2nd Directorate was Engineering and later, Training and Facilities

1969–unknown	Klishov
1976–1979	Major-General Georgiy Stepanovich Shonin
1979–1987	Unknown
1987 Sep–1989 Feb	Colonel Yuri Nikolayevich Glazkov
1989 Mar–1992 Oct	Colonel Valeriy Ilyich Rozhdestvenskiy
1992 Oct–1997 Nov	V.K. Gotvald
1997 Nov–2000 Jul	Colonel Gennadiy Mikhailovich Manakov
2000 Jul–unknown	B.A. Naumov

Sometime after the turn of the century, this department was merged with the 4th Directorate (Medical)

2003–date	Colonel Valeriy V. Morgun

3rd Directorate was formed in 1976 to work on Buran, then Training in 1994, and was then given responsibility for survival training, including the management of the Hydrolab.

1976–unknown	Unknown
1989–1995	Colonel Yuri Vladimirovich Romanenko
1995–1996	Colonel Vladimir Georgiyevich Titov

1996–2003	Colonel Nikolay Sergeyevich Grekov
2003–date	Colonel Yuri Pavlovich Gidzenko

4th Directorate was formed in 2004 to oversee Aircraft Operations. (See 70th Air Wing table)

Deputy Chief: The Political Department

1960 Mar 19–1963 Feb 14	Colonel Nikolay Fedorovich Nikeryasov
1963 Feb 19–1964 Dec 24	Nazar Martemiyanovich Trofimov
1964 Dec 24–1972 Sep 22	Colonel Ivan Makarovich Kryshkevich
1972 Sep 22–1974 Jun 16	Major-General Boris Vasiliyevich Matosov
1974 Jun 16–1978 Jan 24	Major-General Ivan Ivanovich Vaganov
1978 Jan 24–1991 Sep 12	Major-General Pyotr Ivanovich Klimuk
	Post abolished after Soviet coup

Deputy Chief: The Head of Staff

1963 Mar–1969	Colonel Grigori Gerasimovich Maslennikov
1969 Jun 6–1971 Oct 28	Major-General Nikolay Pavlovich Pashkov
1971 Nov 11–1976 Feb 5	Major-General Yuri Aleksandrovich Vasilevskiy
1976 May 31–1987 Dec 23	Major-General Vladilen Maksimovich Rumyantsev
1987 Dec 23–1994 Dec	Major-General Yevgeniy Grigoriyevich Dyatlov
1995 Jun–2000	Major-General Nikolay V. Popov
2000–unknown	Aleksandr Nikolayevich Egorov
current	Major-General V. Shchemyakin

Deputy Chief: Flying and Space Training

1963 Dec–1968 Mar 27	Colonel Yuri Alekseyevich Gagarin
1968 Jul–1972 Feb	Colonel Andriyan Grigoriyevich Nikolayev
1972 Feb–1992	Major-General Aleksey Arkhipovich Leonov
	Post abolished and responsibilities passed to 1st Deputy Director

Deputy Chief: Medical and Biological Training

1960 Mar–1963 Feb	Colonel Vladimir Vasiliyevich Kovalev
1963 Feb 8–1969 Mar 21	Grigoriy Fedulovich Khlebnikov
1969 Jul–1971 Jul 2	Major-General Vladimir Alekseyevich Popov
1972 Jul 7–1974 Jan 7	Colonel Konstantin Georgievich Pushchin
1974 Jan–1989	Unknown
1989–1996	Major-General Stanislav Bugrov
	This Department was redesignated the 4th Directorate Medical
1996–2001	Colonel Valeriy V. Morgun
	This was then merged with the 2nd Directorate in 2001/2002

Deputy Director: Science, Research and Testing

1969 Mar 21–1974	Colonel Nikolay Dimitriyyevich Samsonov
1974–1976 May	Major-General Vladilen Maksimovich Rumyantsev
1976 May–1978 Jan	Major-General Yuri Aleksandrovich Afanasyev
1978 Jan 25–1989 Mar	Major-General Pavel Romanovich Popovich
1989 Mar 9–1992 Aug	Colonel Yuri Nikolayevich Glazkov
1992 Apr–1999	Colonel B.I. Kryuchkov
1999–date	Colonel Aleksandr Nikolayevich Egorov

Commander of 70th Air Wing (Special destination) Based at Chkalovskiy AF Base

1960–1965	Colonel Vladimir Sergeyevich Seryogin
1965–1966	M.I. Lavrov
1966–1967	A.M. Ustenko

Deputy Chief: Flying Training

1960 May 6–1961 Feb 8	Colonel Evstafiy Yevseyevich Tselikin
1961–1963	I.A. Azbievich
1963–1969	Unknown
1969 Aug 22–1975 Jan 6	Colonel Gennadiy Aleksandrovich Alekseyev

The regiment was formed and attached to the training centre on 24 Feb 1967 and later renamed after Seryogin

1967–1968	Colonel Vladimir Sergeyevich Seryogin
1968–1969	Colonel Gennadiy Aleksandrovich Alekseyev
1969–1978	M.I. Lavrov
1978–1988	A.G. Shchedrov
1988–1994	E.V. Cherednichenko
1995–date	V.A. Platonov

Following reorganisation it became the 4th Directorate Aircraft Operations in 2003 or 2004.

REFERENCES

1 Rex Hall and Bert Vis interview with Nikolay Mikhailovich Kopylov, Star City, August 2004.
2 Nikolay Kamanin, 'Skrytyi kosmos: kniga vtoraya 1964–1966', Infortekst, Moscow 1997; entry for 3 January 1964.
3 Ref 2, entry for 20 September 1965
4 Ref 2, entry for 2 September 1966
5 Nikolay Kamanin, 'Skrytyi kosmos: kniga vtoraya 1967–1968', Infortekst, Moscow 1999; entry for 4 January 1967
6 Ref 5, entries for 1 and 2 April, and 4 May 1968
7 Ref 5, entries for 11 and 13 May 1968
8 Ref 5, entry for 21 June 1968
9 CC CPSU and Council of Ministers Resolution No. 932–331 of 28 November 1968, signed by Leonid Brezhnev and Aleksey Kosygin
10 Nikolay Kamanin, 'Skrytyi kosmos: kniga vtoraya 1969–1978', Infortekst, Moscow 2001; entry for 21 June 1968
11 Hall and Vis interview with Gennadiy Ivanovich Sokolov, TsPK and Star City construction overseer from 1969 onwards, August 2004
12 Aviation Week & Space Technology, 25 June 1973, pp. 18–21
13 Ref 5, entry for 29 July 1968
14 Russian Government Resolution No. 478 of 15 May 1995
15 Interfax-AVN military news agency, Moscow, in English, 4 October 2004

Simulators

At the height of the space race of the 1960s, work was conducted on many manned programmes. As these programmes called for the development and construction of quite a number of simulators, the government issued a decree that formalised a reorganisation of the Flight Research Institute on 21 August 1967.

From then on, several subdivisions that worked on simulators were united and started operating under the name SOKB-LII (Specialised Experimental Design Bureau – Flight Research Institute). Among its responsibilities was the development of cockpit consoles for manned spacecraft, and the construction of simulators for cosmonauts to train in. Its first Chief Designer was Sergey Grigoryevich Darevskiy, who had been working on spacecraft simulators from the very beginning of the manned programme.

In 1971, SOKB-LII separated from LII itself to become an independent organisation, while still remaining the prime contractor for the development of spacecraft simulators. In 1983, SOKB-LII merged with another branch of LII and was renamed NII-AO (Flight Research Institute – Aviation Equipment).

Finally, in 1997, a new design bureau called SOKB-KT (Specialised Experimental Design Bureau – Space Technology) was set up within NII-AO. This bureau is described as being the successor to SOKB-LII and is responsible for the development of control panels for manned spacecraft and simulators.

THE FIRST SIMULATORS

In 1960, a prototype of what would later become the first spacecraft simulator was developed and built in the Ministry of Aviation Industry's Flight Research Institute (LII) in the city of Zhukovskiy, southeast of Moscow. A team of engineers led by Darevskiy developed and built a stand that was used to perfect the design and layout of the Vostok cockpit control panel. Shortly after it had been constructed, Kamanin had this stand incorporated into a full-size Vostok mock-up and upon its completion in the autumn of 1960, it was dubbed simulator TDK-1[1]. Although it served its purpose well, only Yuri Gagarin and German Titov ever trained on it in Zhukovskiy.

By then, Kamanin had ordered an improved simulator to be built. This second one, called TDK-2, was designed by Darevskiy's team together with engineers from the Cosmonaut Training Centre, and built by the same laboratory in LII as TDK-1.

After it had been completed, it was placed in the new centre. The cosmonauts who flew the later Vostok missions, and their back-ups, all trained on this simulator, which represented a real Descent Module. It provided the cosmonauts who trained on it with visual information about flight modes, different systems, assemblies and onboard equipment operation. Furthermore, the external visual environment was simulated with the help of an optical visor called 'Vzor'. Optical imitation of the visual environment helped the cosmonauts to see the reaction of the vehicle to the controls, which created the illusion of real space flight.[2]

One of the first instructors to work with the TDK simulators was Valentin Varlamov, who had been one of the cosmonauts selected with Yuri Gagarin. In March 1961 however, Varlamov had been permanently grounded and removed from the cosmonaut team after sustaining a serious back injury during an off-duty accident.

Reportedly, a third Vostok simulator, designated TDK-3KA, was constructed. The timeframe for its use was given as 1963–1964,[3] which suggests that it was never actually used for cosmonaut training because shortly after the mission that was flown by Valentina Tereshkova in June 1963, it was decided to end the Vostok programme.

The difficult 1960s

Experience gained in actual space flight (as well as in training) during the Vostok programme was used to improve the simulators that were being developed for the Voskhod programme. The first all-round simulator for the space programme that was developed was the TDK-3KV, which was used to train Vladimir Komarov, Konstantin Feoktistov and Boris Yegorov for their mission on Voskhod 1. With the MN-14 computer that was an integral part of the TDK-3KV, it had become possible to simulate orbital manoeuvring and manual entry dynamics. The visual environment imitator, the most complex part of the simulator, was designed by TsKB Geofizika and 'Arsenal', both defence industry enterprises.[2]

A second new simulator, TDK-3KD, was built for training Pavel Belyayev and Aleksey Leonov, the crew of Voskhod 2 that was to perform the first extra-vehicular activity (EVA).

Unfortunately, there is a lot of confusion over the Voskhod simulators. Although the official histories of the Cosmonaut Training Centre clearly state that both simulators were built in 1964 and 1965 respectively and were used for the training of the Voskhod 1 and Voskhod 2 crews, Kamanin's diaries suggest that they were not. On 26 June 1965, Kamanin wrote that 'the Voskhod simulator didn't even exist on paper yet.'[4] The following August however, in another entry, he mentioned that a decision had been taken that ordered LII to have the TDK-3KV simulator ready by October of 1965, and the next one, TDK-3KD, in the first quarter of 1966.[5] It is unknown if this time schedule was met, as the first mention that the TDK-3KV was actually available is from the entry of 10 January 1966. These entries would suggest that the simulators Kamanin mentions are not the same ones as those used for training the Voskhod 1 and Voskhod 2 crews, since Voskhod 1 flew in October 1964 and Voskhod 2 in March of 1965.

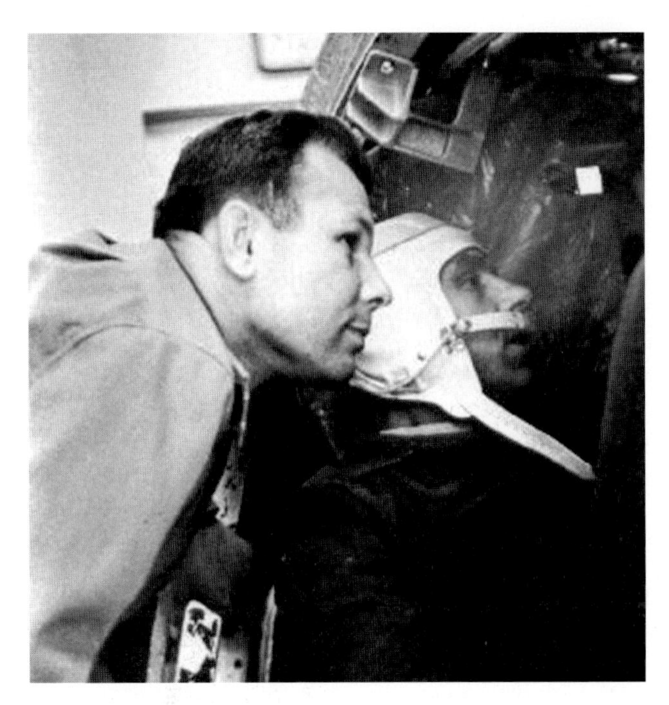

Yuri Gagarin (left) and Vladimir Komarov discuss the upcoming Soyuz 1 and 2 docking mission in the Soyuz simulator

In January 1966, a series of follow-on missions was still scheduled in the Voskhod programme. Voskhod 3 was to be a long-duration flight, with the creation of artificial gravity as one of the objectives; Voskhod 4 was to be flown by two female cosmonauts, one of which would conduct an EVA; and Voskhod 5 was scheduled to be a life sciences mission.

To add to the confusion, the TDK-3KV simulator was reportedly[3] used for both Voskhod 1 and Voskhod 2, while TDK-3KD was said to have been meant for training the crew of Voskhod 3, and possibly subsequent flights. Even if the suggestion that the TDK-3KD was indeed meant as the training tool for Voskhod 3 is correct, it is interesting to see that Kamanin states it had to be delivered in the first quarter of 1966, since that was also the time the mission itself was planned. It would have given the crew very little time to actually use the simulator. Whatever the case, it can be concluded from the diaries that the simulator for Voskhod 3 was available in mid-December 1965. This suggests that he was talking about TDK-3KV.

Early Soyuz and the lunar programme

Apart from the Voskhod simulators, new ones had to be developed for the follow-on programmes, such as Soyuz. One of these was being built at a bureau called TsNII-30 in the town of Noginsk, north-east of Moscow, for the 7K-9K-11K programme, an early concept for a manned circumlunar flight. The mission called for a rocket stage, designated 9K, that would be launched without its load of fuel. It would then be followed by several tanker craft, designated 11K, that would fuel the 9K. Once that had all been completed, the manned 7K spacecraft would be launched to link up with the 9K, after which the complex would be sent on its way to the Moon.

In his diaries, Nikolay Kamanin mentions that the 7K-9K-11K simulator was nearing completion. It was said to consist of a full-scale 7K mock-up and $^{1}/_{30}$ scale models of both the 9K and 11K craft. When training on the simulator, a cosmonaut would have seen the 9K and 11K as they would look in orbit, with the use of a television system or an optical visor.[6] In the end, however, the 7K-9K-11K concept never flew in space and it is not known whether any cosmonaut ever actually trained on the simulator, or whether it was even fully completed.

By then, the plans for manned space flight were more ambitious than ever. This was evident by the fact that twenty new cosmonauts were selected in late 1965, with a note that the Central Committee of the CPSU and the Minister of Defence had given permission to select up to forty. All these programmes with new spacecraft also called for the necessary new simulators, but their development and construction apparently didn't keep pace with the other developments. In his diary, Kamanin stated that in order to prevent further delays in the lunar programme, it was essential to order and develop simulators.[7] It seems that he managed to convince his superiors of this, since only two months later, the Ministry of General Machine Building and other ministries were ordered to 'within two weeks... submit for approval by the Military Industrial Complex a schedule for the development, construction and delivery of multi-purpose and specialised 7K-L1 simulators within a timeframe that will ensure the preparation of cosmonauts for missions on 7K-L1 spacecraft.'[8] The design bureau that would be responsible for building the simulators would be the Flight Research Institute of the Ministry of Aviation Industry (LII-MAP).

Things didn't go smoothly however. In July 1967, Kamanin wrote that the Soyuz simulator had not been functional for three months already, blaming TsKBEM and particularly Mishin and his deputy Tsybin. In addition, the L-1 simulator was not yet finished and was expected to be delivered only in September, and the L-3 simulator was said to be still on the drawing board.[9] The reason for these delays, according to Kamanin, was indolence on the part of Mishin and his design bureau. On 31 August, he noted that while the military Soyuz variant, Soyuz VI, was supposed to make its maiden flight in four months, no simulator had been delivered yet. Kozlov, the designer of Soyuz VI, had simply stated that providing the necessary training equipment for the ship was not his responsibility. It made Kamanin realise even more that a new design bureau (led by Darevskiy) had to be founded that would be responsible solely for producing simulators.

By October, TsPK had finally received an L-1 simulator (TDK-F91), but it didn't resemble the actual spacecraft. This was instigated by the fact that Mishin had sent a letter to Darevskiy, in which he said that the L-1 simulator did not have to be equipped with the automatic systems of the L-1 spacecraft, essentially telling Darevskiy to build a simulator that wouldn't resemble the actual craft. A complaint went out to Mishin, but apparently to no avail, as in November, Kamanin was still complaining that the simulator was 'raw' and causing many problems.

Two weeks later, on 29 November, Kamanin was fed up with the situation and wrote a letter to the Central Committee of the CPSU in which he described his problems in obtaining adequate simulators for training the crews that were to fly. He still didn't have a simulator for the L-3 lunar landing programme and Mishin was trying to keep the Air Force out of this programme. Mishin had indicated that he wanted TsKBEM to build a simulator that only needed to meet his requirements, and he didn't need, or want, any input from the Air Force. Of course, Kamanin opposed these plans. He demanded that the Air Force's input would be part of the specifications for the L-3 training equipment. Furthermore, he stated that if two simulators were built, one of them would be placed in TsPK, and should only one be built, that would also have to come to TsPK.

Kamanin saw his problems mounting. The first flight in the L-3 programme was scheduled for April 1968 and training had not even begun yet, because a suitable simulator was still not available. Kamanin saw this as Mishin's fault – he only wanted simulators that met the requirements of TsKBEM, and all he really wanted was to send TsKBEM's cosmonauts to the Moon. Finally, Kamanin was fed up with it and complained to Sergey Afanasyev, the Minister of General Machine Building and Mishin's superior. He said that a choice had to be made: either TsKBEM should get full responsibility, or the programme should come to TsPK with a list of candidates for training, a training programme, and help in the construction of the necessary simulators. It was a risky move, but Afanasyev said he agreed with Kamanin and added that no one supported Mishin's plans to build simulators only for his own bureau. TsPK needed them more and he gave Mishin a direct order to settle the matter.[10]

The next day, during a meeting without Afanasyev present, Mishin once more attempted to turn things to his advantage. It was found that he had changed the specifications for the simulators on order without consultation with the Air Force, and had unilaterally cancelled the contracts for some of the simulators, including a device described as the 'turboflier' and a V-10 helicopter that was to be equipped with an L-3 lunar lander cockpit. Kamanin told the others at the meeting that these actions by Mishin were unacceptable, and they agreed, promising to correct things as soon as possible.[11] As a result, between 12 and 15 December 1967, TsPK, the Ministry of General Machine Building and TsKBEM sat down together and talked about a list of simulators, cosmonauts to be trained for L-3 and the training programme they would follow.

On 16 January 1968, Afanasyev visited TsPK in order to get an in-depth view of the frictions between the Air Force and TsKBEM. He toured the centre and made short runs in the Volga and L-1 simulators. According to Kamanin, he was quite satisfied with what was being accomplished in TsPK, and expressed that opinion more than once. He took the side of the Air Force when it came to the disagreements over the degree to which the simulators should resemble the actual spacecraft. Reportedly, Mishin still said that he would refuse to give the Air Force the hardware it wanted, stating that cosmonauts could train at the design bureau's facilities. It all ended in an open altercation between Afanasyev and Mishin, with Mishin openly saying that he refused to carry out his minister's orders. Afanasyev said that he had not realised before that things were in as bad a state as he was seeing now, and vowed to do something about it. He added that he would return to TsPK for a follow-up visit to see for himself if things had improved.[12]

Still, a little progress seemed to have been made. On 30 January 1968, Kamanin visited the Institute for Aviation and Space Medicine to inspect the Volchok trainer that had been installed in a centrifuge, for training for entry into the Earth's atmosphere when returning from the Moon. He concluded that although the L-1 cabin, several other pieces of training equipment, and the instructor's console were in place, it was far from ready for serious training because many critical parts, like cabin instrumentation, had not been installed yet. By March, little or nothing had improved, however, even though on 6 February, Kamanin and Mishin had agreed on deadlines for the Soyuz and L-1 simulators to be ready for use.[13]

In March, Kamanin informed the cosmonauts who were to fly the first Soyuz missions of the delays that were being encountered and said that the simulator was not to be expected before early April at best. In addition, an order went out to provide finances for the Kiev Aviation Institute to build an L-3 simulator for lunar landing training.[14]

By November, the need for adequate simulators was getting pressing. With missions planned to take place in the near future, a fully operational Volchok was needed in particular, but it was clear that even with the simulators installed, the problems wouldn't be over. Kamanin had calculated that he would need some two million roubles and an additional staff of some 30–40 people to have them installed and to operate them.

In the end however, matters would become less problematic, at least as far as simulators were concerned, because the L-1 and L-3 programmes slipped more and more and eventually were cancelled altogether. Soyuz VI already had been axed around February 1968.

SOYUZ AND SALYUT

The Soyuz programme was the first to see training simulators that were comparable with what NASA had available in the USA. From the beginning, the programme was aimed at docking operations, so besides training facilities for the spacecraft itself, a docking simulator was also needed.

When a small group of American aerospace journalists was shown around the cosmonaut training facilities in Star City in June 1973, one of the buildings they visited housed two Volga docking simulators. Initially, these simulators were used to practice Soyuz-Soyuz dockings, but later they were also used for training on Soyuz-Salyut dockings.

The simulators consisted of two Soyuz spacecraft docking simulators, situated side by side in the building, while twin miniaturised Soyuz docking targets were suspended from 15-metre long tracks. One of these targets could be viewed using a television system, while the other could be viewed through the Soyuz periscope system. Crew members were trained to work with both systems.

A manned Soyuz-Soyuz link up was planned for only three missions: Soyuz 1 and

Valeriy Kubasov (left) and Aleksey Leonov in training for the Soyuz 11 mission. Shortly before launch, they, and third crew member Pyotr Kolodin, were replaced by Georgiy Dobrovolskiy, Vladislav Volkov and Viktor Paysatev, who were subsequently killed during re-entry

Part of the Volga simulator consisted of models of the Salyut space station (left) and the Soyuz transport ship. A television and periscope system simulated the view from the spacecraft during final approach

2, Soyuz 4 and 5, and Soyuz 7 and 8. In training, the crews of both spacecraft would occupy the Volga's Soyuz cabins at the same time, and were given the illusion that they could see the other Soyuz craft approach.

For Soyuz 2 and 3, during which only one spacecraft would be occupied, by Georgiy Beregovoy, only one cabin of the simulator was occupied. This was also the case in later simulations, when the Salyut space station replaced the second Soyuz in the exercise. From then on, the second Soyuz docking target was replaced by that of the Salyut. During their visit, the American journalists were told that the system could not be used to simulate docking operations for the Apollo-Soyuz Test Project.

Volga was the name for the simulator that had officially been designated TDK-7K. As had been the case with the earlier simulators, Volga had been built by the Flight Research Institute (LII), which in August 1967, combined a number of subdivisions and formed a specialised design bureau called SOKB-LII. Two additional Soyuz simulators, TDK-1S and TDK-2S, were built.

When cosmonauts had to start training for missions on the Salyut 1 space station, TsPK did not have a Salyut simulator available. In order to keep things moving, the various control posts of the station were put inside the L-1 simulator, which allowed the crews to train on controlling the station's orientation. In the meantime, management tried to ensure that a mock-up or simulator would be available as soon as possible. On 30 September 1970, Kamanin noted in his diary that he had visited the Khrunichev plant in Moscow for an inspection of the mock-up of the DOS/Salyut space station. During his visit, he was told that the mock-up would be delivered to TsPK on 20 October of that year. It seems that this date was met, as in November, the Soyuz 10 crew was reported to be using it for training.[15]

Erected in Korpus D, the first simulator building within TsPK, it is unclear to what degree it was a copy of the actual station. While several sources have indicated that it was 'just an empty hull',[16] it has also been said that Salyut 1's interior was completely reproduced, adding that it was possible to link up scientific equipment and computers, to have radio contact with cosmonauts that were training, and to 'prepare them for photographing and filming activities'. Why a simulator was needed for the latter activities was not explained, however.[17]

A view of Korpus 1 of the Engineering and Simulator Building. Behind the Salyut 6 simulator, the Soyuz 2 Descent Module can be seen

TKS/Almaz

Simultaneously with TsKBEM's civilian DOS/Salyut space station programme, the military Almaz space station, along with a transport ship called TKS (Transportniy Korabl Snabzheniya), was being developed by Vladimir Chelomey's OKB-52 in the city of Reutov, a short distance away from Star City. Several simulators were developed by a group of engineers that was led by E.Ye. Zhernov. The group was part of Section 42, the design bureau's cosmonaut department.

TKS and Almaz simulators were put up in Reutov and cosmonauts from the Air Force's Almaz cosmonaut group travelled to Reutov on a regular basis to train there. Reportedly, Chelomey's order to create the simulator for his own design bureau was given only two months before the launch, in April 1973, of Salyut 2, the official designation of the first Almaz station.[18] The simulator would not only be used to train crews for missions on Almaz, but also had a broad range of functions. It was powered up from the moment an Almaz station was launched to the moment it re-entered the atmosphere. In addition, engineers would use the simulator to determine how best to maintain the station and repair any malfunctioning systems.

In 1999, the TDK-F74 mock-up of the TKS Descent Module was still stored behind a curtain in the KTOK hall that also houses three Buran simulators, the simulators for Mir's Spektr and Priroda modules, and the Russian ISS modules Zarya and Zvezda

During flights, 'shadow crews' would stay onboard the simulator the entire time the station in orbit was manned. During important phases of the flight, there would even be radio contact between the crews in Reutov and in orbit.

It is unclear what Almaz training equipment there was in TsPK. It appears that some kind of device was up and running in November 1970, when Kamanin wrote in his diary that it was 'in full use for crew training'. In his 2 December entry however, he mentions problems with the simulator, but it is not clear if he meant that it didn't work as well as it should, or that he meant that it was still not delivered. Later that month, Kamanin met with Darevskiy and they estimated that only 25–30 per cent of the total training needs of the cosmonauts were being met. In order to improve the situation, Kamanin told Darevskiy to devote 75 per cent of his efforts on the Almaz simulator, 20 per cent on DOS/Salyut, and not more than 5 per cent on L-3. Of course, TsKBEM's chief designer, Mishin, number one competitor of Chelomey, wanted no time spent on Almaz at all, 70 per cent on DOS/Salyut, and 30 per cent on L-3.

On 6 May 1971, Kamanin wrote that SOKB-LII was to deliver the Almaz simulator on 1 December 1971, but that they couldn't guarantee it would include equipment that had to be delivered by a range of other ministries. When delivered, the Almaz simulator, designated TDK-F71, was placed at Chkalovskiy Air Base. The reason for this is unknown, although it may have been to keep it hidden, even from civilian cosmonauts and engineers that came to TsPK for training sessions. No other details about this simulator are known, and no photographs of it have ever been released.

It is known however that a crew, consisting of Air Force cosmonauts Gennadiy Sarafanov and Vladimir Preobrazhenskiy, and OKB-52 engineer cosmonaut Valeriy Romanov, conducted an eight-day test onboard the simulator from 20–28 November 1979. Two photographs, showing the three cosmonauts in Sokol space suits and with beards, have surfaced in recent years. They were taken minutes after they had emerged from the simulator upon completion of the test. Reportedly, at least one more crew performed a similar test in Chkalovskiy sometime between 1979 and 1982, although no further details are known, other than that it lasted about two weeks and that the crew consisted of Georgiy Shonin and Yuri Glazkov from TsPK, and Valeriy Makrushin from OKB-52.[19]

In TsPK, training equipment for the Almaz programme was put up in Korpus 1A of the Engineering and Simulator Building. The Almaz simulator hardware was called 'Irtysh', and it appears as if this was not a complete station like the one in Chkalovskiy, but only parts of the onboard systems. Besides Irtysh, there were the TDK-F77 simulator for the TKS Functional Cargo Block, and the TDK-F74, a full-scale mock-up of the TKS-VA landing module which reportedly was operational between 1975 and 1982.[3] In 1999, visitors to TsPK saw the TDK-F74 in storage in the KTOK.

A fourth training device was reportedly constructed by the centre's own simulator department. 'Sirius-77', as it was called, was intended for practicing approach and docking of the TKS spacecraft with the Almaz station.[20]

Cosmonauts Gennadiy Sarafanov, Vladimir Preobrazhenskiy and Valeriy Romanov pose for photos moments after exiting the TDK-F71 Almaz simulator at Chkalovskiy Air Base, after an eight-day test that lasted from 20-28 November 1979

ASTP

When the crews for ASTP started training for their mission, in addition to the Salyut, there were two Soyuz simulators in Korpus 1. Photographs released by NASA in June 1974 show a third, complete, Soyuz vehicle, which presumably was an exhibition model rather than a training device.

While the TDK-7K simulator had been relocated from Korpus D, a new one, designated TDK-7M, had been manufactured by SOKB-LII especially for ASTP training. On the NASA photographs, both simulators can be seen equipped with the Androgynous Peripheral Assembly System (APAS-75), the docking system that was developed for ASTP.

The Salyut 4 mock-up in Korpus 1

Post-ASTP

In preparation for missions onboard Salyut 4, TsPK received a full-scale mock-up of the station itself, and three dedicated simulators, called 'Amur', 'Baikal', and 'Dvina'. Amur and Baikal were the station's astro-orientation and astronavigation control posts.[20] Later, a fourth dedicated simulator, 'Kama', was manufactured by specialists from TsPK's Experimental Plant. Kama was meant for training in operating the orbital station's main control post.

The next DOS station, Salyut 6, was launched on 29 September 1977. In order to allow cosmonauts to be trained for missions to the new station, the Don-17K simulator was delivered to TsPK on 28 December 1976. Although it was no more than a training mock-up initially, it was soon upgraded to the status of fully-fledged simulator.

The next station, Salyut 7, had been built as the back-up for its predecessor and was almost identical to it when it was launched on 19 April 1982. Therefore, only minor modifications were necessary before the existing simulator could be used to train crews for flights to that station. All in all, a total of 58 prime and back-up crews trained on the Salyut 6/7 simulator in the ten years it was used.[17]

In the meantime, the old Soyuz spacecraft was being replaced by the first modification, Soyuz T. In order to prepare the crews, a new simulator called TDK-7ST was built by SOKB-LII and delivered to the Cosmonaut Training Centre in 1979.

As well as this new Soyuz simulator, a second and third one were ordered from OKTB Orbita in Novocherkassk. Called 'Don-732' (it must be noted that while all official publications speak of 'Don-732', the simulator itself had 'Don-732M' painted on it until at least 2001. In 2003, this had been replaced by 'Don-Soyuz-TM'), the first of these simulators was specially built to have cosmonauts train for both automatic and manual approach and docking of the Soyuz T with the Salyut station. The other one was named 'Pilot-732' and was a specialised dynamic

A view of Korpus 1, around 1980. Behind the Soyuz simulator in the foreground is the Salyut 6 simulator

simulator to train for manually controlled re-entry on the basis of the TsF-7 centrifuge profile.

Buran

Buran, the Soviet Space Shuttle programme, called for a new range of simulators and the facilities to house them. Training of the cosmonauts for orbital flights would take place in TsPK, but unlike previous programmes, Buran also called for specific flight training for the approach and landing phases of missions. The test programme for these phases was carried out under the direction of the Ministry of Aviation Industry's Flight Research Centre (MAP-LII) in the city of Zhukovskiy, east of Moscow.

In 1977, a group of test pilots was selected and underwent basic cosmonaut training in TsPK. Four of them later performed approach and landing tests on a

The Don-732M simulator

Buran analogue called BTS-2. Training for these tests took place in a Buran simulator that was located at NPO Molniya, the prime contractor for Buran. This simulator, called PDST, was built by SOKB-LII's successor, NII-AO (Scientific Research Institute for Aviation Equipment). The pilots were also able to train on a Tupolev Tu-154 aircraft that was specially modified to be able to fly Buran descent patterns.

According to NII-AO information, they built two more Buran simulators; TDK-F35 and PDST2. These were used from 1982 to 1991 and from 1987 until 1990 respectively[3] and were situated in the KTOK in TsPK. Up to the present day however, there remain three Buran simulators in the KTOK. Who built the third one is unclear.

KTOK was large enough to house a full-scale mock-up of the Buran orbiter, but that was never delivered. In addition, construction of a separate building was begun, for training crews to work with Buran's manipulator system and for training on an orbiter that could be placed in a vertical position. But this training hardware also was never delivered because of the cancellation of the Buran programme and the building was never finished.

Soyuz modifications

Soyuz T 15 was the last flight of Soyuz T. In 1986, the Soviets moved to a new modification of the Soyuz spacecraft called Soyuz TM. A new simulator, designated TDK-7ST2, was therefore set up in Korpus 1A, built by NII-AO in close

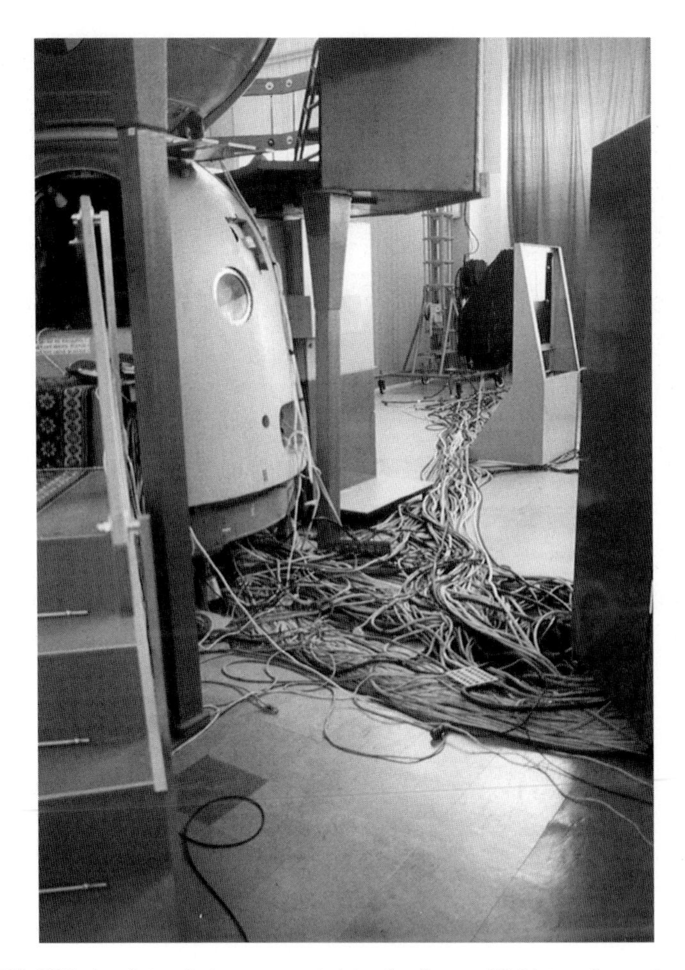

The TDK-7ST simulator being converted to the Soyuz TMA configuration, in 1998

cooperation with engineers from TsPK's simulator department. Commissioned on 4 April 1986, the first training session on TDK-7ST2 was conducted by Vladimir Titov and Aleksandr Serebrov, who at that time were scheduled to fly a long duration mission to the Mir station.

When Soyuz T was phased out, the Don-732 simulator became obsolete, and OKTB Orbita was given the task of modifying it to Soyuz TM configuration. After the reconfiguration, it was renamed Don-Soyuz-TM.

Since Soyuz TM served as the Soviet/Russian crew transportation ship from 1987 until the last in the series, Soyuz TM 34, returned to Earth on 10 November 2002, TDK-7ST2 was used very frequently. On 14 February 2002, the 4,000th training run was performed by the ISS-6 back-up crew at the time, Salizhan Sharipov, Carlos Noriega and Don Pettit. By that time however, it was reportedly becoming problematic to get spare parts for the simulator, as these were no longer being produced.[1]

The Soyuz TM simulator TDK-7TS2

A new player in the field

In November 1993, a new company was founded to build simulators for cosmonaut training. Its name was 'Tsentra Trenazhorostroyeniya' or TsT (Centre for Simulator Building). Two months later, it opened an affiliate in Novocherkassk, into which the OKTB Orbita was incorporated. Orbita had been building simulators for a number of years and had been the prime contractor for Don-732, Don-Soyuz-TM and Don-17K.

THE MIR PROGRAMME

On 20 February 1986, the Mir station was launched from Baykonur. While the transition of training from Salyut 6 to Salyut 7 had been fairly easy since both stations were almost identical, Mir was a new generation space station, which called for new simulators.

Control console for the TDK-7ST2 simulator

The Mir training hall in 2003. In the background is the Vykhod 2 EVA training device

Initially, it was possible to place the simulator for the Mir base block, called Don-17KS, in Korpus 1 of the Engineering and Simulator Building, next to the Salyut simulator. When the first add-on module, 'Kvant', was planned, its simulator (Don-37KE) was attached to the Mir without much problem. However, the modules following Kvant were so large that the Salyut simulator had to be removed in order to make room. The first was Don-77KSD, the simulator of the 'Kvant-2' module. It was followed by Don-77KST ('Kristall').

In 1995 and 1996, two more modules were added to the Mir complex; Spektr and Priroda. Simulators for these modules were put up in TsPK, but there was no more room available in the Engineering and Simulator Building. Korpus 1, the hall which contained the Mir, Kvant, Kvant-2, Kristall and TORU simulators, was filled to capacity, so a new location was found in the KTOK. Even with the three Buran simulators already *in situ*, there was more than enough room left, so it became the home for Don-77KSO (Spektr) and 'Navazhdeniye' (Priroda).

As well as these large simulators, TsT also built Don-GP (Glavnyy Post), a special trainer for operating the main control post of the Mir base block.

A further simulator, called 'Teleoperator' but better known as 'TORU', was set up in Korpus 1 of the Engineering and Simulator Building. It was a replica of the remote control station on board Mir, and was used to train crews in controlling the approach and docking of unmanned Progress supply ships.

After the Mir programme had ended, it looked as if Korpus 1 was destined to become an extension of the TsPK museum. Training activities had been replaced by tour groups and official guests to TsPK being shown around the Mir modules. But in 2001, construction of a new training device began. Called 'Vykhod 2' (Exit 2), this

The Don-77KSO ('Spektr') simulator in the KTOK, October 1999. Barely visible behind it is the 'Priroda' simulator, 'Navazhdeniye'

The 'Teleoperator', or TORU trainer in Korpus 1. Later, Vykhod 2 was set up in the open space in front of TORU

structure could be used by cosmonauts wearing Orlan EVA suits to practice egress and ingress to the International Space Station. After it was finished in 2003, walls were put up between Vykhod 2 and the Mir complex, to ensure that cosmonauts could train in peace and quiet while tour groups were walking around the Mir complex only metres away. Vykhod 2 was built by engineers from TsPK's Experimental Plant in cooperation with TsT. Since TsT describes this trainer as an upgrade, it seems that it was an upgrade of Vykhod 1.[21]

The International Space Station

By this point, the Russians were preparing for the ISS programme as well. Plans called for the first two modules to be Russian; the Functional Cargo Block 'Zarya' and the Service Module 'Zvezda'. An order to build simulators of both modules went out to two prime contractors, TsT in Moscow and NII-AO. Development took place in close cooperation with RKK Energiya and Khrunichev, the prime contractor for both modules.

In the night of 3–4 March 1998, the Zarya simulator was moved from Khrunichev's plant in Moscow to TsPK. On 17 July 1998, it was followed by Zvezda. Both were placed in the KTOK, next to Spektr and Priroda. It seems that both were delivered in a half-finished state, since the final outfitting of the simulators took place in TsPK between mid-1998 and early 1999.[17] Plans for simulators of the Universal Docking Module (which was to be delivered in 1999) and for the Science Power Platform did not materialise.

Both the Zarya and the Zvezda simulators were first used on 23 August 1999, by the ISS-1 prime and back-up crews. Since all ISS Expedition crews, Soyuz 'Taxi' crews, and Space Shuttle crews train on them, it wasn't long before the 1,000th training session took place, on 12 February 2002.

Shortly after the International Space Station had become operational, the Russians began using a new transport craft, Soyuz TMA. Of course, a new simulator was needed to practice operating this new Soyuz modification, so the old STK-7ST simulator was modified from the Soyuz T layout to that of the Soyuz TMA. Work began in 1998, but it was not deemed necessary to move the simulator. Reconfiguration was carried out in Korpus 1A and when it was finished, the simulator was renamed TDK-7ST3. The first training session was conducted on 28 August 2001 by NASA astronaut Don Pettit.

During the transition from Soyuz TM to Soyuz TMA, a TMA Descent Module was 'incorporated' into the Don-Soyuz-TM simulator, enabling both the last Soyuz TM crews and the first Soyuz TMA crews to train for manual dockings. The Russians now had one Orbital Module and two Descent Module simulators next to each other. The first training on the TMA Descent Module was conducted by ISS-5 crew members Valeriy Korzun and Sergey Treshchev on 5 March 2002.

While Pilot-732 was used to train crews for manually controlled re-entry based on the TsF-7 profile, a new simulator was ordered for similar training based on the TsF-18 profile. Although it is not clear where this new simulator is located, it was reported in 2001 that it was being tested.[22]

At present (2005), the TDK-7ST3 is the standard Soyuz simulator that is being used to train the crews for ISS. But developments are on-going and yet another new Soyuz modification (called Soyuz TMM) is being developed by RKK Energiya, although its status is unclear. However, should that craft become operational, a

Control consoles for the Pilot-732 and Don-Soyuz-TM simulators

Pilot-732 is used for practicing approach and manual docking

An overview of Korpus 1A. Don-Soyuz-TM can be seen in the foreground, with Pilot-732 to its right. The Don-7ST3 (Soyuz TMA) simulator is visible in the background

simulator will be needed, and there are already plans to convert the now unused Soyuz TM simulator TDK-7ST2 for that purpose. Once the conversion has been completed, it will be renamed TDK-7ST4.[1]

OTHER SIMULATORS AND MEANS OF TRAINING

Besides these large simulators, TsPK also has a number of smaller ones that were meant for training on specific aspects of Soyuz, Salyut, Mir and ISS. Built by the centre's own simulator department, these were:

- Oka-ARS, to train for manually controlled approach of Soyuz craft
- Yenisey, to train for operating the Soyuz motion control system
- Prichal, for practicing docking of Soyuz with DOS/Salyut orbital stations
- Don-ERA, to train for operating the European Robot Arm (ERA), the ISS manipulator arm
- SPK-Don-21, the purpose of which is unclear, but since SPK is the Russian Manned Manoeuvring Unit, this may have been its training device
- Besides the Hydrolaboratory, the Ilyushins and Vykhod 2, cosmonauts have used the TBK-50 thermal and vacuum chamber which is situated in the space suit design bureau, NPO 'Zvezda', in the town of Tomilino. Here, specific EVA tasks have been practiced in simulated space conditions.
- The training centre also maintains a simulator at the Baykonur Cosmodrome's administrative-technical complex. Built in 1980 by OKTB Orbita, this specialised simulator was called 'Bivni' ('Tusk'). Its purpose was to allow cosmonauts to maintain their skills in manually docking their Soyuz ferry craft, similar to the Don-Soyuz-TM simulator in TsPK. Bivni was later upgraded twice, the modifications being known as Bivni-2 and Bivni-3, in all probability for Soyuz TM and Soyuz TMA.

Parabolic flights

Like NASA in the USA and ESA in western Europe, the Russians use aircraft flying parabolas in order to let cosmonauts train for specific tasks during short periods of zero-gravity. (This is now more correctly referred to as microgravity, but the term zero-gravity remains the more familiar and is used throughout this book.) In the 1960s, a Tupolev Tu-104 was used. This plane was operated by the Gromov Flight Research Institute in Zhukovskiy rather than TsPK. Later, the Ilyushin Il-76 would become the standard plane for these flights and three such planes were added to the fleet of the Seryogin Regiment. Their cabins were modified, with the walls, floors and ceilings being padded to prevent injuries. The planes were designated Il-76MDK and were used to familiarise cosmonauts with zero-g, to train them in moving around in that environment, to don and doff space suits, and to handle large masses. The Il-76s are the largest planes in the world that are used for this kind of training. The cabin measures 14.18 metres long, 3.45 metres wide and 3.40 metres high. The crew usually consists of three pilots, two flight engineers, and eight instructors, who assist

Training photo of the 1970 group in a parabolic aircraft test. The people in the picture are (from the left) unknown, Leonid Popov, Nikolay Fefelov and Vladimir Kozlov

Yuri Gagarin (right) observes Test-Parachutist V.I. Golovin during a training session in the Tupolev Tu-104LL. Golovin is seated in a Vostok ejection seat. On the left is Leading-Specialist L.M. Kitayev-Smyk

a total of twelve cosmonauts or other test subjects. A flight will usually last between an hour and a half and two hours, during which ten to twenty parabolas are performed. Each parabola will result in between 25 and 30 seconds of zero-g.

The plane goes into a dive, developing a speed of 620 kilometres per hour. It then goes into an ascending curve until it reaches an altitude of 7200 metres, when the crew sets the controls to idle and the craft continues to move through inertia along a path resembling a Kepler parabola, during which weightlessness occurs. The aircraft's speed drops to 420 kilometres per hour on the descending branch of the trajectory and the aircraft moves along the curve in horizontal flight. They repeat the cycle every three or four minutes and usually fly ten cycles at a time, but they can do more, depending on the mission to be flown.[23]

Earth Observation

A Tupolev Tu-154 has been modified for specialised observation to train cosmonauts in space navigation and Earth observation of ground- and sea-based objects. They also test space equipment on the aircraft, which is also based at Chkalovskiy and operated by the Seryogin Regiment.

EVA TRAINING FACILITIES

Four years after Yuri Gagarin blazed the trail into space, having called upon more than a year of specialised training, another Soviet cosmonaut, Aleksey Leonov, created history by becoming the first person to venture outside a spacecraft in orbit and walk in space. This new activity required much more specialised preparation – trying to reproduce the effects of zero-gravity while still firmly within the grasp of Earth's gravity. For many years, the Soviets (and the Americans) had studied the possibility of crews working in open space, on the Moon and, eventually, even on Mars. This would all require the development of new equipment and the procedures to support them, as well as techniques to allow potential space walkers to practice their activities on Earth. Both the Soviets and the Americans followed similar paths to prepare their crews for what is officially termed Extra Vehicular Activity (EVA) and more commonly known as spacewalking.[24]

Simulating spacewalking

One of the most effective and sustainable methods of EVA training is to use a large tank of water. Elements of space hardware can be submerged in the tank and test subjects (suitably weighted) can conduct simulated EVA operations. At TsPK, there has been a huge facility available for EVA training for over 25 years and this has allowed cosmonauts to perform regular training sessions over the evolving space station programme. In addition, there are the two Vykhod facilities that cater for 1-g and suspended gravity simulations of procedures and techniques. Combining this with other 1-g facilities for ground run-throughs, the parabolic aircraft flights and sessions in altitude chambers at TsPK, Zvezda and elsewhere, the Russians have been able to train for and complete an impressive log of over 100

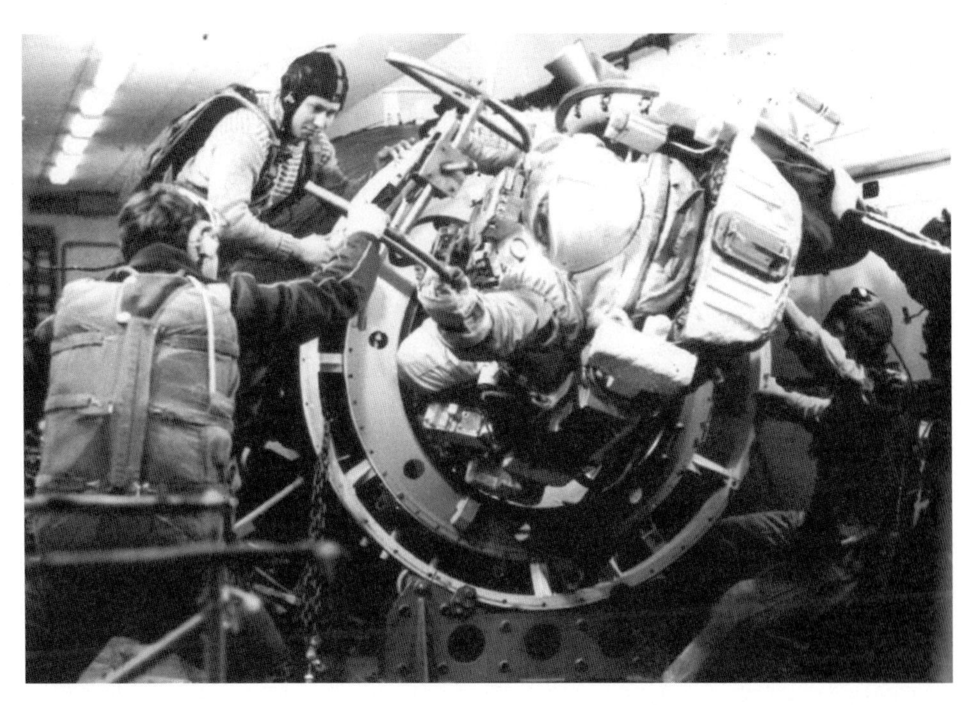

During a short period of zero-g on board an Ilyushin Il-76MDK, a cosmonaut practices egress from the Mir space station while wearing the UPMK manned manoeuvring unit

EVAs, gathering a wealth of experience in long-term EVA operations from space complexes.

Hydrolab (underwater EVA training) Facility

As the emphasis of the Soviet manned space programme shifted from the delayed lunar programme to the development of the long duration space stations in the late 1960s, it became apparent that it would also be possible to perform spacewalks from the stations more frequently than from any previous Soviet manned spacecraft. A suitable, dedicated training facility had to be provided to support such an increased EVA training programme. The current EVA facility, called the Hydrolab, was completed in the early 1980s and features a huge, 23-metre diameter, 12-metre deep water tank filled with 5,000 litres of water maintained at a temperature of around 30°C.[25] Large lamps floodlight the inside of the pool when required and once every 12–18 months, the water is emptied out and the whole facility is cleaned and maintained.

The Hydrolab has been in use since the early 1980s and has supported the EVA training of Salyut 6, Salyut 7, Mir and ISS crews since then. Using an overhead crane system, full-size mock-ups of the Salyut station, modules of the Mir space complex and, more recently, the Russian elements of the ISS, can be fixed on a support frame at the base of the tank for simulated EVA operations. The Hydrolab facility is managed by the Survival Department, whose current Head is former cosmonaut

A training session by cosmonauts Yuri Romanenko and Georgiy Grechko in the old EVA training facility in TsPK, in 1977. From 1980 onwards, EVA training would be conducted in the new Hydrolaboratory

Colonel Yuri Gidzenko. He recently took over from Nikolay Grekov, another former cosmonaut, and Gidzenko is also responsible for both the centrifuge and survival training. The Hydrolab has its own staff of specialists, divers and technicians.

Operationally, a safety and support team of six or seven scuba divers works in the tank, assisting the crew as they follow their test programme. The EVA-suited test subjects descend into the tank by means of a crane, as they are too cumbersome to enter the water themselves. Their activities are recorded by two diver/cameramen, who document their progress as the test proceeds. These films and photos are used for post-simulation evaluation, or are sent up to a resident crew on a station for on-orbit training for new EVA procedures or operations. This has been done on a number of occasions where emergencies have occurred in orbit.

The test crew uses specially adapted suits for underwater work, which have umbilical connections through a mock-up backpack. Weights can be inserted into pockets at the wrist, waist, chest, ankle or back of the suit as required to suspend the test subject at different depths in the tank for the simulated EVA operation that is being performed. The test is monitored by a team of specialists from locations around the exterior of the pool, using view ports to look into the tank. A central control panel is located about halfway down the side of the tank and includes stations for the test conductor, a chief doctor, and an EVA specialist. This team controls the pace and progress of the test being conducted and monitors the parameters of the EVA suits, the medical condition of the cosmonauts, and general safety requirements.

An EVA simulation begins with a briefing for the crew in the conference room located on the second floor of the facility. Here, the principles of the hardware and the EVA time line can be discussed in detail. Mock-up Orlan suits are located in this room for demonstration and illustrative purposes and one wall of this room is adorned with signatures from cosmonauts and astronauts who have visited or trained in the facility, as well as a number of dignitaries who have toured the building.

The planned EVA time line or familiarisation sessions for new equipment are also conducted in these rooms. The EVA crew then receive a medical and proceed to the suiting area near the pool, where the liquid coolant garment, communications cap and inner gloves are donned before they are assisted into the rear hatch door of the

This is the Vykhod 1 facility which allows cosmonauts to get a feeling of weightlessness. It is located on the first floor of the Soyuz hall

Orlan EVA suit. They are supported by the overhead lifting frame for a systems and integrity test and then prepared for immersion in the pool. Once these tests have been completed, the subject is hoisted up and over the pool and then lowered into the water. After release from the framework, support divers move the subject to the prescribed area in the pool or on the mock-up where the test is scheduled. At the end of the test, the procedures are reversed and, after leaving the suit, the test subject takes a hot shower in the nearby changing rooms and then attends a debriefing session in the adjoining rooms, along with the test controllers, doctors, cosmonaut trainers, EVA specialists and other cosmonauts. The Orlan suits are provided by the Zvezda design bureau, but more recently, US (Space Shuttle) space suits have also been stored at the facility and have occasionally been used in EVA simulations within the tank.

This facility is also used for training new cosmonauts in the techniques of EVA; for proficiency training of veteran cosmonauts; for the development of new procedures; evaluation of equipment and hardware; and support for EVA operations being conducted or planned for current missions. Just outside the facility, mock-ups of old hardware are left open to the local weather conditions. The Hydrolab has been extended recently, with a storage area being constructed and linked to the tank itself by a girder rail system. The new building stores space modules, which are craned in and out as required.

Vykhod 1 (EVA 1-g/zero-g) facility

This facility is located on the first floor between the Mir and Soyuz Halls (Korpus 1/1A). It is a facility in which cosmonauts can don Orlan pressure suits and be suspended from the ceiling by pulleys to simulate weightlessness. This is a very small room, with space for only a couple of specialists and a cosmonaut. There is space for only two EVA suits and very limited room to conduct tests and experiments. When not being used to support training, it is a facility that visitors to the centre are taken to and given the opportunity to sample the experience of wearing an EVA suit. Recently, the Russians have developed a second facility, Vykhod 2. It is not clear if this has meant that Vykhod 1 has been closed, or that it will be used to supplement the new facility.

Vykhod 2 (EVA 1-g/zero-g) facility

Located between the Mir mock-up and the TORU simulator in the 'Mir Hall' is the training facility described as Vykhod 2 (or Exit 2, or EVA 2). This was added to the hall in early 2001 to assist in the EVA training of cosmonauts to operate and egress or ingress from the hatches on the ISS. It focuses on the Russian EVA hatches on the Pirs Docking compartment, since that is the only one replicated in actual simulated hardware. (The hatches on the American section are only simulated by two full-sized colour photographs). Two Orlan suits are suspended from overhead cranes and a twin boom suspension system (labelled MOST 1 and MOST 2) allows trainees to reproduce zero-g and movement across a simulated worksite in several directions, independently of each other. The work site includes hand grips and retention devices (foot restraints) and allows practice in moving across the surface of a space station for a few metres.

The Vykhod 2 EVA training device, as seen in June 2003. Shortly afterwards, walls were put up around it so cosmonauts could train on it while guests and tourists visited the nearby Mir simulators

RUSSIAN EVA TRAINING DEVELOPMENT (1964–2004)

Cosmonaut Aleksey Leonov was the first person to perform a spacewalk in March 1965, and at that time, the facilities to prepare for such a pioneering activity were, at best, basic.[26]

Voskhod EVA training: For the first exit into space, training focused on simply operating and wearing the special EVA suit, preparing to exit and enter the airlock, and operating the hatches and the pressure controls. In addition to 1-g familiarisation at the Zvezda simulation facility, training was also conducted in parabolic flights and in altitude chambers (also at Zvezda). There was no Hydrolab training, as it had not been introduced at this time, and the benefits of underwater activities to simulate future spacewalks were only beginning to be evaluated. Leonov has described his own physical training programme to supplement the formal EVA training. In the twelve months prior to the EVA, his training included running over 500 km, cycling over 1,000 km and skiing over 300 km, in addition to hundreds of hours in the gymnasium. Though each cosmonaut still conducts a personal physical training programme to maintain peak fitness, nothing as demanding as this is required for current EVA activities.

Leonov and his colleagues also used the Voskhod 2 simulator, TDK-3KD, for systems training and familiarisation in support of pre- and post-EVA activities. In order to verify the dimensions of the airlock and hatches, the correct and practical location of equipment in the spacecraft, and to evaluate planned crew movements, a wooden airlock mock-up was built and tested in July and August 1964 by test pilot (and future cosmonaut) Sergey N. Anokhin of OKB-1, wearing the Berkut suit mock-up. Also in August 1964, cosmonauts Leonov, Belyayev, Khrunov and Gorbatko completed fit checks of the descent vehicle and airlock. In developing the first EVA, an extensive programme of tests and simulations was developed, using both Zvezda engineers and cosmonauts. These included 1-g simulations at 'sea level' conditions, the use of thermal vacuum chambers at Zvezda, water flotation tests and thermal mock-up tests at OKB-124, and use of the TBK-60 thermal vacuum chamber at the Air Force Scientific Research Institute (GK-NII). These tests were completed by February 1965. Flight tests including short periods of weightlessness were completed on a Tu-104 flying laboratory, with training for the exit conducted at the Gromov Flight Research Institute (LII), under the direction of Leading Engineer E.T. Berezkin.[27] The difficulties Leonov encountered in trying to retrieve the camera prior to entering the airlock during the mission were attributed to the obvious lack of previous EVA experience (this was the world's first spacewalk) and the inability to adequately simulate weightlessness during training on Earth.

Soyuz EVA training: In addition to 1-g simulations, altitude chamber runs and physical training programmes, the cosmonauts assigned to the Soyuz-to-Soyuz transfer completed a number of training flights in the IL-76 Flying Laboratory, flying parabolic curves. These simulations were designed for operations during the Soyuz 1/2 docking flight planned for April 1967, but that was delayed due to the in-flight difficulties with Soyuz 1 prior to the launch of Soyuz 2. The techniques of EVA transfer between Soyuz craft were evaluated for the first time by cosmonaut Valeriy Bykovskiy and engineer (later cosmonaut) Vladimir Aksyonov in 1966, three years before the EVA was achieved during the Soyuz 4/5 docking. A partial mock-up of the Orbital Module hatch was used for exit and entry simulations wearing the Soyuz EVA suit and, in a separate section, a partial mock-up of the linked spacecraft was used for practicing hand-over-hand translation. This was also a demonstration, in part, of the proposed and subsequently cancelled LK/LOK EVA transfer during a manned lunar landing mission. As with the Voskhod 2 EVA experiment, engineers and cosmonauts performed an extensive suit and EVA technique development programme, and this would become a factor in all future development. Thermal testing in the TBK-30 vacuum chamber at Zvezda focused on the suit and EVA life support systems, while activities in a mock-up Orbital Module of the Soyuz were evaluated and practiced in the GK-NII thermal vacuum chamber. In the Tu-104 flying laboratory, parabolic flight profiles were used to test the sequence and method of crew transfer from one mock-up Soyuz OM to another at the LII (Flight Research Institute).

Lunar EVA training: During the 1960s, the Soviets developed their own manned lunar landing programme, with plans to include in-flight EVAs and surface explorations by single cosmonauts. This programme was cancelled in 1974 and very little has been revealed about the EVA training that would have taken place to support the planned operations. Taking the involvement of American NASA astronauts in the development of both the Apollo lunar surface EVAs and the suits as an example, some as yet unidentified cosmonauts must have assisted in 1-g, rig-supported or parabolic ⅙ gravity EVA simulations; the development of procedures and techniques for in-space transfer operations in lunar orbit; fit and function, and exit and entry simulations to determine operational guidelines and surface time lines on the Moon; and simulated surface EVA activities to deploy experiments, collect samples and take photographs.

A number of American astronauts started working on Apollo LM issues in 1963 and continued for the next decade until the end of the programme. They worked with Grumman, the primary constructor of the LM; Bendix, the contractor for the surface experiments; Boeing, the contractor for the Lunar Roving Vehicle; and International Latex Corporation, the contractor for the pressure garments. Reflecting their completely different approach to crew participation, the Soviet 'system' did not allow for detailed crew support in the development of space hardware until later in the programme. In addition, setbacks to the manned lunar programme could be the reason that such involved participation was never really developed. There has been film released of LM ladder descent simulations, but these were probably conducted by test engineers rather than cosmonauts. There does seem to have been a facility (probably located at Zvezda) where 'cosmonauts' were suspended by pulleys and harnesses and then walked around a circular 'terrain' in the open. This facility has been seen on many occasions on film. It consisted of a tall tower with a horizontal arm that could turn 360°. A set of cables were suspended from the end of the arm, which could be attached to retention devices on a specially modified lunar EVA suit. The 'cosmonaut' would be held almost horizontal, with their feet on an inclined bearing wall, allowing them to 'walk' across the smooth surface of the wall with the cables and pulleys suspending up to ⅚ of their Earth weight and leaving the ⅙ 'lunar simulated' weight for evaluating methods of mobility in suits and footwear. Cameras mounted on the top of the tower could then film the action from the side profile to aid the evaluation of walking methods and equipment mobility. Technical tests were also carried out in the TBK-30 vacuum chamber and in the Hydrolab from 1971. Finally, there is nothing to suggest that cosmonauts participated in geological training, something that *all* American astronauts did from 1963 – well before crew assignments and some six years before the first landing.

According to Zvezda records,[28] the Orlan suits intended for weightless EVA during a lunar mission were first manufactured in 1967 after several years of development. Due to the fact that cosmonaut training for EVAs was being developed using the Hydrolab, special versions of the Orlan suit were developed to simulate in-space EVAs. These suits included weights for neutral buoyancy, attachments for a hoisting system to place the cosmonaut in the pool or bring him out, and a simplified life support system provided from the pool side. From 1969, fit and function tests of the Krechet-94 lunar EVA suit began, including work in the mock-up lunar lander located at OKB-1, which tested the shock absorber system and cosmonaut attachment and restraints designed to support the cosmonaut at the point of lunar touchdown on the Moon. Work on a ⅙ test bench located at Zvezda began in 1968

EVA training for Salyut 6, in the old training facility. This picture was probably made during the same training exercise as the one on page 54

and again, simulated short-term weightless (and possibly $\frac{1}{6}$ -g) tests were conducted using a Tu-104 flying laboratory. In addition, vacuum chamber tests were conducted using empty suits, manikins and human test subjects. The work on lunar suit development was suspended in 1972 and cancelled in 1974 with the termination of the N-1/L-3 manned lunar programme.

Mars EVA simulations: A Soviet/Russian desire for the human exploration of Mars has existed for decades, but with no firm programmes, there has been little specific training for such a venture, although many Russians do link the experiences from long-term ground simulations. Simulation facilities for surface exploration of Mars have been supported by Zvezda in Russia and in the US. It is thought that only test-engineers, rather than cosmonauts, have participated in these suit and procedures tests, but this remains an area of future possibilities for EVA training and evaluation.

Space Station EVA training: The use of the Hydrolab at TsPK has supported all space station EVA operations from Soviet/Russian space station elements since the early 1980s. There has also been a programme of 1-g training, air-braking table

A relic from the past. Outside the Hydrolaboratory, the Soyuz Orbital Module that was used for EVA training is now slowly deteriorating in the harsh Russian winters

simulations and parabolic flight tests. The first (unplanned) Russian station-based EVA was performed from Salyut 6 in December 1977 to inspect the forward docking port, following the failure of the Soyuz 25 spacecraft to hard dock with the station a few weeks before. Though rumours indicate there were at least discussions and/or simulations of EVA from earlier stations, none were performed. The 1-g facilities at both TsPK and Zvezda, as well as parabolic aircraft flights, have continued to be used to expand EVA training experience in support of Russian ISS-based EVAs.

To accommodate cosmonaut training in space station operations, Zvezda commissioned a series of training and simulation suits for all variants of the Orlan EVA suits (Orlan D, DM, DMA and M). These included:

Orlan-B (Russian 'V' for Ventilation), used for evaluating fit check work aboard flying laboratories, and ground work. They feature ventilation and positive pressure resources from onboard sources instead of from suit systems. This design also includes a suit enclosure and backpack housing only.

Orlan-GN (a Russian acronym for 'hydraulic weightlessness') is the version of Orlan suit used in the Hydrolab at TsPK. They are almost identical to the flight suits, except that the outer protective garment is not installed and the life support system is supplied via ground sources, not from the backpack, although there is an emergency supply of fifteen minutes air for contingency purposes. There are also weights for positioning the cosmonaut at different depths, and several components have been adapted for use underwater over repeated cycles. When Russia joined the ISS programme in 1993, Zvezda manufactured two Orlan-M-GN suits for use by NASA at the Sonny Carter Water Immersion Facility at Ellington AFB, near JSC in Houston, Texas. These feature interfaces suitable for American adapters and connections.

Orlan-T (training) is the version of the EVA suit used at the Vykhod 2 facility at TsPK and is also used for cosmonaut training in airlock procedures at sea level with out decreasing ambient pressures. This suit was commissioned by the Government Resolution dated 29 September 1985. It is a normal EVA suit with modified ventilation and life support systems. Cosmonauts practice various emergency situations in this suit, including failure of suit component units, leakages and activating suit warning systems. The cosmonauts also use mock-up suits for servicing and maintenance work they have to conduct on orbit to extend the life of operational suits on the space station.

INTERNATIONAL EVA TRAINING AND OPERATIONS

The Hydrolaboratory's own logo

Between 1965 and 2005, forty-nine cosmonauts performed EVA from Soviet or Russian spacecraft and EVA facilities. In addition, a few Russian cosmonauts have used the EVA training facilities at NASA JSC to support Shuttle-based EVAs from Mir and ISS. A number of American, European and Japanese astronauts have also completed EVA training at TsPK, with an increasing number using Orlan pressure garments to conduct EVAs from the Russian elements of ISS. This international expansion of cosmonaut EVA training at TsPK and at foreign training centres in the United States (WET F facilities at JSC in Houston) in Europe (at ESA's Astronaut Training Centre in Cologne) and in Japan (at the Tsukuba Space Centre in Obaraki) will give the cosmonauts greater opportunities to train for a range of cooperative, as well as national, EVA objectives.[29]

Russian STS EVA training and operations

In preparing for flights on the US Shuttle, two cosmonauts (Sergey Krikalev and Vladimir Titov) were the first to complete NASA EVA training at JSC. Though no EVA was performed on their first flights on the US Shuttle, Titov did complete the first international EVA (5 hrs 1 min with Scott Parazynski) from an American Shuttle (Atlantis) on 1 October 1997, while docked to the Mir space complex. As part of his Shuttle-Mir training, Titov had logged 137 hours of US EMU EVA training, while Krikalev had accumulated 24 hours and Yelena Kondakova completed 13 hours of Shuttle EVA training. As part of the STS-106 crew, Yuri Malenchenko conducted a second Shuttle-based Russian EVA from Atlantis (10 September 2000, with Ed Lu) while docked to ISS. Other cosmonauts have received briefings and familiarity training with US Shuttle/ISS-based EVA equipment and procedures.[24]

French Spationaut Salyut EVA training

The first non-Soviet cosmonaut to complete Orlan DMA EVA training with the intention of performing an excursion in space was French spationaut Jean-Loup Chrétien, for his 1988 mission to Mir. He was backed up on that mission by Michel Tognini, who also completed the Orlan DMA EVA training in support of the *Aragatz* mission. Chrétien completed one EVA (5 hrs 57 min) with Aleksandr Volkov on 9 December 1988. Jean-Pierre Haigneré qualified for Orlan M EVA operations and completed one EVA (6 hrs 19 min on 16 April 1999 during the *Perseus* mission) from Mir with Viktor Afanasyev. He was backed-up on this mission by Claudie Andre-Deshays, who also qualified in the Orlan M suit and became the first non-Russian female to train for EVA operations.

EuroMir EVA training

German astronaut Thomas Reiter completed EVA training for Orlan DMA suit operations and logged two periods of EVA outside Mir, on 20 October 1995 (5 hrs 16 min with Sergey Avdeyev) and 8 February 1996 (3 hrs 6 min with Yuri Gidzenko). Reiter's back-up for the EuroMir 95 mission was Swedish astronaut Christer Fuglesang, who also qualified for Orlan DMA EVA operations at TsPK.

American Orlan training and operations

Several American astronauts have also qualified to use Orlan DMA and Orlan M EVA suits. Between them, they have performed three EVAs from Mir and, to date (Apr 2005), twelve EVAs at ISS using Russian equipment and facilities. The NASA Phase 1 Programme Joint Report,[30] included a detailed explanation of qualifying a foreign crew member for Orlan suit operations (from Mir).

Linenger and Foale: During 10–28 June 1996, these two astronauts completed seven theoretical and practical classes (dry) and five sessions in the Hydrolab pool, wearing

Still wearing their special undergarments, NASA astronauts Jerry Linenger and Mike Foale (right) discuss an EVA training session with their instructors (June 1996)

Orlan DMA-GN suits and following standard EVA procedures. Working in the pool with Mir mock-ups (DM, Spektr, and core module mock-ups), the astronauts wore both Orlan DMA-GN and Orlan-M-GN suits, and dimensional mass mechanically operated mock-ups of hardware and EVA systems. The two astronauts also completed two training sessions each in the pool and two practical classes relating to EVA target tasks (instrument attachment and removal). The training teams were Tsibliyev and Linenger (main crew) and Budarin and Foale (back-up crew).

Foale: Ground training for Foale for the unplanned EVA on 6 September (with Anatoliy Solovyov) to inspect the depressurised Spektr module was not completed. He utilised his back-up training to Linenger, his previous US EVA experience and training, and on-orbit revision to prepare for the task.

Wolf and Thomas A: These two astronauts completed training (no dates given in the report) in the pool on Mir mock-ups, using scuba gear and in Orlan DMA and M spacesuits. Scuba training was not conducted, as NASA astronauts qualify for this during their Ascan training programme. When Wolf was assigned to EVA training, his previous experience of working in the EMU at the JSC WET-F was taken into consideration, and his experience in scuba gear during EVA simulations made it possible to reduce the number of simulations Wolf had to perform. Both Wolf and Thomas also completed short-term weightless training in the IL-76MDK flying laboratory aircraft. During standard EVA training operations for the DMA/M spacesuits, as well as the EVA programme and procedures, Wolf and Thomas completed three practical classes each, totalling ten hours. They also completed four checkout submissions wearing scuba gear, and practical training in scuba gear for standard EVA operations (16 hours). For underwater training in the Orlan DMA/M suit for standard EVA operations, Wolf was submerged four times (16 hours). They also completed one flight (4 hours) in the IL-76MDK, where they learned to work with the Orlan suit in weightless conditions, including donning and doffing. (They also took the opportunity to become familiar with the Soyuz Sokol suit in the same way.)

Thomas A and Voss J.S: EVA training for these two astronauts took place between 30 September and 30 November 1997, again using scuba gear and Orlan DMA/M type spacesuits. Again when assigned to the EVA programme, their previous NASA training in scuba gear and EMU space operations was taken into account when planning their Orlan training programme. For theoretical and practical training for standard operations, terminology, tasks, training resources and science hardware, Thomas took nine classes (13 hours) and Voss, ten classes (16 hours). For practical training using Russian scuba gear, Thomas logged three training sessions (9 hours), while Voss completed four training sessions (12 hours). Both astronauts logged four sessions each while wearing the Orlan DMA/M suit (16 hours).

US Astronaut Mir/Orlan EVA training

When Progress M-34 damaged the Spektr module of Mir in the summer of 1997, the need arose for NASA astronauts to qualify in Orlan EVA suits on subsequent residency missions, in the event that further repairs were required. Mike Foale was already aboard the station and had qualified by training as back-up to Linenger. At the time of the incident, Wendy Lawrence was in training to take over from Foale, but it was found that she was too short to qualify in the Orlan suit and had never trained for Shuttle EVA operations. She was replaced by her back-up Dave Wolf, who had to undertake an abbreviated training programme with Andy Thomas. When Wolf replaced Foale on Mir, Andy Thomas trained for the last NASA astronaut residency, with Voss as his back-up. Voss had also undertaken Orlan EVA training specifically for this role and this would help on his ISS assignment three years later. The sizing issue for American astronauts would also have an effect on future Soyuz and EVA training for long duration ISS crews and Soyuz rescue craft qualification. Despite this hurdle, a number of lessons were learned from Shuttle-Mir EVA operations that would apply to the forthcoming ISS EVA training of American (and other) resident crew members that had to qualify for Orlan operations from the Russian segment.

As a result of training Russian-American EVA crews, a number of practical skills were acquired by NASA astronauts:

- Theoretical knowledge and practical skills of working with Russian scuba gear
- Theoretical knowledge and practical skills in donning and removing variants (DMA/M) of the Orlan EVA suit
- Working in Orlan pressure garments in flying laboratory aircraft and the Hydrolab
- Practicing elements of the EVA time line in accordance with flight data files
- Practicing contingency off-nominal situations in accordance with flight data files
- Installation and retrieval of instruments on the exterior of the station

- Working with elements of the pressure garment, including maintenance, checkout, cleaning and use of the EVA tool kit
- Working with mock-ups of space hardware in the Hydrolab
- Working as a team member with Russian training staff, divers and cosmonauts in EVA training scenarios.

ISS Russian Segment EVA training

Training for International Space Station EVA operations has drawn upon years of experience from the Salyut and Mir programmes. From 2001, most of the EVAs conducted by resident crews at ISS have been via the Pirs docking module using Orlan suits. This has been necessary because of the tasks to be performed from the Russian segment by resident crews, the grounding of the Shuttle following the February 2003 Columbia accident, and in-orbit difficulties with the American EMU units which can only be used via the Shuttle airlock or Quest docking module. All Russian cosmonauts are trained on Orlan EVA operations and facilities at ISS and international members of resident crews (up to April 2005, only Americans) are required to qualify in Orlan suits as well as US EMU units. Updated procedures and time lines are implemented as required according to programme requirements, but basic Orlan suit training by international astronauts has been described recently (February 2005) by NASA astronaut Clayton Anderson, who began long duration resident crew training in January 2004. As part of his online journal, he commented about the status of Orlan EVA training for ISS.[31]

Yuri Gidzenko and Bill Shepherd, crewmembers for the first expedition to the International Space Station, conduct EVA training in the Hydrolaboratory, on 14 September 2000

With fellow NASA astronaut Sunita Williams, Anderson has conducted an Orlan training course that will one day lead to an EVA from ISS. The course includes theoretical training on how the suit is designed and works. As part of astronaut training at NASA JSC, the astronauts had become familiar with the current Shuttle EMU suit, airlock and Quest airlock on ISS and, along with the scuba diving training, this experience is useful for quickly becoming acquainted with the Orlan equipment. Anderson wrote about the 'basic skills' required before anyone is allowed to enter the Hydrolab, a facility that is not available in US EVA simulations. According to Anderson, the Russian simulation facility (Vykhod 2) 'provides a suspension system, connected to the suit, which offsets our weight a bit here on Earth, giving us a rough simulation of zero-gravity... at least from the perspective we have from within the suits. In addition, this simulator has a mock-up of the Russian airlock segment (Pirs, or Docking compartment 1) and its requisite panels, as well as a hatch. The combination of all these capabilities allows us to practice some basic space-walking techniques while wearing the Orlan, such as depressurisation (and re-pressurisation) procedures, opening and closing the airlock hatch, use of tethers and most importantly, executing the steps used in the event of equipment malfunctions. In a nutshell, we can do everything we need to prepare us for going out the door.'

JAXA candidate EVA training

As ISS operations expand, so will the training for missions to the facility. For some years, a team of three Japanese astronauts has been conducting ISS training for flights on Shuttle and Soyuz spacecraft to operate Japanese equipment on the station (including the Kibo science facility), as part of a visiting or long duration residency mission. In February 2004, astronauts Satoshi Furukawa, Akihiko Hoshide and Naoko Yamazaki received lectures on the Orlan spacesuit and Russian EVA operations. They participated in EVA training wearing the Orlan training facility (Vykhod 2), where they learned basic procedures for EVA in both emergency and nominal situations with the spacesuit and airlocks. This 1-g training was followed by scuba diving to determine their level of diving skills in advance of later EVA training wearing Orlan suits in the Hydrolab (which still had to be completed at the time of writing – April 2005).[32]

Manned Manoeuvring Units

In February 1990, two Soviet cosmonauts (Aleksandr Viktorenko and Aleksandr Serebrov) flight-tested a UPMK (Equipment for Cosmonaut Transference and Manoeuvring Unit) during tethered EVAs from the Kvant 2 module of Mir. The unit, designated 21KS, was flown by the cosmonauts on 1 and 5 February 1990 before being stored for many years inside the Kvant module. It was finally relocated outside in a support frame, where it remained, unused, until it was destroyed upon the re-entry of the space complex in March 2001. In order to operate this unit in space, special training devices had to be created to simulate the operation of the unit on the ground.

UPMK for Voskhod/Almaz: In the early 1960s, the first cosmonaut manoeuvring unit was developed by Zvezda.[33] The plan, based on a government directive of 27

July 1965, was to test fly it during an additional Voskhod mission that was subsequently cancelled. The project then transferred to the Almaz military space station programme in a directive dated 28 December 1966. During 1966, several test units were produced for developmental work and a developmental prototype test was conducted during 1968, but in 1969, work was terminated when no specific task could be assigned for the unit. Though testing had been completed at Zvezda by staff engineers, it is unclear if any cosmonauts were assigned to help with the project or if any training was accomplished. The unit is located in the museum of the Zvezda design bureau

21KS for Mir & Buran: It was not until 22 March 1984 that a new directive was issued to develop a self-contained manoeuvring unit (21KS) for use on Mir and the Buran shuttle programme. This directive came the month after NASA had demonstrated its MMU from STS 41-B to great success. Once again, Zvezda was the prime contractor and the Ministry of General Engineering Industry and the Ministry of Defence were the 'customers'.[34] Testing of the unit occurred between 1986 and 1989. Special test facilities were established and in November 1988, cosmonauts Serebrov and Viktorenko represented the cosmonaut team in manned tests and evaluation of the system. The two cosmonauts and Zvezda engineers controlled the 21KS by visual evaluations of their attitude and with various angular and linear accelerations and rates to determine the accuracy of 'flying' the vehicle in the test rigs. The cosmonauts tested the attitude control system in automated, manual and contingency modes. Between 1987 and 1989, there were 37 Hydrolab tests of the unit, 32 flights in the flying laboratory aircraft and a large number (unspecified) of cosmonaut training sessions. These included sessions in the Zvezda Aerostatic Support training devices, the Polosa training device and the Don simulator at TsPK. The flight unit was launched aboard Kvant 2 on 26 November 1989. Using a safety tether in the event of a system malfunction and the inability of the Mir complex to manoeuvre and pick up the stranded cosmonaut (as the Shuttle was able to accomplish if required), Serebrov flew the unit for forty minutes to a distance of 38 metres from the hatch on 1 February 1990. Viktorenko completed a 93-minute flight to 45 metres from the station on 5 February 1990. Despite a flawless performance and the cosmonauts' praise for the unit, it was never used again.

SAFER & ISS: With Russia joining the ISS programme, the expansion of EVA activities required improved EVA safety over the more traditional tethers used for the past 25 years on Salyut and Mir stations. As a result, from 1998, Zvezda began the development of a Russian Simplified Aide For EVA Rescue (SAFER), a smaller MMU that could be used for the safe return of a crew member to the station. It was designed to be attached to the back of the Orlan M suit. Though developed separately from the NASA EMU-SAFER but with initial financial support from the American agency, as much duplication was incorporated into the Russian unit as possible to ease crew training for the units. Russian SAFER training is integrated into cosmonaut Orlan M training for ISS flight assignments. It has been passed as flight ready and is awaiting transportation to ISS on a shuttle mission.[35]

Table 2 Cosmonaut International EVA Log

Sequence	*Date*	*Russian*	*International*	*Country*	*Mission*	*Spacecraft*	*EVA Suit*	*Duration*
1	1988 Dec 9	Volkov, A.	Chrétien	France (CNES)	EO-4 Mir/Aragatz	Mir Node	Orlan DMA	5 hr 57 min
2	1995 Oct 20	Avdeyev	Reiter	Germany (ESA)	EO-20 Mir/EuroMir 95	Mir/Kvant 2	Orlan DMA	5 hr 16 min
3	1996 Feb 8	Gidzenko	Reiter	Germany (ESA)	EO-20 Mir/EuroMir 95	Mir/Kvant 2	Orlan DMA	3 hr 06 min
4	1997 Apr 29	Tsibliyev	Linenger	America (NASA)	EO-23 Mir/NASA-4	Mir/Kvant 2	Orlan M	4 hr 48 min
5	1997 Sep 6	Solovyov, A.	Foale	America (NASA)	EO-24 Mir/NASA-5	Mir/Kvant 2	Orlan M	6 hr 00 min
6	1997 Oct 1	Titov, V.	Parazynski	America (NASA)	STS-86/Shuttle Mir 7	OV-104 Atlantis	STS/EMU	5 hr 01 min
7	1998 Jan 14	Solovyov, A.	Wolf	America (NASA)	EO-24 Mir/NASA-6	Mir/Kvant 2	Orlan M	6 hr 38 min
8	1999 Apr 16	Afanasyev	Haigneré	France (CNES)	EO-26 Mir/Peresus	Mir/Kvant 2	Orlan M	6 hr 19 min
9	2000 Sep 10	Malenchenko	Lu	America (NASA)	STS-106 (ISS)	OV-104 Atlantis	STS/EMU	6 hr 14 min
10	2001 Jun 8	Usachev	Voss J.S.	America (NASA)	EO-2/ISS	ISS/Zvezda Node	Orlan	0 hr 19 min (IVA)
11	2001 Nov 12	Dezhurov	Culbertson	America (NASA)	EO-3/ISS	ISS/Pirs	Orlan M	5 hr 04 min
12	2002 Jan 14	Onufriyenko	Walz	America (NASA)	EO-4/ISS	ISS/Pirs	Orlan M	6 hr 03 min
13	2002 Jan 25	Onufriyenko	Bursch	America (NASA)	EO-4/ISS	ISS/Pirs	Orlan M	5 hr 59 min
14	2002 Aug 26	Korzun	Whitson (F)	America (NASA)	EO-5/ISS	ISS/Pirs	Orlan M	4 hr 25 min
15	2004 Feb 26	Kaleri	Foale	America (NASA)	EO-8/ISS	ISS/Pirs	Orlan M	3 hr 55 min
16	2004 Jun 24*	Padalka	Fincke	America (NASA)	EO-9/ISS	ISS/Pirs	Orlan M	0 hr 14 min (terminated)
17	2004 Jun 30	Padalka	Fincke	America (NASA)	EO-9/ISS	ISS/Pirs	Orlan M	5 hr 40 min
18	2004 Aug 3	Padalka	Fincke	America (NASA)	EO-9/ISS	ISS/Pirs	Orlan M	4 hr 30 min
19	2004 Sep 3	Padalka	Fincke	America (NASA)	EO-9/ISS	ISS/Pirs	Orlan M	5 hr 21 min
20	2005 Jan 26	Sharipov	Chiao	America (NASA)	EO-10/ISS	ISS/Pirs	Orlan M	5 hr 28 min
21	2005 Mar 28	Sharipov	Chiao	America (NASA)	EO-10/ISS	ISS/Pirs	Orlan M	4 hr 30 min
							Total	100 hr 47 min

* This EVA was cut short due to a problem with Fincke's primary oxygen tank in his pressure suit. Mission managers decided to reschedule the EVA for 30 June.

(F) = Female

REFERENCES

1 Boris Yesin, 'Zemnye sobratya, kosmicheskikh korabley', Novosti Kosmonavtiki No. 4 (231), 2002, pp. 52–53
2 'RGNII-Tsentr Podgotovki Kosmonavtov imeni Yu. A. Gagarina', Moscow, Kladez-Buks, 2002, page 125
3 Institute of Aircraft Development (NII-AO) website http://www.niiao.ru/space/sokb_-hi.htm
4 N. Kamanin, 'Skrytyi kosmos: kniga vtoraya 1964–1966', Infortekst, Moscow 1997; entry for 26 June 1965
5 Ref 4, entry for 13 August 1965
6 Ref 4, entry for 26 February 1964
7 Ref 4, entry for 17 February 1966
8 Resolution # 101 of 27 April 1966
9 N. Kamanin, 'Skrytyi kosmos: kniga vtoraya 1967–1968', Infortekst, Moscow 1999; entry for 2 August 1967
10 Ref 9, entry for 2 December 1967
11 Ref 9, entry for 3 December 1967
12 Ref 9, entry for 17 January 1968
13 Ref 9, entries for 30 January and 6 February 1968
14 Ref 9, entry for 31 March 1968
15 N. Kamanin, 'Skrytyi kosmos: kniga vtoraya 1969–1978', Infortekst, Moscow 2001; entry for 17 November 1970
16 Rex Hall and Bert Vis interview with Nikolay Mikhailovich Kopylov, Star City, August 2004
17 B.M. Yesin, 'Trenazhory rossiyskogo segementa KKS v TsPK im. Yu. A. Gagarina', Novosti Kosmonavtiki 17–18/1998, pp. 58–59
18 V.A. Romanov, 'The 'Analogue': the ground-based double of Almaz', Tribuna, NPO-Mashinostroyeniya's (formerly OKB-52) in-house newspaper, unknown date
19 Rex Hall and Bert Vis interview with Valeriy Makrushin, Star City, 11 April 2001
20 Gagarin Cosmonaut Training Centre website http://www.gctc.ru/eng/facility/default.htm
21 'Tsentra Trenazhorostroyeniya' (Center for Simulator Building) website http://www.asrdc.tpark.ru
22 'RGNII-Tsentr Podgotovki Kosmonavtov imeni Yu. A. Gagarina', Moscow, Kladez-Buks, 2002, page 129
23 Zero-gravity Training Exercise for Space Mission Candidates, Red Star, 26 July 1991
24 For a detailed account of the techniques and development of EVA see Walking in Space, David J. Shayler, Praxis/Springer-Verlag, 2004
25 Astro Info Service notes, Bert Vis notes, from a personal tour of the Hydrolaboratory, June 2003
26 The Rocket Men, Springer-Praxis 2001; Soyuz: A Universal Spacecraft, Springer-Praxis 2003, both by Rex Hall and David J., Shayler; Russian Spacesuits, Isaak P. Abramov and Å. Ingemaar Skoog, Springer-Praxis, 2003; and Walking in Space David J. Shayler, Springer-Praxis 2004
27 Russian Spacesuits pp 59–79, previously cited
28 Russian Spacesuits pp 99–124 previously cited
29 Walking in Space, D. Shayler, previously cited
30 NASA SP-1999–6109, January 1999, in English

31 ISS Expedition Journal – Training, Chapter 11 February 2005, by Clayton Anderson, http://space flight.nasa.gov/station/crew/andersonjournals/training11.html
32 JAXA ISS Astronaut Activity Report, February 2004, http://iss.sfo.jaxa.jp/astro/report/2004/0402_e.html
33 Russian Spacesuits pp 193–197 previously cited
34 Russian Spacesuits pp 197–206 previously cited
35 Meeting between S. Abramov and Rex Hall, Zvezda Design Bureau, Moscow, 16 April 2002

Other national and international facilities

From 1960 until the early 1990s, all the training of Soviet and Russian cosmonauts had taken place at TsPK, or at other Russian and east European facilities, such as contractors, design bureaus and academic institutions. Apart from a few short training trips to the United States for the Apollo-Soyuz Test Project, there was no formal foreign cosmonaut training programme until the creation of Shuttle-Mir. Due to the nature of that programme, the cosmonauts would have to become acquainted with the American Space Shuttle, its systems, facilities and launch and landing systems, with training sessions held at JSC in Houston and KSC in Florida. In addition, as a precursor to ISS, training on partner components in Europe, Canada and Japan has seen an expansion of cosmonaut space flight training around the world, expanding the scope of experience and cooperation like never before. A brief summary of each of these 'foreign' training facilities is detailed here.

RUSSIA

The primary training facility for Russian cosmonauts (and foreign candidates) is the Cosmonaut Training Centre named for Yuri Gagarin (TsPK). However, not all the training equipment and facilities can be located on one site, so several locations around Moscow are used to support specific elements of cosmonaut training or for gaining hands-on experience.

Energiya

Formerly Korolyov's OKB-1 design bureau, Energiya continues to be the leading design bureau for the Russian space programme and the source for many of the country's engineer cosmonauts. Working for years in various departments, many Energiya engineers support the training and preparation of cosmonauts by testing and evaluating space hardware, preparing flight documentation and methodology and taking cosmonaut training classes. Many of these engineers have never become formal cosmonauts, but others were selected to join the Energiya cosmonaut team to utilise their experience in the design bureau directly on space missions. In addition to a number of classrooms and lecture facilities, cosmonauts have the opportunity to use a Soyuz simulator for generic familiarity training and for evaluating upgrades. In the Soyuz and Progress Assembly room, the cosmonaut flight crew conducts fit and function activities in the near-complete OM and DM of the Soyuz. Elsewhere, the complex houses the Soyuz and Progress production-lines, but even though these facilities are on the same site, Energiya is a vast complex and even some of the most experienced Energiya cosmonauts have never ventured into some of the spacecraft production areas. In the Dynamics Testing Station, spacecraft are cordoned off to complete a programme of tests on new systems or upgrades. For example, during one test in June 2003, rookie cosmonaut Sergey Revin was evaluating Soyuz TMA hand controls, supervised by veteran Energiya cosmonaut Pavel Vinogradov. Though not part of a formal crew training programme, this experience acting as 'test-subjects' adds to the cosmonaut's experience and between flight assignment activities. No obvious pictures have been released of this simulator, which is also used to give engineers first-hand experience of space flight conditions.

TsUP Mission Control – Moscow

Located near to Energiya is the leading Russian mission control centre, some 10 km north-east of Moscow and 10 km north-west of TsPK in the town of Korolyov. (This was formerly the town of Kaliningrad, which was itself formerly called Podlipki. Many still talk about Podlipki and the railway station on the Moscow-Monino line still bears this name.) The centre is known as the Tsentr Upravleniye Polyotom (the Centre for the Control of Flight), abbreviated to TsUP (pronounced 'tsoop'). Many cosmonauts also use the call sign 'Moskva' (Moscow), in much the same way as US astronauts refer to their mission control at JSC as 'Houston'. Though not the original flight control centre, this facility, one of the major research facilities of the RSA Central Research Institute for Machine Building (TsNIIMash), was created in 1970 and became operational from September 1973 for the flight of Soyuz 12. It has been used since then (along with a duplicate room designed for the cancelled Buran Shuttle programme) for all Soviet/Russian manned space flights. Within the five rows of consoles (24 in all), teams of specialists monitor the systems, function and progress of each mission. For many years the position of Flight Director has been filled by former cosmonauts:

Table 3 Soviet/Russian Flight Directors

Date	*Director*
1961–1964	Sergey Pavlovich Korolyov
1964–1965	Unknown
1966–1968	P.A Aradzhanov
	Boris Yevseyevich Chertok
1968–1973	P.A.Aradzhanov
	Yakov Isayevich Tregub
1973–1982	Aleksey StanislavovichYeliseyev
1982–1986	Valeriy Viktorovich Ryumin
1986–1988	V.G.Kravets, in charge, Buran
1988–unknown	Vladimir Alekseyevich Solovyov

One position on the flight control team usually filled by cosmonauts (normally unflown, but also veterans) is spacecraft communicator (called 'Glavniy Operator' at TsUP and Capcom in the NASA programme). In the early programme, these positions were also found on Soviet tracking ships or remote stations outside of the Soviet Union, prior to the commissioning of the larger TsUP facility. In 1975, one of these 'remote Capcom' positions was held by Valeriy Illarionov, an unflown cosmonaut, at NASA MCC-Houston during the ASTP mission. In receiving an assignment as Capcom, the cosmonaut needs to be familiar with all aspects of the flight, hardware and experiments, as well as with the crew, acting as point of contact between the cosmonauts in orbit and the mission control and support room staff on the ground. This is valuable experience in progressing to assignment to a flight crew. A number of former cosmonauts from the Air Force selections have worked at MCC after their retirement from the team. The Russians do not restrict who actually talks with the crew on orbit.

In recent years, as well as establishing offices at Star City to coordinate their own astronauts' training, NASA and ESA have also established small 'mission controls' in adjoining rooms to the main control point. This enables them to talk to their own astronauts, monitor their health and pass on requests from researchers. These are called Mission Support Rooms (MSR).

IMBP

From early 1960, the staff of the Institute for Medical and Biological Problems (IMBP) supervised the isolation training of cosmonauts, observing their body language and gestures. From March 1960, at the Central Aviation Institute of Medicine, a programme of isolation training and tests was devised for the selection of cosmonauts, and from 1963, a new type of programme was introduced. In October 1963, a series of long-term 'missions' was devised in conjunction with the Aviation Institute, including low pressure simulations and total isolation training, remaining in facilities for 10–15 days (female candidates were tested for up to seven days). This training featured reversing the day/night cycle, a programme of four hours work followed by four hours of rest, and sessions in the altitude chambers at the IMBP.[1]

On 23 March 1961 one of the cosmonaut trainees, Senior Lieutenant Valentin Bondarenko, was tired and lacking concentration at the end of a ten-day run in the isolation chamber. He casually tossed an alcohol-soaked swab that he had used to clean off the adhesive used to fix medical sensors to his body across the chamber. The swab dropped on a heating ring and, in an oxygen-rich environment, caused a flash fire to erupt in the chamber, engulfing him and leaving him with 90% burns. He died from his injuries just eight hours later, barely three weeks prior to Vostok 1.[2] This tragic accident brought home the fact that training for a flight into space could be as dangerous as the flight itself. The low pressure chambers took a long time to open and, after this incident, they were used for isolation tests only. A different chamber with added safety features was used for low-pressure simulations.

A 'typical' test started at 08:00 hours with ablutions, hygiene and breakfast. Then a programme of 'methods' was conducted, which included a German-devised experiment of black and red chutes, down which balls were fed to test the candidate's

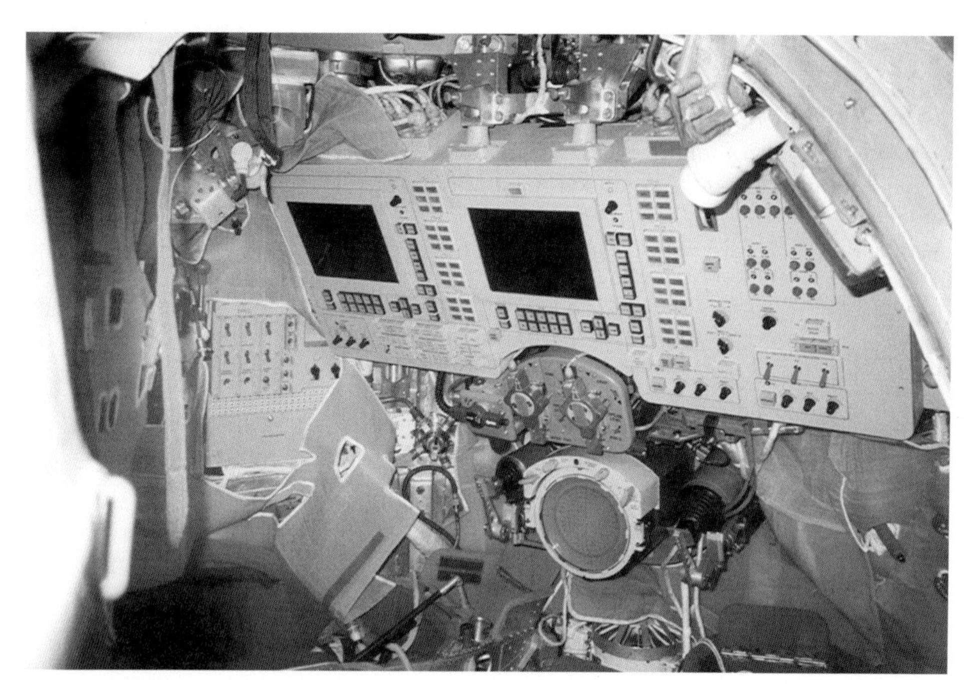

Inside the new Soyuz TMA simulator which is located in the Soyuz hall at the training centre

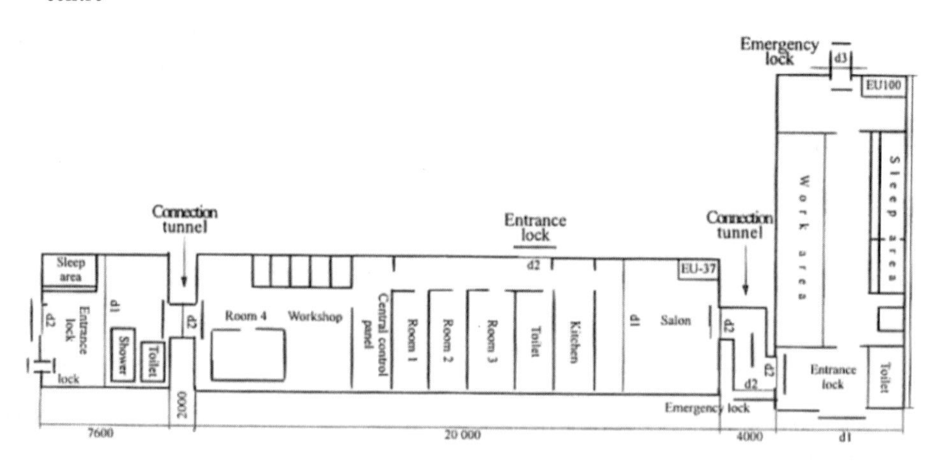

Layout of the Ground Based Experiment Complex, the long duration isolation facility in the IMBP, as it was used for the SFINCSS-99 test in 1999-2000. (Courtesy IMBP)

reactions and demonstrate mental agility. Some of tests were also set with music in the background. The cosmonauts were taught how to place electrodes on their own bodies and how to prepare their calorie controlled diet inside the chamber. Each meal was marked as part of the nutritional studies, so the candidates had to report what was not eaten (the most unpopular item on the menu was pureed carrots!). Urine and solid waste was collected for mass measurement. The lights were dimmed, but never switched off completely, to further confuse the candidate as to what time it was, and a 24-hour monitoring operation was completed by a team of 8–9 researchers – including psychologists – outside the chambers.

A record of the candidate's time in the facility was taken, along with notes on their reactions and items they took inside to relieve the boredom. For example, Tereshkova took poetry to read, while Ponomaryova wrote a book, some drew pictures and Popovich sang. Much later, Valeriy Polyakov, who worked at the institute, undertook similar isolation experiments, performing memory experiments about his dreams. Valeriy Bykovskiy was the first to volunteer for these arduous selection and training procedures, for which he received a prominent bravery award. The details of chamber usage were revealed briefly to the authors by Dr. Irina Ponomareva[1], and although the authors were only able to record the details of nine of the twenty candidates, nineteen of them completed the test:

1960. The first six were:

Apr 6–16	Bykovskiy
May 23- Jun 2	Volynov
Jun 12–22	Popovich
Jul	Leonov
Jul 26-Aug 5	Gagarin
	Nikolayev

We were unable to record the rest of the order for the 1960 selection until Bondarenko in March 1961. He was the seventeenth of the group to do the test and was followed a few weeks later by Zaikin and Filatyev. Varlamov did not participate in this phase of the training, following his accident and removal from the team

The women of the 1962 selection took their isolation tests in the following sequence:

Solovyova started her test on 13 April 1962
Tereshkova
Ponomaryova
Kuznetsova
Yorkina started her test on 15 June 1962

The training did continue after this, with a group of testers conducting more extreme tests, including some with no clocks available.

At IMBP, there are facilities to support medical-biological investigations and physical conditioning of the cosmonauts. In addition to the medical monitoring of cosmonauts during flight, the facility also houses an isolation simulator (used for cosmonaut selections), a nine-metre centrifuge and a special long duration facility known as the Ground Based Experiment Complex, consisting of either a 100-cubic-metre (EU-100) or 200-cubic-metre (EU-37) chamber, which can be used for space simulations lasting many months. These chambers have been used to conduct a number of national and international isolation tests in support of space station

The facility on the grounds of the IMBP in Moscow that houses the Ground Based Experiment Complex

IMBP's 9-metre centrifuge. (Courtesy Mark Shuttleworth)

missions and in planning for manned Mars missions. Recently, simulations of ISS residency missions have been completed with a number of international participants from ISS partner countries (Europe, Japan and Canada). The major simulations over the years have included:

Date	*Purpose*
1967 Nov 5–1968 Nov 5	A year-long simulation performed by a 'crew' of three men
1970–1973	Various three-person tests lasting between 20 and 50 days each
1974–1998	During the intervening years, crews completed various short simulations
1999 Jul 1–2000 Apr 14	SFINCSS-99 consisted of a core crew of four, completing a 240-day simulation, plus four visiting crews totalling 21 test subjects (20 male, 1 female, including 15 Russian, 3 Japanese, 1 German, 1 Canadian and 1 French). SFINCSS stands for Simulation of a Flight of International Crew on a Space Station
Planned	Mars 500 – a simulation of a flight to Mars

IMBP has also conducted a number of hyperkinesis experiments over many years, in some cases lasting over 120 days. The record is over 370 days, conducted by a 'crew' of nine men during 1996. Another facility used is the altitude chamber, GBK-63. This chamber is designed to conduct long-term continuous scientific investigations with humans and equipment. It has a volume of 5.4 cubic metres and can accommodate two persons. It is used by cosmonauts for qualification tests and experiments. The extent of the facilities available is not fully understood.[3]

Khrunichev

At the Khrunichev facility, there was a full-size mock-up of the Mir space station complex. It was used for public relations purposes and there is no evidence of formal training occurring in this mock-up.

OKB-52

To support training for the Almaz ('Diamond') military space station missions, a simulator of the station was located at OKB-52 in Reutov. This was used for real-time simulations of Almaz missions. There was also a Transport Logistics Spacecraft (TKS) simulator used for flight crew training, consisting of a mock-up of the crew capsule which combined with the Functional Cargo Block (FGB) to form the TKS ferry craft.

Zvezda

The Research Development and Production Enterprise (RD&PE) Zvezda, in Tomilino, south-west of Moscow, houses several facilities to aid and support training and familiarisation with space equipment and procedures, with specific emphasis on pressure garments, EVA and escape methods. At the facility (in a separate building to the facility's museum holding numerous displays of pressure suits and ejection and survival equipment) are two vacuum chambers, one of 30 cubic metres (used to test the Krechet lunar suit) and one of 50 cubic metres in an adjacent hall. As Zvezda produces the Orlan EVA suit, these chambers are used to test the integrity of each suit and for the early training of cosmonauts in vacuum conditions while wearing them. Two test subjects can be accommodated in the 50-cubic-metre chamber, where they are observed by a test crew from consoles outside using TV cameras. This ten-person crew includes suit technicians and engineers, a doctor, a communications officer and a test crew leader. Nominal and emergency procedures are simulated in the chamber in 1-g. Long duration familiarisations are also possible in these chambers, to get used to working inside the suit over many hours. Using a 15-cm step up, physical exertion in the suit can also be evaluated and can help to condition the cosmonaut to working within the restrictions of the suit. This simulation work is additional to work at the Hydrolab at TsPK.[4] The first simulator used for EVA training may have been located at Zvezda for suit and exit/entry evaluation tests. It consisted of a flown Vostok capsule (Vostok 3) converted to a Voskhod configuration and fitted with an airlock for practicing airlock exit and entry.

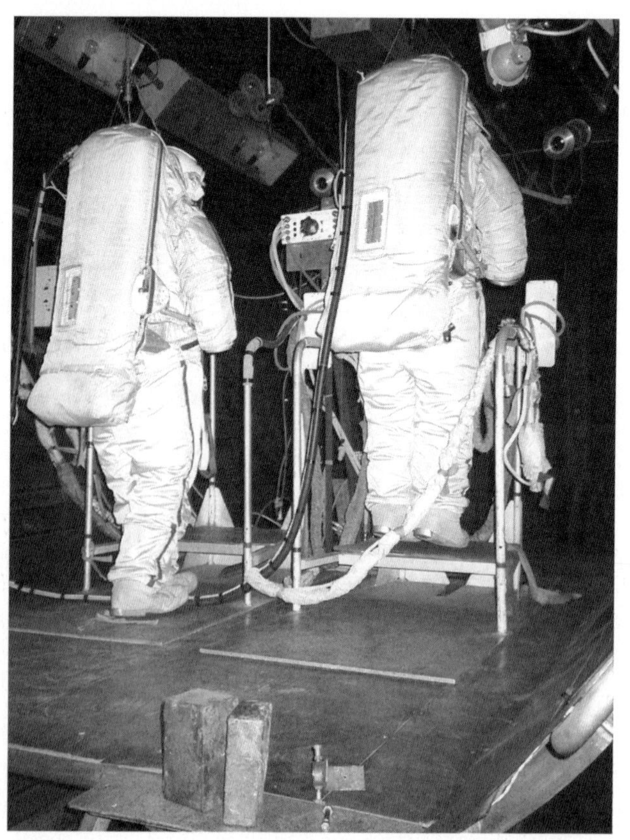

Two Orlan EVA suits in the 50 cubic metre vacuum chamber at RD&PE Zvezda, in Tomilino

Zvezda has manufactured all the pressure garments used by cosmonauts in the Soviet and Russian programmes and it is here that the cosmonauts are fitted and tested in the Soyuz Sokol pressure garment used for launch and entry phases of the mission. Fitting begins by putting on underwear, after which the cosmonaut is measured by a team of five engineers and then seated in a plastic mock-up of a Soyuz seat, where even more measurements are taken. Then, in a second mock-up of a Soyuz 'Kazbek' seat (resembling a bathtub), the cosmonaut's body is covered in a fast drying plaster to create a mould for a form-fitting custom seat liner. After initial couch fitting, the cosmonaut is fitted for a Sokol suit and then the measurements for a comfortable fit in the launch position while wearing the suit are recorded for future reference. Prior to launch on each mission, the cosmonauts visit Zvezda for adjustments to their suits and seats, remaining in the flight posture for two hours. The results from this test allow adjustments or refinements for safety and comfort. The cosmonauts then carry out a programme of tests in the vacuum chamber wearing their Sokol suits.

There is also a magnetic-cushion simulator (or minimum friction floor facility) for EVA training, which was used extensively for MMU operations. During the late 1960s and early 1970s, it was also the site of the lunar $\frac{1}{6}$ gravity walking simulator rig, although this has not been in use for many years, and seems not to be in place any more. At least two US astronauts, George 'Pinky' Nelson and Jerry Ross, have checked out this facility in the early 1990s, wearing an Orlan suit.

RD&PE Zvezda's minimum friction floor training facility

It is not clear what other cosmonaut-related training is conducted at Zvezda. Many simulations of suit development and operational systems checks are conducted in weightless conditions on aircraft under the control of the Cosmonaut Training Centre.

Chkalovskiy Air Base

In Houston, US astronauts use the nearby Ellington Field that houses the T 38 aircraft they use to fly to different facilities and centres across the United States. At the Cape, the nearby Patrick Air Force Base or Shuttle Landing Facility can be used for the same purposes. Near the Cosmonaut Training Centre is the Chkalovskiy Air Base, which has housed MiG 15s and MiG 17s, as well as the Czech L-29 and, currently, L-39 aircraft for the use of the cosmonauts.

Interkosmos facilities

During the Interkosmos programme, cosmonauts travelled to East European research institutes for briefings and familiarisation training for the joint experiment programmes flown on each Interkosmos flight. This type of international familiarisation training began with ASTP in 1973–1975 and the Soyuz 22 mission, in which an East German Zeiss camera was flown for flight certification for later Salyut second generation spacecraft.

Baykonur cosmodrome

Final checks of the Sokol pressure garments are completed at the facilities at Baykonur a few days prior to launch. There is very little 'training' completed at the launch site, as a recent interview with Dutch ESA astronaut André Kuipers confirms.[5] Kuipers (who flew to ISS aboard a Soyuz in 2004) explained that his crew went to Baykonur twice. After the first visit, they returned to TsPK for a few days for the official goodbyes, after which they returned to Baykonur.

At the launch complex, they engaged in the official and unofficial traditions, such as flag raising, but were otherwise kept busy with physical exercise (sports) and with physical preparation for the mission. This included sessions on the tilting table, lying with their heads in a position lower than their bodies to create a condition of blood flow to their heads. The crew also had final sessions with scientists, and went through various launch and in-flight procedures.

There are three checkout sessions in the actual spacecraft when it is still in the MIK. One is performed by the back-up crew and the others by the prime crew. The first of the prime crew sessions takes place in coveralls and for the second, the cosmonauts don their Sokol suits. During these sessions, the position of the cargo is checked to see if everything has been put in the correct place and to see if the crew can move around in the cramped vehicle.

It is also during this session that the crew does egress training, but Kuipers stressed that this was post-landing egress, not emergency egress on the launch pad. The springs that push the seats upward seconds before landing were activated and the crew had to get out of the craft on their own. Kuipers explained that this was very difficult, as the activation of the springs had caused his seat to move upwards

Cosmonaut Vladimir Kozlov (left) relaxes a little after an isolation test

and his face was now only some twenty centimetres away from the instrument panel, with no way to push the seat back to its original position. After what he thought was about fifteen minutes, he managed to get out.

Kuipers also said that emergency egress was not really an option. Egress has to take place through the same hatch as ingress, unlike the Shuttle, where there is an emergency egress possibility through the cockpit windows. Being in the left seat (where Kuipers was) meant that he would be the last one to get out, but there is also nowhere to get out to. The service tower with the elevator is already gone, and the possibility a Shuttle crew has – to walk through the white room and gantry to the service tower where they can use the slidewire system – is not an option for Soyuz.

Kuipers added that they were also warned a number of times that when certain warning lights illuminate, they had better brace themselves, as that meant that activation of the Launch Escape System was imminent '... and when that system is activated, with its 20-g acceleration, you want to be in your seat!'

No photographer was present during his ingress, nor did Kuipers know if one ever had been. However, he knew that Mark Shuttleworth had carried a small camera and had taken some pictures himself, which he had seen. The Russians had been rather annoyed at this because they strongly oppose carrying loose objects in the launch phase. If the item had not been secured or had been dislodged during a high-g phase, it might have caused damage. Kuipers said that he had also carried a private camera, but he had not been able to get to it until he was in flight.

In the summer of 2004, a short survival training session was conducted at Baykonur. The exercise lasted seven days in all, with one day of preparation followed by a two-day session for each of three 'crews'. The first crew consisted of instructors, while the other two crews were made up of cosmonauts (Yuri Lonchakov, Maksim Surayev and Oleg Kononenko on the first, and Aleksandr Skvortsov Jnr, Mikhail Korniyenko and Konstantin Valkov on the second). Summer in Baykonur is very hot.

EUROPE

European astronauts have been training at TsPK since the mid 1970s, initially from eastern bloc countries, then from west European states and now from the European Space Agency. In recognition of this expanded cooperation in human space flight programmes with Russia and America, ESA created a European Astronaut Centre (EAC) in Cologne, Germany in 1990, with the primary responsibility of selecting, preparing and supporting European astronaut operations. ESA is responsible for training all ISS astronauts on European station elements (Columbus science module, the Automated Transfer Vehicle, and European-supplied experiments and research facilities).[6]

EAC – Cologne

Located at the EAC are the following training devices that will also be used by cosmonaut members of residential ISS crews.[7]

Columbus Trainer Europe (COL-TRE) is a simulator supporting hands-on training for controlling and monitoring all systems and subsystems of the science module.

Columbus Mock-Up (COL-MU) contains all the system components with mechanical interfaces that are designed for crew operation, which includes possible or planned in-flight replacement hardware.

ATV Crew Trainer (ACT Lite) provides simulations of both ATV and ISS characteristics, allowing crew members to train for ATV approach, docking, undocking and separation. Safety features to command the ATV to hold its approach, abort the profile and guide it back to a safe distance to prevent collision are also included. This is a high fidelity simulator of the docking profile, which simulates the video image of an approaching ATV monitored by the crew aboard ISS, and overlays numerical data either on the video screen, or on a separate display in the station's Service Module. The training staff can insert a wide range of flight profiles, from normal to a variety of malfunctions and off-nominal scenarios. Commands from instructors and responses from the crew are recorded for post-flight evaluation.

ATV and Service Module Mock-Up (ATV/SMMU) is a crew training facility for logistics transfer or emergency situations involving the unmanned cargo carrier. This is a full-size mechanical replica of the Russian Zvezda Service Module, with an attached ATV integrated cargo carrier (the pressurised element of the ATV vehicle). This allows crews to train on cargo, gas and water transfer, stowage and inventory management and, where required, in-flight maintenance issues.

ATV training In accordance with the Expedition Training Requirements Integration Panel, the first crews to prepare for ATV operations as part of their training programme were the ISS-12 prime (Bill McArthur and Valeriy Tokarev) and back-up crews (Jeff Williams and Aleksandr Lazutkin). They were assigned to complete two weeks of training each at EAC during March-May 2005, with training dates for ESA astronaut Thomas Reiter and his back-up Leopold Eyharts to be determined later. Training programme development and support work has also been provided by cosmonauts Sergey Krikalev and Valeriy Ryumin. Crew training focuses on four areas: ATV Core Block (5 hours), rendezvous and docking, undocking and departure (31 hours), ATV attached phase operations (15 hours) and ATV Emergency training (1 hour). This programme is repeated during 'complex training' at TsPK later in the training cycle. An onboard simulator is being developed to provide on-orbit refresher training. Three training dry runs at EAC were completed between July 2003 and April 2004, involving Russian, American and ESA crew members, and two special training runs began at TsPK in May 2004. Facilities at TsPK include the teleoperation training facility for stand-alone simulations of ATV-to-ISS rendezvous and docking, undocking and departure, and a Russian Segment Training Facility (RSTF-ATV) for simulated rendezvous and docking, undocking and attached phase operations (refuelling, orbital control). This is also used for integrated training. At TsPK, the crews complete ISS (ATV) prerequisite training on emergency equipment, the Russian docking system and Progress vehicles. They then complete ATV Training Period 1, where they train on the systems in the Russian segment for ATV operations, such as computer and communications equipment. Finally, ATV Training Period 2 covers rendezvous, undocking and departure proficiency skills and simulations.

Columbus training ESA figures indicate that ISS crew members are trained at three different qualification levels: User, Operator or Specialist. Each training profile is designed to match the required qualification level:

- *User:* Safe living aboard ISS by utilising systems/operations/payloads to achieve individual mission success
- *Operator:* Same as User level, but with frequent nominal and maintenance task training, off-nominal task (quick response) training and higher payload capability training
- *Specialist:* Both User and Operator training, plus infrequent nominal and maintenance tasks, increment specific tasks and repairs.

For ESA elements at EAC this training requires:

Qualification	*ATV*	*Columbus*
User	8 hours	25 hours
Operator	49 hours	82 hours (+ 39 hours for initial activation and checkout activities ISS 1E (first European) mission)
Specialist	59 hours	128 hours (+ 49 hours for 1E mission)

This does not include training at TsPK.

UNITED STATES OF AMERICA

Following the limited training sessions during the Apollo-Soyuz Test Project in the early 1970s, no Russian cosmonaut training was held in America until the creation of the Shuttle-Mir programme two decades later. The majority of this training was held at NASA facilities in Houston and in Florida.

JSC – Houston, Texas

This has been the home of the American astronaut corps since 1962. Cosmonauts assigned to Shuttle-Mir and ISS crews train here on fixed-base or motion simulators, practicing the ascent and entry phases of the Shuttle mission profiles either as a Shuttle crew member or as a returning Mir/ISS resident crew member. There are also practical and theoretical classes in the facilities of the Shuttle orbiter (food selection, meal preparation systems, waste management, sleeping, stowage), as well as emergency crew escape procedures in 1-g mock-ups, before training at the Sonny Carter water tank, where water egress training from the Shuttle is completed. This facility is also used for EVA training sessions. In addition to the facilities in Building 9 (Shuttle and ISS mock-ups) and Building 5 (ISS and Shuttle simulators), the cosmonauts have also visited nearby off-site institutions such as Boeing and United Space Alliance, who are major sub-contractors that provide elements of crew equipment (EVA suits, escape suits, mission peculiar equipment), and Spacehab – contractor for the Shuttle augmentation module. As the Shuttle vehicle is phased out of the American programme, Russian cosmonaut training will focus more on ISS simulators, and possibly elements of the Crew Exploration Vehicle currently under development to replace the Shuttle by 2010–2014. Though several Russian cosmonauts have participated in American Mission Specialist training, none have yet been assigned as full Astronaut Candidates, unlike several Japanese, Canadian and European astronauts (and one Brazilian astronaut). It will be interesting to see whether Russian cosmonaut candidates complete a NASA Ascan training programme (and equally, whether NASA Ascans complete a cosmonaut candidate training programme) prior to qualification as official astronauts (or cosmonauts).

As for NASA astronauts training on Soyuz, former Chief of the Space Flight Training Division of the Mission Operations Directorate, Frank E. Hughes, was able to acquire a mock-up for training purposes in Houston. In 1998, the Kansas Cosmosphere Space Museum in Hutchinson, Kansas came across an original, complete Soyuz T in a marketplace in Thailand. When it was shipped back to Kansas, Hughes managed to procure funds to have a mock-up made from it which was used for training purposes in Houston. When NASA finished with the mock-up, it was given to the Kansas Cosmosphere.[8]

KSC – Florida

The primary launch site of the NASA Shuttle system is also used for some cosmonaut training, primarily in the procedures for escaping from the launch pad in the case of an emergency. Here, the cosmonauts are briefed in evacuation systems, including the slide wire baskets and the personnel carrier which the cosmonauts have to qualify to drive as part of their training (which many find great fun!) Briefings on the facilities and procedures at the Cape are also part of the cosmonauts' preparation for a launch on the Shuttle, and participation in crew Countdown Demonstration Tests and fit and function sessions aid in familiarisation with the actual vehicle to be flown in orbit.

CANADA

Canada is the prime contractor of the Shuttle Robot Arm (Canadarm) and the Space Station Mobile Servicing Systems consisting of three main elements: the Space Station RMS (Canadarm2), the Mobile Base System (MBS) and a mobile platform with a two-armed robot called the Special Purpose Dexterous Manipulator (SPDM). These elements were installed on the ISS from 2001 and training for their operation is conducted at the Canadian Space Agency's headquarters in Saint-Hubert, Quebec, using the Mobile Servicing System Operations and Training Simulator (MOTS).

MSS System Training With the installation of the MSS system, training of resident crew members was included in their programme of preparations. All astronauts and cosmonauts assigned to a resident crew aboard ISS are trained to operate the MSS and complete a two-week intensive training programme at the CSA facilities. Initially, from 1998, this training focused on Canadarm2, but since December 2001, it has expanded to include training on the operation of the MBS. A course on the SPDM is also part of the training programme. A three-quarter scale mock-up of the US Destiny Lab contains the Robotic Workstation (RWS), featuring a display and control panel, a portable computer system for displays, two hand controllers handling the translation and rotation of the Latching End Effector (LEE), and three monitors displaying the views from the three cameras located on Canadarm2 and other locations outside the ISS. The trainer also features the Canadian Space Vision System (SVS), which pinpoints the exact location and movement of payloads, presenting accurate distances and orientation.

Theoretical training is followed by practical sessions held at the CSA headquarters. Candidate crew members complete training in the Multimedia Learning Centre, the MSS Operations Training Simulator, and by Virtual Reality systems. After completing the two-week training course, the crew members are certified as Mobile Service System Robotics Operators (MRO) and are awarded a certificate and an embroidered emblem crest, with two embroidered wings featuring an illustration of the SSRMS arm. This training is very intensive, with two five-day sessions and some of the days lasting between ten and twelve hours.

A special course for ground staff (controllers, Capcom, Flight Directors and crew support astronauts / cosmonauts, as well as instructors) has been developed by the CSA. Tailored for each group, this course instructs the ground team on the systems operating in space, offering additional support and awareness to flight operations. Several cosmonauts in support roles have also completed this training course.

On-orbit training Experience with other systems indicated difficulties in training with a system months before operating it in space. Developed in cooperation between the Canadians and Russians, the the on-orbit MSS training System for Maintaining and Monitoring Performance (SMP) comprises a small laptop computer and two hand controllers. This allows the resident crew to practice capturing free-flying objects before attempting the real thing. The software was developed by CSA, using Russian expertise of onboard system integration and experience from the Russian TORU system used on Mir. An SMP training system prototype was delivered to the ISS by Progress M-47 in February 2003 and the first person to use it on orbit was cosmonaut Yuri Malenchenko, Commander of the ISS-7 crew. In order to prevent deterioration of SSRMS skills, the SMP prototype was used on ISS to train crew members in specific robotic operator tasks until their skills reached a 'stable' level. This training will be used as a baseline for comparison with future training sessions. After their last training session, the subjects conducted performance tests at different levels that were designed to assess how their skills had changed and performed 'recovery training' to return to the stable level again. A control group of six cosmonauts are performing experiments at TsPK in cooperation with six CSA instructors and engineers. These cosmonauts have all completed MSS training or ISS missions. There has also been an ISS on-orbit test programme planned (stated as ISS-7 to 10 crews), as part of a two-year experiment programme prior to its becoming a qualified and operational programme on the station for free-flying satellite capture.[9]

Survival training There have been discussions between the CSA and RSA over the possibility of landing a Soyuz crew on Canadian territory. However, it appears that this has come to nothing, mainly over a question of money, not least the fact that the Canadians would have to pay the Russians for permission to land a Soyuz in their territory. However, in cooperation with Russia and America, a joint emergency landing exercise was conducted in the Gelenndzhik Bay on the Black Sea, during September 2003. The exercise was aimed at coordinating the aviation and naval rescue forces of the three countries to recover a 'crew from the ISS' in a simulated water rescue scenario. One of the trio of cosmonauts was evacuated from the emergency splashdown site to a hospital; the second was taken aboard one of the recovery ships to receive medical aid; and the third cosmonaut was immediately returned to the shore.[10]

At least one cosmonaut has undergone joint training in the Canadian Arctic under the supervision of the Canadian Armed Service. Dimitriy Kondratyev participated in a programme of Arctic Survival Training with an ESA astronaut, a Canadian astronaut and three NASA astronauts, focusing on psychological training in preparation for coping with higher stress levels encountered during long space station missions and based on NASA experiences during Shuttle-Mir. The concept is based on a winter exercise developed by Canadian armed forces and combines elements of Air Force winter survival training and Canadian Army leader / winter warfare training. The programme begins with three days of 'static training' at Valcartier Garrison, Quebec, learning how to use the equipment, perform daily routines and learn safety procedures. The fourth day sees the team taken by helicopter out to the field training site for completion of the programme, spending five nights and six days alone and hauling their equipment to each night's campsite. At the completion of the sixth day in the field, the helicopter returns the team back to Valcartier Garrison for debriefing, rest, a welcome shower, medical examinations and a sleep in a proper bed, before departing for Houston for further debriefing. This programme is expected to expand over the coming years and will feature more cosmonauts from Russia in the teams.[11]

JAPAN

To support their own participation in the International Space Station, the Japanese space agency, JAXA (formerly NASDA), has created a Space Station Integration and Promotion Centre (SSIPC) at the Tsukuba Space Centre (TKSC), Obaraki prefecture, Japan. Though no formal training of international crews has taken place due to the Japanese elements (primarily the Kibo research module) still awaiting launch, a number of cosmonauts have visited the centre and Japanese astronauts have participated in Soyuz training as part of their ISS preparation programme. Astronauts Furukawa, Hoshide and Yamazaki completed their Soyuz training from July-September 2003 and January-March 2004 to qualify them as Flight Engineers. They completed study courses on various sub-systems such as propulsion, docking, and attitude control, and were also fitted for Sokol suits. There was also centrifuge training and a physical training programme, as well as eight-hour days in two, three-month periods, receiving lectures, conducting simulations and completing verbal tests for each of these systems.[12]

Tsukuba Training Centre

As well as developing and operating the Kibo element and supporting the experiments performed onboard the facility, the Tsukuba Centre is used to support crew training. This will eventually include some training by Russian cosmonauts.[13]

Space Station Test Building includes the Kibo pressurised module trainer, capable of training crew members to handle racks, connections and associated electrics and plumbing. Rack configurations are trained for using the Kibo Experiment Logistics Pressurised Section Trainer (ELM-PS Trainer); control system simulation training is handled by the Kibo Systems Trainer, using the JEM (Japanese Experiment Module) Control Processor (JCP) and System Laptop Terminal (SLT). A Kibo Airlock Trainer is used to train the crew on the mechanisms and controls of experiment support tables and hatches. There are also simulation facilities for the manipulator main arm and small arm systems.

Astronaut Training Facility is used to train Japanese national astronaut candidates and includes an isolation chamber, hyperbaric chamber, health care facility, space medicine research, vestibular function research and bed rest study facilities. Though it is not aimed at foreign crew members, it does feature elements of the training of

NASA astronauts and Russian cosmonauts and cooperative programmes are feasible.

Weightless Environmental Test Building is a 16-metre diameter, 10.5-metre deep water immersion facility that can hold a full-size mock-up of Kibo to aid EVA simulations. This could be used to train international partner crew members (including cosmonauts) on Kibo EVA operations, as well as Japanese astronauts.

COSMONAUT TRAINING GOES INTERNATIONAL

For over thirty years, the training of Russia's cosmonauts was largely conducted in secret and inside the Cosmonaut Training Centre, well away from foreign view unless specially invited. With the desire to offer commercial deals to foreign countries and participation in the ISS programme, the controlled opening up of the TsPK centre was also supplemented by the need to send cosmonauts abroad to train on foreign space equipment and procedures. This not only challenged the technical skills of the cosmonauts and Russian training staff, but also their personal and team working philosophy. This expansion not only enriches the Russian training programme, but also broadens the experience and depth of knowledge of those who participate, and in turn supports a greater return from each mission.

REFERENCES

1 Personal visit to IMBP facilities by the authors 17 June 2003, information provided during an interview with Dr Irina P Ponomareva, who was the Chief researcher and member of the Russian Tsiolkovskiy Academy of Sciences State Research Centre for IMBP.
2 The Rocket Men, Rex Hall, David Shayler, Springer-Praxis 2001 pp 131–132
3 Simulation of Extended Isolation: Advances and Problems, Ed V.M. Baranov, Associate of the Russian Academy of Medical Sciences, Moscow 2001
4 Information provided during a tour of Zvezda facilities by David J Shayler and Bert Vis, 16 June 2003
5 Telephone interview with André Kuipers by Bert Vis, May 2005
6 ESA magazine On Station, No 18, November 2004, various articles
7 Astronaut Training Centre for the European ISS contributions Columbus module and ATV, Peter Elchler *et al*, paper presented at the 54th IAC Congress, Bremen, Germany, 29 September–3 October 2003; ESA On Station, Issue no 18 November 2004
8 Interview with Frank E. Hughes, former Chief of the Space Flight Training Division of the Mission Operations Directorate (JSC-DT), Moscow, 10 April 1998
9 Canadian Space Agency website, http://www.space.gc.ca/asc/eng/astronauts/osm_system
10 ITAR-TASS news release 29 May 2003
11 Astronaut Team Training In Canada, CSA Backgrounders, http://www.space.gc.ca/asc/eng/media/backgounders/2004/0122.asp
12 Japanese Space Agency (formerly NASDA, now JAXA) website, http:iss.sfo.jaxa.jp/Astro/report/2003/0307_e.html and http:iss.sfo.jaxa.jp/Astro/report/2004/0402_e.html
13 Japanese Space Agency website, http://iss.sfo.jaxa.jp/ssip/index_e.html

Survival training

Survival training is overseen by the Third Directorate of the Cosmonaut Training Centre. This department is currently headed by Colonel Yuri Gidzenko, a veteran of three space flights, and it has previously been headed by both flown and unflown cosmonauts, including Yevgeniy Khludeyev, Yuri Malyshev and Nikolay Grekov. The department occupies one of the buildings in the training centre and survival training is conducted in cooperation with the Federal Aerospace Search and Rescue Administration. The training enables the crews to acquire the skills needed in the event of an off-nominal landing (or splashdown).

THE ROLE OF THE THIRD DIRECTORATE

Survival training is an important part of the basic (OKP) training of all cosmonauts. It is repeated on a regular basis during a cosmonaut's career, and when they are assigned to a new crew, they also complete a new cycle of such training. Every Soyuz craft carries a Granat-6 (Pomegranate) survival pack, which includes a 'Forel' (Trout) hydro-suit – a one-piece orange nylon flotation suit with attached rubber soled feet and a hood trimmed with 'CCCP'. The suit contains a 'Neva' inflatable collar with an emergency mouthpiece, emergency beacon and a signal device on the shoulder. It also has rubberised cuffs, Velcro-close pockets on the legs (with ten pairs of small rings on the legs and eight pairs of grommets on the boots), and a pair of brown jersey mittens with separate thumb and index finger stalls, with watertight cuffs and adjustable orange nylon wrist straps. There is also a TZK-14 cold weather suit, with a royal blue nylon zip front anorak with attached mittens. This has two slash pockets with contrasting zips and a draw closed waist. Also included is a wool knit balaclava, a lined wool knit cap with button flaps, wool gloves, one pair of shearling socks and one pair of nylon over boots, elasticised at the top with Velcro-close at the heels. There are three other orange nylon packages in the pack. These contain survival equipment including a large canteen, a soft flask, dried food, a medical kit, a frying pan, signals and flares, a machete (which also doubles as the shoulder rest of the rifle/shotgun), a Makarov pistol with cartridges (TP-82m), a foraging bag, fishing tackle, and metal wire garrottes for use as a saw as well as for hunting. The combination of the 'Forel' suit and thermal suit is intended to keep the wearer alive for up to twelve hours, if needed, in water of 2°C, with an ambient air

Survival equipment hanging in an office in the 3rd Directorate department

temperature of -10°C (14°F). Coupled with the shelter of the descent craft, it is hoped that the clothing and supplies could support a cosmonaut for up to three days in conditions of severe cold.[1]

The package weighs around 32.5 kg and is located in two triangular carrying cases that wedge snugly between the cosmonauts' seats. The package is produced by the Zvezda Production Association. The first kits, called NAZ (portable emergency kit), were produced by Zvezda in 1960–61 and were carried on Vostok craft.[2] After the problems encountered during the ballistic return of the TMA capsule containing the ISS-6 expedition crew of Bowersox, Pettit and Budarin, a satellite phone system has since been added to the kit.

Survival training is divided into a number of different aspects:

- Winter training
- Mountain training

- Desert training
- Swamp training
- Sea recovery training
- Parachute jumping
- Zero-g aircraft

Other types of emergency situation are covered with the use of simulators and the centrifuge. This training is covered elsewhere in this book, as is the use of zero-g aircraft.

Winter survival training

There are examples from the 1960s of Soviet and Russian cosmonauts undergoing simulations in winter conditions. Pictures of these simulations show cosmonauts training for Voskhod and Soyuz missions with their craft on its side in the snow. In the 1980s and 1990s, cosmonauts were regularly taken to Vorkuta in the Arctic for their winter survival training. A Soyuz capsule was set up on the snow and the cosmonauts were expected to stay there for two days. Before this, they would have already undergone some training to prepare them for the prospect of a Soyuz being stranded in deep snow on return to Earth. At Vorkuta, they had to build an igloo to provide shelter, as conditions in the Soyuz capsule itself are extreme. They also constructed a tepee, using the parachute and an 'A' frame hut built from trees they had cut down. The parachute could also be cut up to act as cover.

The specialists in charge of this training are examining the psychological qualities of the crew, as well as their physical shape, but the Russians have planned Soyuz emergency landing sites in both Siberia and Canada, where these conditions are most likely to be encountered. The cosmonauts are medically examined every evening.

This training mirrors a real situation encountered by the Voskhod 2 crew, who made an emergency landing in the North Urals where the temperature was twenty degrees below zero and with half a metre of snow on the ground. It took ground services two days to find the crew, and this was by ski teams as the helicopters could not fly due to the weather conditions. These extreme conditions are illustrated in an interview with Sergey Bedziouk, a candidate to join the Energiya cosmonaut team in the late 1970s and a member of the design bureau's flight test department. 'We tested the light variant of cosmonaut equipment after an emergency landing under tundra conditions. The temperature was minus 65 degrees Celsius and the equipment was very light, especially with regard to clothes. Using a knife we were forced to make multi-layered covers out of the landing vehicle's drogue chute and build an igloo in the morning. Food was another problem. All the water froze immediately and the device for warming it up had a hole in the middle of it. The Institute of Psychology of the Academy of Sciences conducted an experiment called 'continuous staying awake regime'. We had to work continuously for three to four days without closing our eyes for even a minute.'[3]

Currently, crews going to the ISS are left in a remote part of the grounds of Star City, but the Russians are also continuing to develop new systems for winter rescue. In early 2004, this was covered in a BBC monitoring report. 'Russian athletes flew

Survival training in 1970. (from left) Cosmonauts Vladimir Kozlov, Yuri Romanenko, Boris Volynov (the group training commander) and Anatoliy Berezovoy

Training in the winter of 1970

into the Antarctic today to make the first parachute jump from a super-light aircraft in the history of the icy continent.' The expedition's leader, Aleksandr Begak, told ITAR-TASS by telephone from the Russian research station of Novolazarevskaya; 'The members of the expedition, who flew into the Antarctic, have started testing new means enabling cosmonauts to survive in extreme conditions and are in preparation for parachute jumps from a super-light aircraft.

'Expedition members intend, for the first time in the Antarctic's history, to fly what they call a paracraft (in Russian, a paralet) – a super-light vehicle with soft wings and a 32-horsepower petrol engine. The flight will start the moment the wind becomes a little milder. At the moment, the wind speed is up to 15 metres per second. The area around the station is sunny and the temperature is minus 14 degrees Centigrade. Expedition members will perform a few parachute jumps from the paracraft. They will be jumping from the height of 1,200 metres, wearing oxygen masks and heat-shielding suits from the cosmonaut survival kit.'

Valeriy Trunov, head of the Cosmonaut Training Centre's survival department, who also took part in the expedition, told ITAR-TASS, 'The modified heat-shielding suit from the Granat-6 portable cosmonaut survival kit was designed by Star City specialists together with Russian scientists. In this suit, it is possible to spend 72 hours at a 60 degree temperature with minimal mobility.' Trunov also commented that the suits the cosmonauts wear during landing are 'fairly light and do not save from cold. In order to survive a landing in a remote area in winter, heat-shielding

Training in the winter of 1990. (from left) Talgat Musabayev, who was training for TM 13, and German astronauts Reinhold Ewald (centre) and Klaus Flade

Winter training camp

suits are included in the survival kit.' Under Russian safety rules, rescuers must spot the landing capsule and reach the landing site within three days.

In addition, Trunov said that a new Russian remote medical control system, which has no analogue in the world, would be tested in the Antarctic's extreme conditions. 'This is a special belt with sensors and a radio transmitter, whereby all information on a cosmonaut's state of health can be bounced off a satellite for at least 48 hours. In the event of an emergency landing or descent in an unintended area, doctors will be able to continuously monitor the cosmonaut's state of health.'[4]

Desert training and the heat chamber

Cosmonauts are exposed to extended testing in a heat chamber based at the Cosmonaut Training Centre. This simulates desert conditions, as well as testing the cosmonaut's mental condition. The chamber is approximately five metres by three metres and also doubles as the isolation chamber. The tests are monitored with the aid of instruments and observation (via a window from the control room) by staff from the medical department of the training centre. The purpose of the training is to test the cosmonaut's ability to work in humidity and in high temperature conditions. It also brings to light deviations in the cosmonaut's behaviour or character which cannot be detected in any other manner. In the 1960s, the testing regime was very rigorous, with subjects spending many days in the chamber. They had to endure temperatures of about 40°C, adapt to changes in the gas composition inside the chamber and endure temperatures of up to 80°C while wearing fur lined flight clothing.

Desert training

An article written in 1991 describes such a test, undertaken by cosmonaut Valeriy V. Illarionov. The air temperature was 60°C and a humidity level of thirty per cent was maintained inside the chamber. A powerful fan blew hot air into the chamber. Illarionov had pickups attached to his body and an electro-thermocouple was placed under his tongue. The test was fairly short, lasting only a few hours, and was observed through a window in the wall and monitored by doctors from the staff of TsPK.

Cosmonauts also now have to undertake a 24-hour survival test in the desert. This type of training started in the late 1970s with a capsule that was put into the desert; the same type of capsule that is used in sea and snow training. The cosmonauts had to spend time in the capsule, before exiting and constructing a tent and hammock using the parachute material. Their supplies included a radio, weapons, a flashlight, food and water. They also learned to build a fire to combat the cold nights. The site they used in the 1980s was near the town of Mary, which is in the Kara Kum desert in Turkmenistan. Since the break up of the Soviet Union, cosmonauts have conducted this type of training at the Baykonur Cosmodrome.[5]

Mountain training

Cosmonauts have also simulated landing in mountainous areas. This training was similar to that undertaken in deserts and in the snow and again proved useful during

Mountain training

an actual mission when Vasiliy Lazarev and Oleg Makarov landed in the Altai Mountains in April 1975. The Buran cosmonauts did their training in the mountains of the Pamirs, from the city of Frunze to Issyk-Kul, along with some colleagues from Energiya. The trek on foot lasted seven days. The Buran group also conducted a mountain ski trip in the Dombai region.

Swamp training

This type of simulation uses the same type of capsule, called Ocean, as they do for sea training. It is not clear whether every crew undertakes this training, nor its exact location.

Sea recovery training

Sea survival training is organised at the Special Survival Centre for Air Force flight personnel, located twenty kilometres from Anapa.

Though the Soyuz is designed to come down on land, the Russians simulate the possibility and dangers of a sea landing. A capsule with a cosmonaut crew inside is dropped by crane from a ship called the '*Sevan*' into the Black Sea near Sochi. The crew learns how to don their orange 'Forel' survival suits and then exit the craft and get into an inflatable dinghy while inflating the suits. During this test, the crew is monitored by divers and specialists in the sea and in boats. After two hours bobbing

Swamp training

around on the Black Sea in the Soyuz and donning this extreme weather and water survival gear, the cosmonauts are very close to heat exhaustion. Pulse rates can reach 180 or higher and many cosmonauts suffer from motion sickness. Another option for this training is that the crew can be recovered by helicopter. In this case, they use a blue painted Soyuz capsule called Ocean and there is also an orange version with a dolphin painted on the side. The crew also practices how to keep together in water while awaiting rescue, using a water tank on the deck of the ship.

Before the craft is put into the water, the crew will simulate all their actions on the deck of the ship using the same capsule. This is called 'dry training' and the craft is shaded by the parachute. Some of the Americans have done this test in the Hydrolab at Star City before simulating it in the Black Sea. Again, this training has proven invaluable in a real situation, as Soyuz 23 landed in high winds at night, in the snow in Lake Tengiz in Kazakhstan. The Soyuz sank and the crew had to be rescued by some very brave helicopter crews. The lunar version of Soyuz, called Zond, was recovered from the Pacific Ocean after it returned from lunar orbit. This recovery was monitored by an Australian Air Force plane, much to the annoyance of the ship's crew. The lunar training group undertook major sea landing training in preparation for their anticipated flights in Zond spacecraft.

Sadly, at the end of one test, cosmonaut candidate Sergey Vozovikov was killed when he got caught up in fisherman's nets and was drowned. There were attempts to save him, but to no avail. He had decided to use some of the equipment from the NAZ (portable emergency supplies) to go fishing to mark the end of a successful test.[6]

OKB-52 cosmonaut candidate Vladimir Gevorkyan undergoing sea training

In 1963, when Tereshkova was undergoing her sea recovery training for her Vostok mission, the sea was very calm, so the Soviet commander hired a number of motor boats to pass very close to the craft at full throttle, which made the sea very choppy and made Tereshkova sea sick.

Sea recovery training can be combined with mountain trekking and climbing, as this site is very close to hand.

Parachute jumping

All Soviet and Russian cosmonauts undergo extensive courses in specialist parachute training. Parachute training is an important psychological aspect of the training programme, helping to prepare them to deal with the dangerous nature of their job as a cosmonaut. Many cosmonauts have gained the Instructor Parachutist award as a result of this training, which includes free-fall jumps as well as those from helicopters. Several of the cosmonauts have done over 500 jumps.

During the first phase of parachute training between 1960 and 1963 (overseen by Colonel Nikolay Nikitin), the training was done in cycles of a few jumps at a time. Table 4 shows the parachute jumps undertaken by cosmonaut trainee Valentin Filatyev in the period 1960–1963. It illustrates the cycles and development of the parachute training. Cosmonauts seem to spend five days to a week doing two jumps a day and most of the 1960 selection would have done this sort of schedule, although after Gagarin flew on Vostok, he was banned from doing jumps and only resumed in 1964 after a long campaign.

On 28 May 1963, the cosmonauts' parachute instructor Nikolay Nikitin was killed after his parachute tangled in the air with that of another member of the drop

Table 4 Parachute Jumps by Filatyev during 1960–1963

Date	*Aircraft*	*First Jump Altitude*	*Delay*	*Second Jump Altitude*	*Delay*
1960 Apr 18	Antonov AN-2	800 m	none	800 m	none
1960 Apr 19	Antonov AN-2	800 m	none	800 m	none
1960 Apr 21	Antonov AN-2	800 m	none		
1960 Apr 22	Antonov AN-2	800 m	none	800 m	none
1960 Apr 23	Antonov AN-2	800 m	none		
1960 Apr 23	Antonov AN-2	800 m	none	800 m	none
1960 Apr 25	Antonov AN-2	1,000 m	5 sec		
1960 Apr 26	Antonov AN-2	1,100 m	10 sec		
1960 May 3	Lisunov LI-2	1,300 m	15 sec	1,300 m	15 sec
1960 May 4	Lisunov LI-2	1,300 m	15 sec		
1960 May 7	Lisunov LI-2	1,300 m	15 sec	1,300 m	15 sec
1960 May 8	Lisunov LI-2	1,600 m	20 sec	1,600 m	20 sec
1960 May 9	Lisunov LI-2	1,600 m	20 sec	1,600 m	20 sec
1960 May 10	Lisunov LI-2	1,600 m	20 sec	1,600 m	20 sec
1960 May 11	Lisunov LI-2	1,800 m	25 sec	2,100 m	30 sec
1960 May 13	Mil MI-4	2,100 m	30 sec	2,600 m	40 sec
1960 May 14	Ilyushin IL-14	3,000 m	50 sec	2,100 m	30 sec
1960 Aug 5	Mil MI-4	1,100 m	10 sec	1,300 m	15 sec
1960 Aug 6	Ilyushin IL-14	1,800 m	25 sec	3,000 m	50 sec
1960 Aug 11	Ilyushin IL-14	1,000 m	none		
1960 Sep 2	Ilyushin IL-14	2,100 m	none		
1961 Mar 9	Ilyushin IL-14	1,000 m	5 sec		
1961 Mar 16	Ilyushin IL-14	1,000 m	5 sec	1,000 m	5 sec
1961 Sep 28	Antonov AN-2	1,300 m	15 sec	1,300 m	15 sec
1961 Sep 29	Antonov AN-2	1,600 m	20 sec	1,600 m	none
1961 Sep 30	Mil MI-4	1,000 m	5 sec		
1961 Oct 2	Antonov AN-2	1,300 m	15 sec	1,300 m	15 sec
1961 Oct 4	Mil MI-4	1,000 m	none		
1961 Oct 5	Mil MI-4	1,600 m	20 sec		
1962 Aug 30	Ilyushin IL-14	1,000 m	5 sec	1,100 m	10 sec
1962 Aug 31	Ilyushin IL-14	1,100 m	10 sec	1,100 m	10 sec
1962 Sep 1	Ilyushin IL-14	1,100 m	10 sec	1,600 m	10 sec
1962 Sep 2	Ilyushin IL-14	1,600 m	20 sec	1,600 m	20 sec
1962 Sep 3	Ilyushin IL-14	2,100 m	30 sec	2,100 m	30 sec
1962 Sep 4	Ilyushin IL-14	1,600 m	20 sec	1,600 m	20 sec
1962 Sep 5	Ilyushin IL-14	3,000 m	50 sec	3,700 m	70 sec
1962 Sep 6	Ilyushin IL-14	3,700 m	70 sec		
1963 Feb 12	Antonov AN-2	1,300 m	15 sec	1,300 m	15 sec
1963 Feb 13	Antonov AN-2	1,600 m	20 sec	1,600 m	20 sec
1963 Feb 14	Antonov AN-2	2,100 m	30 sec	2,100 m	30 sec
1963 Feb 15	Antonov AN-2	2,100 m	30 sec		

Taken from Filatyev's parachute record, Star City, August 2004

Parachute training in the 1960s

group, Aleksey Novikov. The testing was being conducted at Chkalovskiy Air Force Base, where Novikov, who was a pilot, was based. He was also killed in the accident. The news of Nikitin's death caused an atmosphere of gloom within the team. Nikitin's funeral was scheduled for 30 May, and Kamanin was very worried about the effect of his death on the female cosmonauts' nerves, particularly with the launch of Vostok 5 and 6 only days away. Nikitin, who was one of the Soviet Union's top parachutists, had supervised the parachute training of the cosmonauts from the very beginning. He is buried in a military cemetery within the air base. He is also seen in the front row of the famous Sochi group photograph, reflecting the importance of his role within the early training for Vostok.

There is a specific department within the training centre that is responsible for overseeing this aspect of training. Nikitin was its first head until his death in 1963. When he was killed, his role was taken over by two cosmonauts who had considerable parachuting experience. They were Major Gennadiy Kolesnikov, who joined the cosmonaut team in 1965, and Irina Solovyova who, prior to joining the team in 1962, had been a member of the Soviet parachute team and a world champion in the sport. In 1967, the role of Department Head was reinstated and Sergey Aleksandrovich Kisilev (who is married to Solovyova) was appointed from 1967 till 1988. He was also one of the Soviet Union's top parachutists, reflecting the continued importance of this role. Kisilev and his wife both still live at Star City. He was succeeded by Colonel Viktor Ren (1989–95) and Colonel Valeriy Trunov (1996–8). Trunov was succeeded by Colonel O.G. Pushkar, who seems to have taken over in 1998, although Trunov resumed the role possibly a year later. He is now Deputy Commander of the Third Directorate, reporting to Colonel Yuri Gidzenko. The department with responsibility for the parachute training is Number 32.

In 1991, the Russians were using an airfield in the Crimea for these tests. This was also used for their sea training and in some instances, they were combined. The

Parachute training 2004. (back row) Cosmonauts Skripochka, Valkov (training supervisors), Candidates Aymakhanov (Kazakhstan), Serov and Zhukov. (front row) Ryazanskiy and Aimbetov (Kazakhstan). (Courtesy Novosti Kosmonavtiki)

cosmonauts jumped from both helicopters and aircraft using the D-1–5U parachute. Irina Solovyova, a psychologist as well as the first back-up to Tereshkova, has said, 'The reason for parachute training was that it helped cosmonauts to control their emotions at the launch stage and to prepare them psychologically for encounters with danger.' Solovyova has made over 5,000 jumps and is one of the most experienced parachutists in Russia.[7]

The latest selection of cosmonauts, recruited in 2003, has completed their first parachute training cycles at the Russian airfield in Tambov. It was supervised by Air Force cosmonaut Konstantin Valkov, who has completed over 500 jumps himself, and Energiya cosmonaut Oleg Skripochka. They used an Mi 8 helicopter and seem to have undertaken a similar cycle of jumps as the 1960 group. Their training has been extensively covered in recent issues of *Novosti Kosmonavtiki.*[8]

REFERENCES

1 Sotheby's catalogue, December 1993, Russian Space History sale
2 Sotheby's catalogue, March 1996, Russian Space History sale
3 Thorny Road to the Stars, by Aleksander Zheleznyakov, Spaceflight **44** September 2002; Mark Shuttleworth's diary, on his website African in Space, www.firstafricaninspace.com
4 ITAR-TASS news agency, Moscow, in Russian, 0735 GMT 12 Feb 2004
5 Private interview with Vasiliy Tsibliyev by Bert Vis in 1992; Specialist training schedule for Buran cosmonauts. Desert Training Exercise of Journalist Candidates for Space Mission, Red Star, 6 November 1991
6 Cosmonaut candidate drowned during training exercise, TRUD, 6 August 1993
7 Goals, Exercise of Cosmonauts Special Parachute Training Programme, Red Star, 27 July 1991
8 Novosti Kosmonavtiki: OKP Training review 2004, issues 4, 8, 9 and 10

One of TsPK's buses, used for transporting groups of cosmonauts and other centre personnel

The entrance hall of the Headquarters and Administration Building. On the wall at the back are portraits of all commanders of the Cosmonaut Training Centre, including the present one, Lt.-General Vasiliy Tsibliyev

The 'White Room' in the main administrative block. This is the main meeting room in the centre and here, it is holding a senior staff meeting. The front row includes Aleksandrov (representing Energiya), Morgun, Afanasyev, Yegarov, Grekov and Mayboroda

The doorway to the experiment building, which was the first training hall in the centre

A Tupolev Tu-134 jet that is used for transporting crews and TsPK officials

A cooling structure within the training centre

Korpus 3, home of the medical department and the TsF-7 centrifuge

The central heating boiler complex. The year '1973' has been put up on the smokestack by using different coloured bricks

The wing of the Profilactorium, which is the home to the NASA and ESA offices in Star City, and to astronauts training for their missions

Korpus 3A, the TsF-18 centrifuge building

The Swedish-built TsF-18 centrifuge

Korpus 3A under construction in the early 1970s

A view of the back of the Engineering and Simulator Building. On the left, Korpus 1 is visible

The office of the 1987 cosmonaut selection group, as it looked in 1994. These offices are located in Korpus 2

In 1981, a new wing was added to Korpus 2, which would become the home of the planetarium. It's dome is just visible in the picture

The 'White Room' in the Headquarters and Staff Building is used for official functions, such as the awarding of certificates to mark the completion of basic cosmonaut training, or the passing of exams to complete flight training

The rear of the wing of Korpus 2 that houses the planetarium

The KTOK, initially built for the Buran program, is now the principal training facility for the Russian section of the International Space Station. Simulators of the Zarya and Zvezda modules are located in the hall on the left

Mir training modules are placed on the floor of the Hydrolaboratory. The floor can be lifted above the water level for replacing modules, or for cleaning and maintenance

The door between the Hydrolaboratory and the Annex. Training modules can be hoisted from one building to the other by cranes

The large training hall in the KTOK was meant for a full-scale Buran mock-up, but is now the home of the Zarya and Zvezda training modules (left). Behind the curtain, there is still a fixed-base and a motion-base simulator of Buran

The opposite side of the KTOK hall, where the simulators for the Mir modules Spektr and Priroda still stand. To their left is a Buran flight deck simulator, with all its wiring disconnected and hanging idle in the cockpit

On the road from the entrance of the training centre to the Headquarters and Staff Building, the granite slabs on this monument commemorate all Soviet spaceflights from Vostok 1 up to and including Soyuz 11

'Soldier's Lake'. When the Hydrolaboratory is emptied, the water is released into the lake through the pipes in the foreground. The lake is a popular swimming spot for the conscripts that serve in TsPK

A view of the Mir hall with the Soyuz 2 capsule, which is part of the display shown to tour groups

After cancellation of the Buran programme, its training facility that was still under construction was left to the elements. It is now slowly deteriorating

The interior of the facility has fallen victim to people spraying graffiti on its walls

Aerial view of TsPK (bottom) and Star City (top) looking west. At the top left of the photo, which was taken in or shortly before 1999, the Chkalovskiy Air Base's runway is barely visible

This statue, which symbolises and is named 'Science', was moved to TsPK from the VDNKh in Moscow

'Technology', is the twin statue of 'Science'

The Cosmonaut Group of the RGNII TsPK

The history of the development of the Vostok programme and the politics behind the agreement to fly a human in space was examined in *The Rocket Men*, published by Springer-Praxis in 2001.[1] This book looks at the selection of the cosmonauts who would fly Soviet and Russian space craft.

In the 1950s, there was a lot of experimental work on biomedical research and its relationship to a first flight in space by a human. There was a long debate about which type of person would be most suited to a cosmonaut adventure into space after a comparatively short course of training. Some argued for the men of the submarine fleet, while others spoke for men with experience of high-altitude activities, such as parachutists and mountaineers, for whom courage and endurance were by-words. But most of those present favoured aviators and the one requirement they thought obligatory would be an engineering background. The scientists fixed upon flyers for their need to think fast and make a decision in a split second – being used to making decisions under stress. A cosmonaut would need to be an all-round

The Air Force cosmonaut team. This picture was taken on its 25th anniversary in 1985

specialist to do the job of pilot and navigator – watching instruments, communicating with Earth, taking notes and pictures, and flying the ship. In choosing the candidates, however, they also had to 'pay attention to spiritual and ethical qualities, ideological views and social attitudes.'[2]

THE FIRST SELECTION

In July 1959, a conference of specialists in aerospace medicine, from the USSR Academy of Sciences, Korolyov's OKB-1, and various scientific research institutions, created the general plan for the selection and training of the first cosmonauts. Preliminary selection of the pilots would be made by the units of the Air Force. Then there would be rigorous clinical and psychological testing in central Moscow. This would be overseen by Major-General Konstantin Fyodorovich Borodin of the Soviet Army Medical Service, who is described as the President of the Central Flight Medical Commission. Another member of the Commission was Colonel A.S. Usanov, Commander of the Central Aviation Scientific-Research Hospital, where the main testing of candidate cosmonauts would be carried out. Another key official was Major-General Aleksandr Nikolayevich Babiychuk, who from 1959 was flag officer doctor on the Soviet Air Force General Staff, reporting to the Commander in Chief of the Air Force, Marshal Konstantin Andreyevich Vershinin.[3]. Other members of the commission included Colonel Yevgeniy Anatoliyevich Karpov, who would become the first Commander of the training centre in January 1960, and Colonel Vladimir Ivanovich Yazdovskiy from the Medical Services, who became the cosmonaut doctor who oversaw the first flight of Yuri Gagarin.

In 1959, 29 candidates passed the tests at the Central Military Scientific Aviation Hospital (TsVNIAG) in Moscow, out of the 154 short-listed. In that era, this was the equivalent of the Medical Commission. These 29 then went before the Credential Committee, the top government committee that passed cosmonauts and gave them the title of Candidate.

The Cosmonaut Training Centre was established, under the command of Karpov, on 11 January 1960. In the structure of the centre, authorisation was given to recruit a group of up to twenty cosmonaut candidates. The first cosmonauts were assigned by order of the Air Force, dated 7 March 1960. They were Lieutenant Aleksey Leonov, Senior Lieutenants Ivan Anikeyev, Valeriy Bykovskiy, Yuri Gagarin, Viktor Gorbatko, Grigoriy Nelyubov, Andriyan Nikolayev, German Titov, Boris Volynov, and Georgiy Shonin, Captain Pavel Popovich, and Engineer Captain Vladimir Komarov. A further order on 9 March 1960 added Senior Lieutenant Yevgeniy Khrunov to the group. Further members of the first selection were added by an order issued on 25 March, when Senior Lieutenants Dimitriy Zaikin and Valentin Filatyev joined the group, and another four candidates – Major Pavel Belyayev and Senior Lieutenants Valentin Bondarenko, Valentin Varlamov and Mars Rafikov – joined on 28 April 1960. The last member to join was Captain Anatoliy Kartashov, by an order issued in June 1960.

Войсковая часть
Штамп части
54876

Форма 4а 9

10 марта 19 60 г.
№ 49

кубинки ПРЕДПИСАНИЕ

Кому Капитану Поповичу П.Р.
(воинское звание,
фамилия и инициалы)

Предлагаю Вам 12 марта 19 60 г. убыть в в/ч 26266 гор. Москва, в распоряжение командира части, для
(куда, в чье распоряжение
прохождения дальнейшей службы
и для какой цели)

Срок прибытия 14 марта 19 60 г.

Основание: Приказ ГК ВВС № 0267 от 7.3.1960 г.

Для проезда выданы воинские перевозочные документы—требования
за №№ ВИ-157538

Начальник штаба
Войсковой части 54876
м. п.
Подписи: гв. подполк. (Карякин)

Линия отреза

Командиру (начальнику)

На Ваш № ____ от ____ 19__ г.

сообщаю, что
(звание, фамилия и инициалы)

____ 19__ г. прибыл

____ 19__ г. Подписи

Г-146656 7.5.56 ЦТ МО—5514

The formal order transferring Pavel Popovich to the team in 1960

The nine who did not pass the selection process have also been identified. They were N.I. Bessmertnyy, B.I. Bochkov, G.A. Bravin, G.K. Inozemtsev, Yevgeniy Karpov (the first commander of the training centre), L.Z. Lisits, V.P. Siderov, I.M. Timokhin and M.A. Yefremenko.[4]

The first selections of the group began their training on 15 March at the Frunze central airfield (the Khodynka airfield) in downtown Moscow. The cosmonauts were, in the main, young and inexperienced pilots, with only Belyayev and Komarov having graduated from a higher education institute. Much of the early training was boring and repetitive and also involved a lot of physical exercises. Leonov has described it as if they were training for the Olympics. But it quickly became clear that it would not be possible to train all twenty of the candidates to the highest standards necessary to undertake a space mission. Instead, it was decided to create a group of 'immediate preparedness' – six candidates selected for consideration for the first Vostok flights. This group system was used by space planners for many years to prepare crews to a peak for their mission. The first group of six cosmonauts were identified to do the accelerated training programme under the command of Col. Mark Gallay, a very experienced pilot and a Hero of the Soviet Union. The six were

German Titov doing physical training on a swing in the early 1960s

Popovich, Titov, Gagarin, Nikolayev, Varlamov and Kartashov.[5] They were given the nickname 'The Lilies', reflecting perhaps a touch of jealousy. It is based on the 'Lilies of the field'.

Their training was tough and soon two candidates were disqualified due to medical problems. Kartashov, suffered pin point bleeding around his spine while undertaking a centrifuge test of 8-g and Varlamov suffered a neck injury while diving during a swimming outing at a lake. Neither recovered their flight status and both stood down from the team. They were replaced in the top six group by Bykovskiy and Nelyubov.[6] By October, the group was in full training.

On 6 January 1961, they took their state exams and were awarded the title of Pilot Cosmonaut. The exams were very exacting, but Gagarin, Titov, Nikolayev and Popovich were awarded excellent marks. The group met with Vershinin, who commended them and said that the mission was close. After this process, the State Commission ranked the candidates, with Gagarin, Titov and Nelyubov receiving the top marks. In his dairies, Kamanin agreed that these three stood out, but also commented that Nikolayev was quiet and Bykovskiy was introspective and withdrawn. He called Popovich a 'mystery' but added no further explanation.

These two images show Yuri Gagarin undergoing centrifuge tests in the Central Clinical Institute prior to Vostok 1. (Courtesy Dr. Kapitolina Sidorova)

The 1960 group did suffer casualties. Varlamov left on 16 March 1961 under order of the Air Force 0321, following the problems after his diving accident. He stayed on at Star City, working in the training centre until his death in 1980. Bondarenko, the youngest member of the team, was killed as a result of the injuries he sustained in his isolation chamber test (see page 76). Kartashov also stood down from the team, on 7 April 1961, under order of the Air Force number 0462. He left Star City but remained in the Air Force and became a test pilot. He currently lives in Kiev in the Ukraine.

There were further losses to the group for disciplinary reasons. Mars Rafikov was dismissed on 24 March 1962 under order of the Air Force number 060. His dismissal was due to family problems as well as a pending divorce and he had also gone absent without leave along with another cosmonaut. This was at a time when Kamanin was very concerned about the behaviour of the cosmonauts following a number of accidents and drinking allegations that had come to the attention of the cosmonaut leaders. On 17 April 1963, three cosmonauts were dismissed for an incident while drunk, when they got involved with the police and were arrested. They were given an opportunity to offer their apologies, but Nelyubov refused. Leonov gave his version of the incident. 'Anikeyev, Filatyev and Nelyubov were out at a restaurant late one night drinking and got into trouble coming back to the centre. In effect, they disrupted their and others' training programmes. All of us had agreed on certain rules of behaviour and the penalty for violating them was expulsion from the group, so a vote was taken and they were expelled.' This was confirmed by order of the Air Force number 089.[7]

After waiting nearly ten years to fly, the last unflown member of the group, Dimitriy Zaikin, was medically disqualified under Air Force order number 01075, dated 25 October 1969. He had failed a centrifuge test some weeks before. The identities of the first twenty cosmonauts were not formally disclosed until a set of articles in '*Izvestia*' in 1986 by Yaroslav Golovanov. This was the first time that the names and fates of the eight who had not flown were made public. Prior to this, they had been known by their nicknames.

The formal welcome of Yuri Gagarin to Star City in April 1961. To his left is Yevgeniy Karpov, the first commander of the Training Centre

The first female selection

The exact reasons why a female flight came about are obscure, but the timescale and events seem to be as follows: Towards the end of 1961, Sergey Korolyov wrote to Kamanin (who at that time was the assistant to the Deputy Chief Commander of the Air Force in charge of cosmonaut training) arguing that, in the near future, sixty cosmonauts of various specialties would be needed, including women. At that point, there were only seventeen pilots in training from the Gagarin selection. The group in training for the 1962 Vostok flights had lost Gagarin and Titov, who were on public relations duties, although Komarov and Volynov had replaced them. Korolyov's suggestion also reflected his desire to fly civilian engineers and doctors into space in an era when, within Soviet space circles, there was an optimism about the advance of cosmonautics and the speed of development. Competition with the Americans was very important and the Soviet Union had gained an advantage with the orbital flights of Gagarin and Titov in 1961. At that time, American women were lobbying to get into the Mercury programme, especially Jerri Cobb, who was a first class pilot. These developments were being monitored in the Soviet Union.

WOMEN COSMONAUTS ARRIVE AT STAR CITY

The decision to select women was made at the highest levels, including the Central Committee of the Communist Party. In Kamanin's dairy on 24 October 1961, he wrote, 'After Gagarin's flight, I persuaded (the Air Force Chief Commander) Vershinin, (the Chief Designer) Korolyov, and (the President of the Academy of Sciences) Keldysh to organise the selection of a small group of women. This project, however, is moving with great difficulty. In my view, it is necessary to prepare women for space flights mainly for the following reasons:

- Without doubt women will fly into space and it is therefore necessary to start preparations for women's flight now.
- Under no circumstances can one allow the first woman in space to be an American. This would hurt the patriotic feelings of Soviet women.
- The first Soviet woman cosmonaut will become an agitator for Communism as great as Gagarin and Titov.[8]

Following the debate about who should be recruited for the 1960 selection and the connection with aviation, in 1960 they selected only combat pilots, despite the reservations of Korolyov. For the female candidates, it was decided that they should have a sports parachute background, as they would have to land by parachute from their Vostok capsule. Parachute jumping is a skill that they would not have time to teach potential candidates, so they would look to recruit the candidates from aviation clubs. Kamanin contacted the Central Committee of the Voluntary Association of the Army, Aviation and the Navy to assist him. He asked for details of 200 candidates, but after the initial screening in the sports clubs of the European part of the Soviet Union, only 58 met the criteria. On 18 January 1962, Kamanin saw 23 of the 58 candidates. He felt that none were fully qualified, but he needed women who were young, physically fit and had had flight or parachute training for at least six months. By 28 February 1962, the list had been reduced to only seven. Kamanin ranked Irina Solovyova, Valentina Tereshkova and Tatyana Kuznetsova as the best, followed by Yefremova and Vera Kvasova (of whom nothing is known) and Ludmilla Solovova (who was a sports pilot), and then Marina Sokolova (who was an instructor pilot). After a very detailed screening process, five were identified as potential candidates. Four of them came from parachute jumping but had different levels of expertise. Solovyova, for example, was a Master of Sport with over 800 jumps and was a member of the Soviet parachute team at the World Championships. The fifth candidate, Valentina Ponomaryova, came from sports aviation and had done only eight jumps. She was a third category jumper, but a very skilled pilot.[9] Ponomaryova was not included in the first screening, but was encouraged to apply late by Mstislav Keldysh, the President of the Academy of Sciences. Little is known about Kuznetsova before her selection, aged 20.[10]

The order for the admission of the women candidates was given on 12 March 1962 (order number 67), admitting Solovyova, Kuznetsova and Tereshkova to the team. It was followed on 3 April 1962 by a second order (order number 92) admitting Yerkina and Ponomaryova. They reported for training at the centre immediately

Meeting of the State Commission admitting the 1962 women into the cosmonaut team. On the left are Karpov and Gagarin; on the right are the women candidates

and were enrolled as privates in the Soviet Air Force, as they now belonged to a military unit. They found the transition to military discipline very difficult and their commanders had problems dealing with them. The training centre was really only offices, so they commuted into Moscow for training and lived in a rehabilitation centre. They went through the same training as the men, but although there was a lot of opposition to the prospect of a women's flight from the first group, this did not stop them helping the women in their training and, according to Ponomaryova, 'teaching us how to deceive physicians and how to pass tests easier.' Like the men's group, they had specialists from Korolyov's design bureau available to teach them about spacecraft systems.

When they completed their training and passed their state exams in December 1962, Kamanin asked them if they wanted to become regular officers in the Air Force. They consulted with their male colleagues and decided it was necessary to join the staff of the Air Force. Ponomaryova said in an interview, 'It was necessary to be like everybody else.' This was an important decision, because when they wanted to get rid of the women after Tereshkova's flight, the fact that they were Air Force officers was a significant obstacle to their transfer or removal from the team. The female flight was delayed due to the construction of the spacecraft and the fact that their adapted flight suits were not ready, so their training was extended. The mission did not occur until June 1963.[11]

In his diary entry of 27 December 1962, Kamanin wrote, 'A decree ordering the training of sixty cosmonauts has been laying around, and suddenly the leadership wants to enforce it. Fifteen new trainee male cosmonauts and fifteen women are to be recruited – an overall total of twenty by the end of 1962 and forty by the end of 1963. And crews are to be formed and trained, even though there are no spacecraft being built for the missions.' This reflected desires within the Soviet leadership for ambitious space plans.

In the summer of 1962, the Vostok training group was reduced to four when Nelyubov failed a centrifuge test. He was replaced by Komarov, who in turn was replaced by Volynov when he developed a heart problem. In late 1962, twenty-five applicants passed the selection process and the medical tests at the TsVNIAG for entry into the Air Force cosmonaut team.

MEDICAL SELECTION OF COSMONAUTS

One of the best accounts of the process of medical selection was given by Colonel Eduard Buinovskiy, who was part of the 1963 selection, in an article in *Spaceflight* magazine called 'A dream which almost came true'. This has been reproduced here with his permission. He has recently released his life story in a new book published in Russia in the autumn of 2004. Buinovskiy was asked to apply for the next Air Force selection on 9 June 1962 by his unit commander. This seemed to be the route for many applications, as the call had gone out to military units for potential candidates.

'Soon I was called to a clinic for medical check-up. The Military Air Force Clinic was situated in a big grey stone building in the centre of Moscow. There I came to find out just one thing: whether my health was good and I was fit enough to carry on with further tests and experiments. There were so many candidates that the doctors were operating like assembly line workers: 'Open, close, turn, bend – fit/not fit – Next!' For some 'potential heroes', the process was so quick that after thorough examination by several doctors, the poor guys found themselves back on the street not fully understanding what had happened and why they had not been chosen to be cosmonauts. I was lucky this time and passed all the examinations, though there were several moments when I dreaded that it would be the end of it.

'The second stage of the examination was carried out in the Central Scientific and Research Aviation Hospital, situated in an old mansion in one of Moscow's parks. The hospital was mainly used for conducting routine medical check-ups of jet pilots, so the doctors were highly professional and knew their job very well. But this time they were facing a difficult task – to select people capable of working under the space conditions, not knowing specific norms and criteria. The short-term flights by Gagarin and German Titov gave too little knowledge to work out an objective methodology for selecting future space pioneers.

'Each of us received a sheet of paper with the list of tests and procedures we were to undergo while staying in hospital. I still keep that paper – there were 25 types of examinations listed in it. Some of them were familiar, like surgeon, ophthalmologist, psychologist; others sounded rather mysterious: pressure chamber for 'diving', vibrostand, radical speeding centrifuge, etc. Tests like that were routine for professional pilots, but not for us, representatives of other professions. So we were kind of scared of such tests and did not expect a positive outcome. There were also examinations, which seemed to be funny and not serious in the beginning. For instance, you enter a room and see an ordinary swing, like the one they have at children's parks. A pretty nurse tells you to sit on it and starts swaying you. It feels good in the beginning – you laugh, joke and tease the nurse. But then after 10, 15, 20 minutes of swinging, you forget all your jokes and start looking for a bucket – normally it was there in the corner modestly waiting for its time to come. Then there was another test, which also seemed to be a game in the beginning but in the end we were crawling out of the room literally on all fours. Another attractive-looking nurse tenderly ties you to a table. You lie silently and motionless for about 30–40 minutes. Then suddenly the table spins 45 degrees, and you find yourself suspended head over heels, and continue hanging like this for a long, long time. We jokingly called this test 'Gestapo tortures'. But being young and strong we somehow managed to survive all the 'tortures'.

'During the short time I stayed in the hospital, I turned into an expert in medicine! I knew exactly why the numerous blood tests were taken, when it was better to undergo a certain examination – in the morning or in the evening – what products I should eat and what to avoid in order to get positive results. Every day, we raced from one examination room to another, had our sheets of paper filled in with test results, told each other about our feelings and experiences, and even found time for joking and friendly teasing. Medical selection board was the only part of that long, hard way towards space glory when relations between the applicants were even and friendly, without envy and intrigues against each other. At least, I didn't perceive my roommates as competitors standing in my way. All of us were glad to see somebody's success and sincerely sad when somebody had to leave us.

'I was among the lucky ones who stayed till the very end. The second stage of the selection was over. Then we were told 'Wait!' The waiting period lasted for about two months. It was so difficult and tiresome – to wait while your fate was being decided somewhere. In order not to miss the crucial call I tried to be close to the phone all the time. I even forgot all my dates with girls and rushed straight home after work.'

New cosmonauts

'Finally, on 8 January 1963, I faced the Mandate Commission headed by General Kamanin, the cosmonaut corps commander. I answered several questions and then heard the verdict: 'Senior Lieutenant Buinovskiy, you are enlisted in the cosmonaut detachment of the Air Force!' Should I describe what I went through and the storm of emotions I felt when I left the room where the Commission had its sitting? This was one of the brightest, most unforgettable moments of my life. The order for the assignment of the group was made on 10 January 1963 (Air Force order number 14).

'On 25 January 1963, Vitaliy Zholobov and myself (we became friends in the hospital) arrived at the mysterious Centre for Cosmonaut Training. At that time there was no Zvyozdnyy Gorodok, just a few two- or three-storey buildings in a wonderful pine tree forest. We joined the team of other newcomers; all in all there were 15 of us. In fact, ours was the third team in the detachment; Gagarin's was the first one, then the group of girls, and then us. But our detachment was a unique one – for the first time it was decided to incorporate aviation engineers and representatives of the rocket forces (engineers included) into the training programme. So our 'international' team consisted of seven engineers (Engineer Lt-Colonel Lev Demin, Engineer-Major Yuri Artyukhin, Engineer-Captain Pyotr Kolodin, and Senior Engineer-Lieutenants Eduard Buinovskiy, Vitaliy Zholobov, Vladislav Gulyayev and Eduard Kugno) and eight pilots (Lt-Colonel Vladimir Shatalov, Majors Aleksey Gubarev, Anatoliy Filipchenko, Georgiy Dobrovolskiy, Anatoliy Kuklin, and Lev Vorobyov, and Captains Anatoliy Voronov and Aleksandr Matinchenko). I am aware that today only a few of the above names sound familiar to the general public – not all of us managed to complete the training and only seven of those who did were lucky to participate in space missions. But at that time, almost forty years ago, we were just a group of 15 young men selected to become heroes.

'The hotel we were to live in had a small, cosy canteen, where we got acquainted with the guys from the first team. The girls, led by Valentina Tereshkova, were actively playing the roles of mediators. Later on, at dinner, in an easy, almost family atmosphere, we met the four heroes (Yuri Gagarin, German Titov, Andriyan Nikolayev, and Pavel Popovich). My dream was about to come true – I was to become a cosmonaut and fly into space.'

This is a graphic first hand account of what was a very new process for selection,

The original prime crew of Soyuz 13 taking a break from the flight simulator. (from left) Lev Vorobyov (Commander) and Valeriy Yazdovskiy (Flight Engineer)

and one that was very different from the previous ones. The trainees include one from VMF Navy Aviation, two from the PVO Air Defense, four from the RVSN Strategic Rocket Forces, and eight from the VVS Air Force. All were transferred to the Air Force. The identity of the six candidates who failed in their application is also known. They were Boris Belousov (who reapplied successfully in 1965), Valentin Sidorenko, Korotkov, Suvorov, Georgiy Beregovoy (who was added to the selection in 1964) and Georgiy Katys (a civilian scientist who was selected for Voskhod and Soyuz training at a later date).[4]

The new group were older and more experienced than the majority of the first selection and had all been to, and graduated from, military academies. Some had held command positions and were of similar and even higher rank than the first selection. This led to real tension within the growing team. In an interview in June 1990, Leonov stated, 'They were interested in rank, titles and position and we were pushed aside.' There was a lot of resentment between the two groups and the senior men chafed at taking orders from younger men like Gagarin.[7]

On 1 February 1963, there were thirty-five cosmonauts in training, divided into six groups.

- Group 1: Four female cosmonauts (Solovyova, Ponomaryova, Tereshkova, Yerkina) in final training for two simultaneous flights in March 1963
- Group 2: Three male cosmonauts (Komarov, Bykovskiy, Volynov) in training for two or three individual flights of over five days duration in the second half of 1963

- Group 3: Four flown cosmonauts (Gagarin, Titov, Nikolayev, Popovich) in academic training but also heavily occupied with public relations tasks
- Group 4: Six cosmonauts from the first group – not trained for Vostok and available for Vostok or Soyuz flights from 1964 onwards (Nelyubov, Shonin, Khrunov, Zaikin, Gorbatko, Filatyev)
- Group 5: Seven pilot-cosmonauts, just selected and starting training
- Group 6: Eight engineer-cosmonauts just started training.

In September 1963, a group of eight cosmonauts (Belyayev, Komarov, Shonin, Khrunov, Zaikin, Leonov, Gorbatko and Volynov) were in training for the 1964 missions. There was also increased pressure from Korolyov on the Air Force to include civilians in the crews for upcoming missions. At the beginning of 1964, the Soviet cosmonaut team had a complement of 33, six of whom had flown in orbit. In 1963, three candidates, including Beregovoy and Sidorenko, had been turned down due to their age. The age limit for candidates was 35, which had been set by the Communist Party's central committee. Marshal Sergey Rudenko, the Minister of Defence, had an idea that a small team of older candidates would give added experience to the team, but the Mandate Commission met at the turn of the year and turned down the request. They cited the 'scepticism by cosmonauts, who had patiently waited in line to make a mission and were now likely to be leapfrogged by these candidates.' Kamanin was overruled by Vershinin and, with reluctance, did not block the decision. Sidorenko did not pass the medical and was not considered, but Colonel Beregovoy, who was already a Hero of the Soviet Union and a top test pilot, was transferred on 25 January 1964 (Air Force order number 072). He was attached to the 1963 selection for training and administration purposes. Later in the year, as plans for the Voskhod programme were being finalised, a number of Air Force candidates were short-listed to work on the programme. They were all Air Force doctors and experimenters and included Colonel Vasiliy Lazarev, a pilot and military doctor, and Vladimir Degtyaryov, an Air Force doctor who specialised in space medicine. He was not assigned to the Voskhod mission, but was selected in 1965 as part of the next formal selection. Lazarev did serve as back-up to the Voskhod 1 mission in 1964, but then returned to his normal duties.

In January 1965, fifteen candidates passed their state examinations, completed their general space training and were given the designation Cosmonaut. There were thirteen from the 1963 group, plus Beregovoy and Kuznetsova. Permission was given to recruit up to forty more cosmonauts, which would, for the first time, include civilian cosmonauts and scientists. (See page 147). They also agreed that the cosmonauts would be between 30 and 32 years old. There were 284 candidates identified from military ranks, of which sixty passed the tests. But it was agreed that a limit of twenty military men would be selected (with the rest being civilians) to join in 1966.

The next selection, the third by the Air Force, was the 'Young Guard' selection, with 22 pilots, engineers and a doctor selected under order of the Air Force number 0942, dated 28 October 1965. Some members of the group were as young as 22. This reflected a view that young pilots would be easier to train as space pilots, and this was seen in the 1960 selection as well. Leonov commented that this was a mistake and certainly by the mid 1980s, the criteria for selection had changed to take into account previous experience. 'We did not need space careerists,' Leonov said.

The new selection consisted of Lieutenants Valeriy Voloshin, Vyacheslav Zudov, Leonid Kizim, Pyotr Klimuk, Anatoliy Fedorov, Aleksandr Skvortsov, Oleg Yakovlev, Vasiliy Shcheglov, Gennadiy Sarafanov, Aleksandr Kramarenko, Ansar Sharafutdinov and Aleksandr Petrushenko (all pilots), and Engineer-Major Boris Belousov, Engineer-Captains Gennadiy Kolesnikov, Eduard Stepanov and Mikhail Lisun, Senior Engineer-Lieutenants Yuri Glazkov, Valeriy Rozhdestvenskiy and Yevgeniy Khludeyev, Navigator-Lieutenant Vitaliy Grishchenko and Sergeant Vladimir Preobrazhenskiy (all engineers). Preobrazhenskiy became the only non-commissioned officer appointed to the team. The doctor was Major of Medical Services Vladimir Degtyaryov. Belousov was older than the agreed criteria, but was allowed to join the team. He was a member of the Rocket Forces and transferred to the Air Force.

Degtyaryov withdrew from the selection almost immediately because he did not want to give up his research to train for many years to fly into space. He was replaced immediately by Lt-Colonel Vasiliy Lazarev, who backed up the Voskhod 1

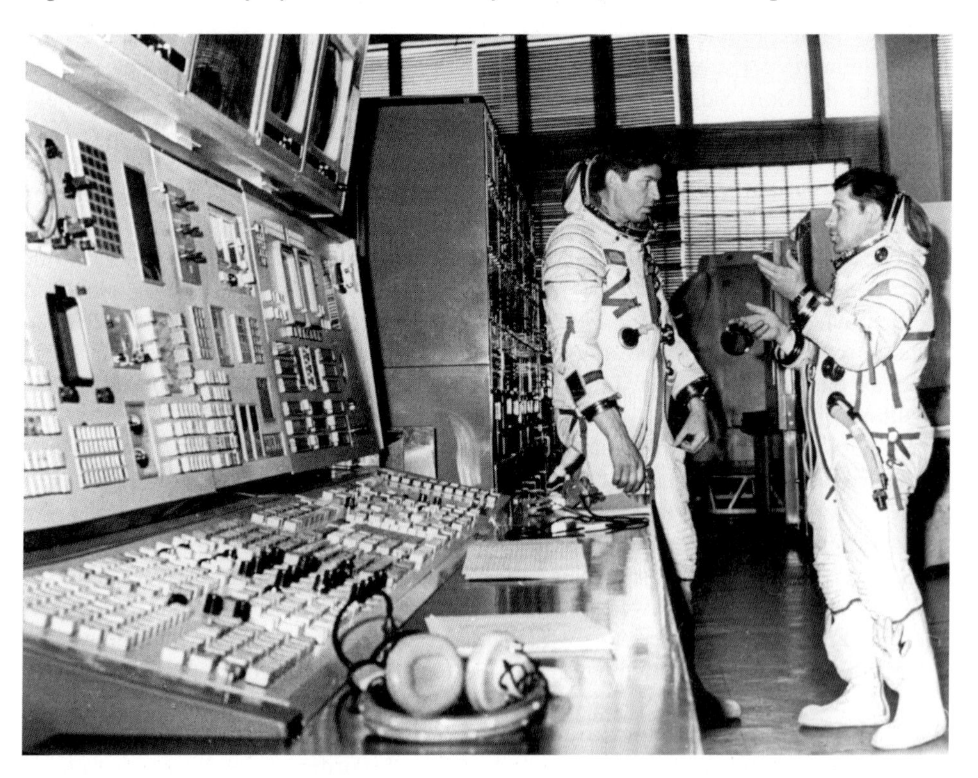

The Soyuz 32 crew in the Salyut 6 simulator hall. On the left is Valeriy Ryumin and on the right is mission commander Vladimir Lyakhov mission and was an experienced aviation doctor and pilot. This came on 17 January 1966, under order of the Air Force number 037 and Lazarev was attached to the 1963 selection for training and administration. During the year, a number of cosmonaut candidates withdrew from the team due to medical and political issues, but the Air Force still had ambitious plans for manned space flight and in 1967, Kamanin called up twelve more pilots and engineers to form the fourth Air Force selection. Seven of these had been candidates for the 1965 selection and had just missed out, so they were not re-interviewed. They had been told to wait for a future call.

The selection process started on 28 February 1967 and the intention was to select twenty candidates. Due to various errors, only thirteen were selected and when the order was signed, only twelve were included. They were Senior Lieutenants Valeriy Beloborodov, Vladimir Kovolyonok, Vladimir Kozelskiy, Vladimir Lyakhov, Yuri Malyshev and Viktor Pisarev (all pilots), and Engineer-Majors Vladimir Alekseyev, Mikhail Burdayev and Nikolay Porvatkin, Engineer-Captains Sergey Gaydukov and Mikhail Sologub, and Senior Engineer-Lieutenant Vladimir Isakov (all engineers). Their transfer came under order of the Air Force number 0282, dated 12 April 1967, for Alekseyev, Burdayev and Porvatkin, who left an Air Force research institute, and under order number 0369, dated 7 May 1967, for the rest of the group. They arrived at Star City soon after the death of Komarov on board Soyuz 1 and there was a very despondent feel to the town. Much of the ambition shown throughout the early 1960s had now gone, in part due to the delays in missions, the death of Korolyov a year before, and the problems inherent in developing new technology. One feature of this selection was the inclusion of three military scientists from the Science Research Institute for Air Defence Forces, because it was anticipated that there would be a need for complex, in-depth experiments for the military use of space. Alekseyev, Burdayev and Porvatkin all became involved in the Soyuz VI and Almaz programmes, but their skills were never called upon in space.

In late July and in September 1967, Kamanin's diaries mention two planning documents which reflected future plans and manning levels with the cosmonaut team and the training centre. The first dealt with the training of civilian cosmonauts and two phases of training were planned; the first phase at MOM institutes and Minzorar, and the second at TsPK and the VVS (Air Force). In addition, fifty new Air Force pilots were identified for space duty in three groups in 1968, 1969, and 1970. They would be ready for the planned large number of 7K-VI and Almaz flights beginning in 1972. Soviet leader Leonid Brezhnev also wanted to see more Voskhod flights.

The second document dealt with the General Staff's space plans, which were impressive. In 1968–1975, they foresaw no less than twenty Almaz space stations, fifty military 7K-VI missions, 200 Soyuz training spacecraft flights and 400 Soyuz space transport flights. This was based on the assumption that the crew of the military space stations would be rotated every fifteen days. That would require 48 transport spacecraft per year, implying not less than thirty ready crews, with three cosmonauts in each crew (this in turn implied that each cosmonaut would fly 1.5 space missions per year). Since supplies would have to be delivered to the stations, that would require another 200 additional transport spacecraft launches. On top of all this was potential for civilian Soyuz flights, L-1, L-3, and various other civilian spacecraft – implying a total of 1,000 launches in the period. This would require 800 Soyuz-class launch vehicles, 100 Protons, and 10 to 12 N1 boosters. The inevitable conclusion for Kamanin was that most of the transport launches should be made by a reusable winged spacecraft, air-launched from an An-22 heavy transport, which was the goal of the *Spiral* project. Based on these plans, Kamanin foresaw a requirement for 400 active cosmonauts, organised in two to three aerospace brigades, supported by ten aviation regiments and including the TsPK training centre – altogether about 20,000 to 25,000 personnel by 1975. The cost of building new aerodromes and facilities alone would be 250 million roubles, all chargeable to the VVS, and the total cost would be expected to run into tens of billions of roubles per year.[12]

On 27 March 1967, the Central Committee and the Council of Ministers adopted a resolution to end the Air Force monopoly on training cosmonauts. This resolution set up the formal basis for the civilian organisations to set up their own teams and four took advantage of this decree, although Kamanin viewed this decision as a mistake. The first groups of civilians were formed on 22 May by the Academy of Sciences and on 27 May by the TsKBEM bureau. Final crew training would still occur at the training centre, but much of their basic work would be done in-house.[13]

In December 1967, the 1965 group took their examinations, but several failed and were removed from the team. Of the eighteen in the group, thirteen scored a '5', four scored '4' and one scored '3'. Belousov, Grishchenko, Skvortsov, Sharafutdinov, and Voloshin – the low scorers – were all dismissed from the cosmonaut corps. Sharafutdinov also had medical problems and was excluded for failing a medical. This, in part, hid problems that were part of the vetting process for cosmonauts. All cosmonauts were vetted by the KGB and some certainly suffered because of their political beliefs, or for criticising the system (which Kamanin was doing using the privacy of his diaries). In fact, Voloshin was not removed and instead received a command assignment, but Grishchenko lost his status because his grandfather lived in the West, Kugno was removed because he would not join the Communist Party, and Belousov's father-in-law had been a translator for the Germans and had served three years in jail. He had his supporters due to his considerable skills, but soon after selection, the KGB withdrew their support. Volynov, one of the 1960 group, had also had his selection to a mission delayed, because his mother was Jewish.[14]

The influence of the authorities on the cosmonauts was best illustrated by the fate of one of the 1963 candidates, Eduard Kugno. In his diaries, Kamanin wrote, 'He is ideologically and morally unstable. Answering a question from the Deputy Head of Political Affairs about why he had not joined the Party, he (Kugno) said. 'I will not join this party of swindlers and lickspittles.' It is unpleasant to hear such things from any Soviet citizen, but from the mouth of a cosmonaut candidate, these words sound like a verdict.' Kugno was dismissed after a year in the team.[15]

COMMAND STRUCTURE REVISED

In September 1960, members of the first selection who had not attained a higher degree were enrolled in the Zhukovskiy Higher Engineering Academy. They completed a correspondence course and were joined by some of the female group a year later. Gagarin, G. Titov, Nikolayev, Leonov, Popovich, Bykovskiy, Khrunov, Gorbatko, Zaikin, Volynov, and Shonin all received their diplomas in early 1968 and Khrunov graduated with honours. Ponomaryova and Solovyova graduated in the second half of 1968, leaving only Tereshkova, Kuznetsova, and Yerkina, who completed their courses in 1969. Soon after gaining this qualification, however, Gagarin was killed in an air crash and his death had a major effect on the Air Force cosmonauts. They lost a leader and someone they all admired. His influence seems to be everywhere, even today, in the approach taken by the command structure of the Air Force and beyond. Since his space flight, he had been protected from harm and forbidden from flying, parachute jumping and other dangerous activities. He did a lot of public relations work until 1963, when he was made Deputy Commander of the training centre. He was returned to flight status in 1966. Gagarin did have problems adjusting to the fame, as did Titov, and there are lots of stories regarding his behaviour. But he was clearly marked for a senior command position in the development of the training centre.

In the early part of 1968, the cosmonauts were organised into three detachments: Nikolayev was in command of the first detachment, which was training for L-3, L-1, and Soyuz flights and it was anticipated that eighteen cosmonauts would be assigned to these missions. Popovich was in command of the second detachment, which was training for Almaz and 7K-VI military space missions. The third detachment was commanded by Nikolay Nikeryasov. This was the 'observer' detachment, consisting mainly of new cosmonauts undergoing basic training. On 3 July 1968, these assignments were amended as the positions of Chiefs of the Cosmonaut Detachments were confirmed and announced. Nikolayev became Deputy Chief of TsPK, replacing Gagarin, with Bykovskiy taking over as Commander of the First Detachment of Cosmonauts. Titov took command of the Second Detachment, with Popovich as his Deputy. According to Kamanin, there were some who were not pleased with these appointments. The General Staff also approved the creation of a fourth training detachment at TsPK, charged with flight, engineering, and experiment development and requiring an additional 200 staff.

On 13 December 1968, Kamanin reviewed the organisational structure of the NII-TsPK Gagarin Centre. There was a commander, three deputies, 700 staff, and twelve MiG 21s for flight training (eight single-seat combat aircraft and four two-seat trainers). The three training streams for the cosmonauts were also still in place (orbital, lunar, and military), reflecting the plans within the system for human space flight. Leonov's L-1 group would complete their training on 20 January 1969 and Kamanin hoped that they would fly a lunar mission a couple of months later.

In January 1970, Belyayev died after a botched operation. This was a blow following the other untimely deaths of Gagarin and Komarov, but the 1970 selection process started on 17 February with 400 candidates, of which 154 were sent to

Moscow for screening. Only 21 passed the medicals and were declared fit enough for the final screening, when they had hoped to appoint up to 30. Kamanin noted that only nine of the sixteen cosmonaut candidates that eventually completed the arduous selection process were cleared by the KGB and Communist Party for acceptance for cosmonaut training. He felt this made the whole time-consuming selection process a waste of time, particularly as the VVS was reluctant to submit officers as cosmonaut candidates, fearing that if they failed the vestibular table tests, they would not only be rejected as cosmonauts, but would be unable to return to flight duty with the Air Force. The result was a final selection that Kamanin considered to be 'dullards, who are not intellectual, or literary, or sports enthusiasts; who are poor readers and not really interested in space flight or cosmonautics.'[16]

The new selections joined the team under order of the Air Force number 0505, dated 27 April 1970. The selection consisted of Captains Anatoliy Berezovoy and Vladimir Dzhanibekov, Senior Lieutenants Anatoliy Dedkov, Yuri Isaulov and Yuri Romanenko, and Lieutenants Vladimir Kozlov and Leonid Popov (all pilots). The two engineers were Engineer-Captain Valeriy Illarionov, who was a member of the Air Force, and Senior Engineer-Lieutenant Nikolay Fefelov, who was transferred from the Rocket Forces.

Kamanin reviewed plans for the Gagarin Centre on 6 June 1970. Within ten years, scientific institutes, housing, and training facilities would have been erected to

The 1970 Air Force selection formal portrait. (front row) Dzhanibekov, Romanenko, Popov, Kozlov, Isaulov, Illarionov and Fefelov. (back row) Berezovoy and Dedkov. This is a very unusual shot, as this was the first time a complete selection group had posed for a group picture

support 500 cosmonauts and Kamanin noted that the five-year plan ending in 1971 called for 140 cosmonauts to be in training. Instead, TsPK had 47 active cosmonauts plus the nine candidates, for which he blamed the low launch rate and the failure of industry to deliver enough spacecraft.

In September 1970, training for space station flights (which would become the norm in future decades) was taking shape, with twelve cosmonauts working on the civilian DOS station programme and 22 working on the military Almaz programme under the command of Popovich. Bykovskiy was still working on lunar missions and the 1970 group was still involved in basic training under the command of Volynov. But all these plans and training schedules were disrupted and put on hold with the death of the Soyuz 11 crew. This led to a two-year break in human flights and also heralded a number of changes in the command structure of the Air Force cosmonauts and the training centre itself.

The first major change was the retirement of Kamanin as head of cosmonaut training and his replacement by Shatalov. This was partly due to having a less than positive relationship with the new head of the Air Force, Pavel Kutakhov. Kutakhov was not interested in space and Kamanin had tried to recommend his deputy, Leonid Goreglyad, to take up the post, but he was overruled. This change was soon followed by Kuznetsov retiring as Commander of Star City and he was replaced by Beregovoy, with Nikolayev named as his deputy. They had been made Generals and were also Heroes of the Soviet Union, so for the first time, all the key jobs were occupied by cosmonauts. This coincided with the development of facilities in the training centre, with new training halls and simulators.

The 1976 Air Force group during a class session. (front row) Vasyutin, Ivanov, Kadenyuk, A. Volkov, Moskalenko and V. Titov. (back row) Saley, Protchenko and A. Solovyov

New pilots and a new direction

There was a gap of two years (from 1971 to September 1973) before another Soviet manned craft flew and there was no real need for more cosmonauts in the team. But in late 1975, they decided to recruit a sixth group of pilots. This selection was made with the Buran Space Shuttle in mind and nine pilots were assigned to the centre on 23 August 1976, under Air Force order 0666. They were Captains Aleksandr Volkov, Leonid Ivanov, Leonid Kadenyuk, Nikolay Moskalenko, Sergey Protchenko, Vladimir Titov, and Yevgeniy Saley, and Senior Lieutenants Anatoliy Solovyov and Vladimir Vasyutin. All were immediately assigned to the Chkalovskiy Test Pilot School at Akhtubinsk for their first year of training, reflecting their need to be test pilots in order to fly the Buran programme. This group did not report to Star City until late 1977 to start their basic training and in 1978 they were joined by two more candidates, Captains Aleksandr Viktorenko and Nikolay Grekov. They were the seventh Air Force group, formed by order of the Air Force number 0374, dated 25 May, and the members of both of these groups were all older than their predecessors. Viktorenko and Grekov also went for test pilot training, graduating in May 1979 before starting basic training. In the summer of 1979, Vasyutin and Titov left the test pilot programme and reported to Star City for Almaz training and Protchenko left due to medical reasons, although he remained a test pilot for the Ministry of Radio Industry. The remainder of the group stayed in Akhtubinsk to improve their flying skills, but Ivanov was killed in an air crash while testing a MiG 23 at Akhtubinsk on 24 October 1980. He is buried at Star City. The remaining trainees commuted between the two locations as their training demanded. They would be the last Air Force candidates for eight years.

Forced retirements

On 26 January 1982, an order was signed which removed a large number of experienced cosmonauts from the team, although a number stayed on in command positions. Those removed included Bykovskiy, Leonov, Nikolayev, Popovich, Artyukhin, Demin, Filipchenko and Glazkov. Isaulov also left due to medical problems. He had been training for a long duration mission on Salyut 7, having been considered for other missions previously. This dismissal order was followed up in 1983 with one that affected the unflown cosmonauts.

In 1983, it was realised that many of the active cosmonauts had little chance of flying in space following the cancellation of a variety of programmes. On 20 April, a number of pilots and engineers were stood down, including Kolodin who, having received a number of prime assignments, had been very unlucky in having the missions cancelled. Alekseyev, Burdayev, Isakov, Kozelskiy, Porvatkin and Dedkov also left, but the ranks of the Air Force team were enhanced when four more of the 1976 selection (Anatoliy Solovyov, Moskalenko, Saley and Aleksandr Volkov) reported full time for Soyuz training. On 22 March Leonid Kadenyuk, a member of the 1976 selection, had left due to his pending divorce, but he continued test flying at Akhtubinsk. In 1988, he joined the team from GKNII VVS, who were training to fly the Buran. He left that programme in 1996 when the training was curtailed and moved to the Ukraine, where he had been born, taking out Ukrainian nationality. He then became the prime candidate for the Ukrainian mission on board the Space

Shuttle and flew as a Mission Specialist on STS-87 in November 1987. He was the only member of the team to fly for one of the new Republics following the break up of the Soviet Union. Two other Russian cosmonauts have been claimed as one of their citizens in space by other Republics, due to their ethnicity. Talgat Musabayev, who is a Kazakh, and Salizhan Sharipov, who was born in Kyrgystan, have received Hero's stars and pilot cosmonaut awards from these now independent countries.

By the time the next selection came about, a new criterion for selection was place. Candidates had to be over 30, but those with test pilot training could be over 35 years old. This led to two groups arriving at the same time, and they also went through a different selection procedure, going before the Mandate or State Commission. In 1987, five more pilots were assigned to the programme, having gone before the Mandate Commission on 26 March 1987. Two of these pilots were much older than the others, Lt-Colonels Vasiliy Tsibliyev and Valeriy Korzun, who were transferred to the team by order number 0622 on 23 July 1987. The three younger pilots were Captains Yuri Malenchenko, Yuri Gidzenko and Vladimir Dezhurov, who were transferred by order 0948 on 6 October 1987. These five made up the eighth Air Force selection, but prior to this group, three test pilots were recruited for the Buran programme in 1985. They were very experienced Air Force test pilots, Colonel Viktor Afanasyev and Lt-Colonels Gennadiy Manakov and Anatoliy Artsebarskiy, who had gone before the Mandate Commission on 2 September 1985. When they completed their basic training, instead of joining the Buran test programme, they were reassigned by an order dated 8 January 1988 to the training centre for Soyuz T training and became the ninth Air Force selection. At the same time, a number of the Air Force Shuttle selections, including Ivan Bachurin and Aleksey Boroday, were asked if they wanted to join the group at Star City. They refused, even though they had done some Soyuz T training.

In 1989, having already gone before the State Commission, three cosmonauts, Yuri Onufriyenko, Gennadiy Padalka and Sergey Krichevskiy, were assigned to the team by order number 0275 on 22 April. Three more candidates, Sergey Zalyotin, Salizhan Sharipov and Sergey Vozovikov, joined the team on 8 August 1990 by order 01142. Vozovikov was subsequently killed in an accident on 11 July 1993, while undertaking sea survival training at Anapa on the Black Sea near the Crimea. Talgat Musabayev, who was originally selected as part of an agreement to fly a Kazakh national, was a civilian airline pilot who was transferred to the Air Force group on 6 March 1991. He became the twelfth Air Force selection, but had been working at the centre since 21 January 1991.

Russian cosmonauts selected

Further expansion of the team occurred when Oleg Kotov, an Air Force doctor, was added to the team on 7 June 1996 by order number 0365. Valeriy Tokarev, who had been an Air Force shuttle test pilot, had expressed the desire to transfer to the Air Force team when the Buran programme closed in 1995. The State Commission agreed this move on 25 July and he joined the team on 16 September 1997. He was a very talented and experienced pilot who was given this opportunity when it became clear that the Buran programme would be cancelled.

The 1997 group of pilots and engineers graduating after completing their OKP training. (back row) Kondratyev, Valkov, S. Volkov, Korniyenko, Skripochka, Moshchenko and Yurchikhin. (front row) Lonchakov, Skvortsov, Surayev and R. Romanenko

In May 1996, permission was given to recruit a new group of pilots. Five were to be selected and medical screening started in October. There were 77 short-listed candidates, with 55 of them taking medicals. The new group was selected in 1997 and in fact, there were eight, plus Tokarev, who were all pilots and joined the team over several months. Major Aleksandr Skvortsov, the son of a 1965 candidate, joined the group on 20 June 1997 under order 0635. He was followed by Senior Lieutenant Maksim Surayev on 24 June 1997, under order number 0676. Five more candidates reported by order number 0162, dated 26 December 1997. They were Senior Lieutenants Konstantin Valkov and Sergey Volkov, Majors Dimitriy Kondratyev and Oleg Moshkin and Captain Roman Romanenko. Volkov and Romanenko are also the sons of former cosmonauts. Major Yuri Lonchakov joined this selection last, arriving on 24 June 1998. This group had passed the State Commission in July 1997 but due to military duties, their entry had been delayed. Moshkin did not pass his basic training (OKP) and left both the cosmonaut group and the Air Force in 1999. The way this group was selected was different than before, with all the candidates introduced to General Pyotr Deynekin, the Commander of the Air Force. They then went before the GMVK, led by Russian Space Agency director Yuri Koptev and, if approved, their order transferring them to the training centre was endorsed by Defence Minister Yuri Rodionov. This reflected the importance of these appointments within the military system, even in this new era of Russian politics.

A major issue with the International Space Station for the Russians is the time that the cosmonauts have to spend training overseas. Many spend months in Houston, and there is now a cosmonaut resident there. They also have to spend time in Canada, where they train on the Shuttle and ISS arms and undergo winter training exercises, and at both the ESA facility in Cologne and the Japanese centre, where the cosmonauts train on the mock-ups of the modules produced by those agencies. All of the current team is learning English and some are very fluent, although some do struggle. This is an additional strain on what was already a demanding job. So far, only Lonchakov has flown in space, on both a Shuttle and a Soyuz. He is currently the commander of the cosmonaut team, having replaced Korzun in late 2003. A number of others are assigned to crews, however, and should fly in the next few years.

One candidate who had been selected by the Strategic Rocket Forces in 1996 transferred to the Air Force on 2 September 1998. He was Lt-Colonel Yuri Shargin, who subsequently transferred back to the Rocket Forces (RVSN) in 2001. On 14 April 1998, civilian Yuri Baturin, who was a political adviser to President Yeltsin on space matters, was attached to the Air Force selection with a rank of Colonel. An additional group started their basic training in 2003, the first that had adapted to a new selection process similar to that of civilian selections. The four candidates were Major Anton Shkaplerov (pilot), who passed the GMK on 12 September 2002, Captain Anatoliy Ivanishin (pilot) and Captain Yevgeniy Tarelkin (test engineer), who passed GMK on 1 March 2002 and commander of the group, Lt-Colonel Aleksandr Samokutyayev, who passed the GMK on 20 January 2003. This selection also included a number of cosmonauts from other departments and organisations, including one civilian, Sergey Zhukov, who has been attached to the Air Force selection. They have all started basic training.

The formal badge of the Air Force Group of Cosmonauts

In 2003 and 2004, a number of Air Force cosmonauts left the team, reflecting a change of order within the group. Gidzenko stood down to take up a number of positions in the training centre and currently heads the survival training department, which is designated the Third Directorate. After completing their long training schedule and missions on the International Space Station, cosmonauts Dezhurov, Onufriyenko and Korzun stood down to take up positions within the command structure of the training centre. Korzun became the Deputy Commander of the training centre with a promotion to the rank of Major-General. Musabayev, who had flown three missions, left to take up a teaching post at the Zhukovskiy Academy, also with a promotion to the rank of Major-General. Sergey Zalyotin left by Air Force order number 560, on 20 September 2004, when he successfully stood as a member of the Duma, the Russian Parliament, for the Tula region. This leaves fourteen cosmonauts under the command of Lonchakov and his deputies Afanasyev and Baturin, plus four candidates who should complete their basic training in 2005. The group includes only seven who have flown in space, although a number of the rookies are slated for future missions to ISS. Two who are currently in training for missions are Lt-Colonel Kondratyev and the newly promoted Lt-Colonel Sergey Volkov.

Since 1960, 127 members of the Air Force have been selected and trained at TsPK. Of these, 56 had flown in space by the end of December 2004 and eleven of the cosmonauts have flown missions which, when combined, have lasted more than one year in total duration.

Table 5 Air Force cosmonauts who went before the State Commission instead of just being transferred by Military Order

	Born	*Medical Commission*	*State Commission*
Artsebarskiy	1956 Sep 9		1985 Sep 2
Afanasyev	1948 Dec 31		1985 Sep 2
Manakov	1950 Jun 1		1985 Sep 2
Gidzenko	1962 Mar 26		1987 Mar 26
Dezhurov	1962 Jul 30		1987 Mar 26
Korzun	1953 Mar 5		1987 Mar 26
Malenchenko	1961 Dec 22		1987 Mar 26
Tsibliyev	1954 Feb 20		1987 Mar 26
Krichevskiy	1955 Jul 9		1989 Jan 25
Onufriyenko	1961 Feb 6		1989 Jan 25
Padalka	1958 Jun 21		1989 Jan 25
Vozovikov	1958 Apr 17		1990 May 11
Zalyotin	1962 Apr 21		1990 May 11
Sharipov	1964 Aug 24		1990 May 11
Shargin	1960 Mar 20		1996 Feb 9
Kotov	1965 Oct 27		1996 Feb 9
Valkov	1971 Nov 11		1997 Jul 28
Volkov	1973 Apr 1		1997 Jul 28

	Born	Medical Commission	State Commission
Kondratyev	1969 Apr 25		1997 Jul 28
Moshkin	1964 Apr 23		1997 Jul 28
Romanenko	1971 Aug 9		1997 Jul 28
Skvortsov	1966 May 6		1997 Jul 28
Surayev	1972 May 24		1997 Jul 28
Tokarev	1952 Oct 29		1997 Jul 28
Lonchakov	1966 Mar 4		1997 Jul 28
Baturin	1949 Jun 12	2000 Sep 27	1997 Sep 15
Kotik*		2001 Aug 29 Air Force	
Tarelkin	1974 Dec 29	2002 Mar 1 Air Force	2003 May 29
Shkaplerov	1972 Feb 20	2002 Sep 12 Air Force	2003 May 29
Samokutyayev	1970 Mar 13	Air Force	2003 May 29
Ivanishin	1969 Jan 15	Air Force	2003 May 29

* Kotik is the only pilot to pass the Medical Commission without being subsequently passed by the State Commission

Table 6 Summary of Air Force Groups

Selection	Year	Number
First	1960	20
Women	1962	5
Second	1963	15
Supplemental	1964	1
Third	1965	23
Fourth	1967	12
Fifth	1970	9
Sixth	1976	9
Seventh	1978	2
Eighth	1987	5
Ninth	1988	3
Tenth	1989	3
Eleventh	1990	3
Twelfth	1991	1
Thirteenth	1996–97	2
Fourteenth	1997–98	10
Fifteenth	2003	4
Total		127

Table 7 The Air Force Cosmonaut Group Commanders

Commander of the Air Force Detachment	
1961 May 25–1963 Dec 20	Gagarin
1963 Dec 20–1968 Jul 7	Nikolayev
1968 Jul 11–1969 Sep 28	Bykovskiy
1969 Mar 3	The Detachment was split into training groups
1969–1970	This structure was under the First Directorate headed by P.I Belyayev
Section 1: Earth Orbital Ships including Soyuz and DOS Salyut	
1969–1971	Shatalov
1971–1973	Leonov
1973–1976	Lazarev
Section 2: Military space programmes Almaz and Soyuz VI	
1969–1973	Shonin
1973–1974	unknown
1974–1976	Artyukhin
Section 3: Lunar programmes including ASTP	
1969–1976	Bykovskiy
Section 4: Aerospace Projects including Spiral	
1969–1970	Titov G
1970–1973	Filipchenko
1973–1976	Khrunov
Student Cosmonauts	
1969–1970	Kuznetsov
1970–1973	Volynov
1973–unknown	Kuklin
1978 Mar 30	The detachment was reformed into a single command structure but was divided into a number of departments. Under the Commander was a deputy, and the departments who each had a head were: Political; Orbital Operations; International; Air and Space; and Candidates.
1978 Mar 30–1982 Jan 20	Leonov
1982 Jan 26–1982 Nov 19	Gorbatko
	In February 1982 the department was reorganised again: 1. Orbital spacecraft and space stations. 2. Multipurpose spacecraft 3. International programmes 4. Researchers 5. GOGU Mission Control 6. Students
1982 Nov 19–1990 Jun 30	Volynov
	In April 1990 sections 3 and 5 merged, as did sections 4 and 6

1990 Jun 30–1998	Volkov A
1995 Jul 24–1996 Nov	Fefelov stood in for Aleksandr Volkov
1998 Jan 15–2003 Nov 4	Korzun. While Korzun trained and flew on ISS, Afanasyev VM deputised
2003 Nov 4–date	Lonchakov

REFERENCES

1 The Rocket Men, Rex Hall and David Shayler, Springer-Praxis, 2001
2 Article 'Cosmonauts' by Col. E. Pyotrov, p 277, 1962. He used a secret name to disguise his true identity. It is likely that he was Col. Yevgeniy Karpov, the first Commander of the training centre
3 Chelovek, Nebo, Kosmos, Major-General A.K. Babiychuk, 1979
4 The Selection of Cosmonauts, Sergey Shamsutdinov, JBIS, **50**, 1997, pp 31
5 The Soviet Cosmonaut team 1960–1971, Rex Hall, JBIS, **36**, 1983, p 468–473
6 Izvestia, 4 April 1986, Background to Manned Space Flight
7 Interview with Leonov by Michael Cassutt, June 1990; Kamanin Diaries entries for 24 March and 27 March 1962
8 Nikolay Kamanin, The Hidden Cosmos and Diaries; 4 books published in Russian, 1995, 1997, 1999, 2001
9 Kamanin Diaries, entries dated 18 January and 28 February 1962
10 Women in Space: Following Valentina, David J Shayler and Ian Moule, Springer-Verlag 2005, pp 44–48
11 History of Recent Science and Technology website http://hrst.mit.edu/hrs/Apollo/soviet/interviews.htm, Interview with Ponomaryova, 17 May 2002
12 Kamanin Diaries, entries dated 31 July and 13 September 1967
13 Kamanin Diaries 1967–1969, Bart Hendrickx, JBIS, Nov/Dec 2000
14 Kamanin Diaries, entry 25 December 1967
15 Kamanin Diaries 1964–1966, Bart Hendrickx JBIS, **51**, 1998, p 413–440
16 Kamanin Diaries, entry 24 March 1970

The major source for this chapter were the books of Nikolay Kamanin, whose diaries were published in four volumes. They covered the period 1960 to 1971, when he retired as head of Cosmonaut Training. '*Skrytyy Kosmos Knigi Chetvyortaya*' *Volume 1* 1995, *Volume 2* 1997, *Volume 3* 1999 and *Volume 4* 2001.They have been covered extensively in the Russian press and have been extensively used in various publications.

Bart Hendrickx did a set of articles summarising the contents in the *Journal of the British Interplanetary Society*; (JBIS) **50**, 1997 pp 33–40, **51**, 1998 pp 413–440, **53**, 2000 pp 384–426, **55**, pp 311–360

Spaceflight 'A dream which almost came true' by Eduard Buinovskiy and Marina Buinovskaya

The Cosmonaut group of RKK Energiya

The design bureau headed by Sergey Korolyov was formed on 26 April 1950, after the old NII was split into two, with Korolyov now heading OKB-1 (Test Construction Bureau No 1), reflecting the importance of rocket design within the Soviet Military complex. OKB-1 was responsible for designing the R-7 booster, Sputnik, planetary craft and the first manned spaceship, 'Vostok'. It was based in Podlipki (which was renamed Kaliningrad in 1926).

ENGINEERS IN SPACE

Korolyov was very keen that civilians should be included in space crews, almost from the time the 1960 Air Force group was selected. He lobbied hard, but opportunities were limited and he would have to break the monopoly of the Air Force. Kamanin in particular was an obstacle to the inclusion of civilians into forthcoming crews and all the various selection and medical commissions were under the control of the military. He did succeed in lowering some of the medical requirements, however. When Voskhod came along, it gave Korolyov a unique opportunity to expand the specialisations within the crew and OKB-1 was asked to provide candidates for the Voskhod mission. Korolyov started the process in March 64 and in late May, a total of eleven candidates were sent for medical screening. The seven that passed were Konstantin Feoktistov, Georgiy Grechko, Valeriy Kubasov, Oleg Makarov, Nikolay Rukavishnikov, Vladislav Volkov and Valeriy Yazdovskiy, who would form the basis of the OKB team for years to come. It was dubbed 'Korolyov's Kindergarten' and Sergey Anokhin, a very skilled test pilot and a Hero of the Soviet Union, was appointed by Korolyov in 1964 to be the head of the team and its selection process. It is likely that some internal selection process was undertaken after his appointment and Anokhin reported to Konstantin Bushuyev, a deputy of Korolyov. Korolyov and Anokhin selected Feoktistov for consideration as a Voskhod crewman, even though they could have included up to three engineers in the candidate group. The inclusion of Feoktistov, who was a leading member of OKB-1 and had been involved in the design of the craft, was due to waivers to the restriction that cosmonauts could only be recruited from the Air Force. His selection meant that a large number of civilian engineers came forward for medical screening.

Korolyov's next opportunity to establish a civilian cosmonaut team came in

September 1965 with the decree authorising Soyuz (7K-OK) construction. The multi-person craft would give scope to a new breed of cosmonaut, the first for the Soviets, and following the selection of the first group of scientist-astronauts by NASA in June 1965. Korolyov never saw his dream of an OKB-1 team fulfilled, as he died in January 1966. His death caused a lot of confusion in the management of OKB-1 while they made decisions which would affect space planning for years. On 6 March 1966, OKB-1 was renamed TsKBEM (Central Construction Bureau of Experimental Machine Building), although Vasiliy Mishin, Korolyov's first deputy director, was not confirmed as his successor until 11 May 1966. On 23 May, TsKBEM Flight Test Department number 731 was founded with the appointment of eight testers. They were not given the title of 'cosmonaut', as Mishin did not wish to offend Kamanin and the Air Force. The eight candidates, many of whom had been identified by Korolyov, were Vladimir Bugrov, Gennadiy Dolgopolov, Valeriy Kubasov, and Aleksey Yeliseyev, plus Anokhin, Grechko, Makarov, and Volkov. The structure of the department is unclear at this point, but Anokhin now reported to a new director called Leonid Kuvashinov, who headed the team for a short time. In 1966, a number of future cosmonauts, such as Vladimir Aksyonov, Aleksandr Aleksandrov (who worked as a technician) and Valentin Lebedev, were working in a 'Flight Test Department'. Mishin received a decree from the State Commission on 15 June 1966 to allow him to include civilians in the crews of the Soyuz craft.

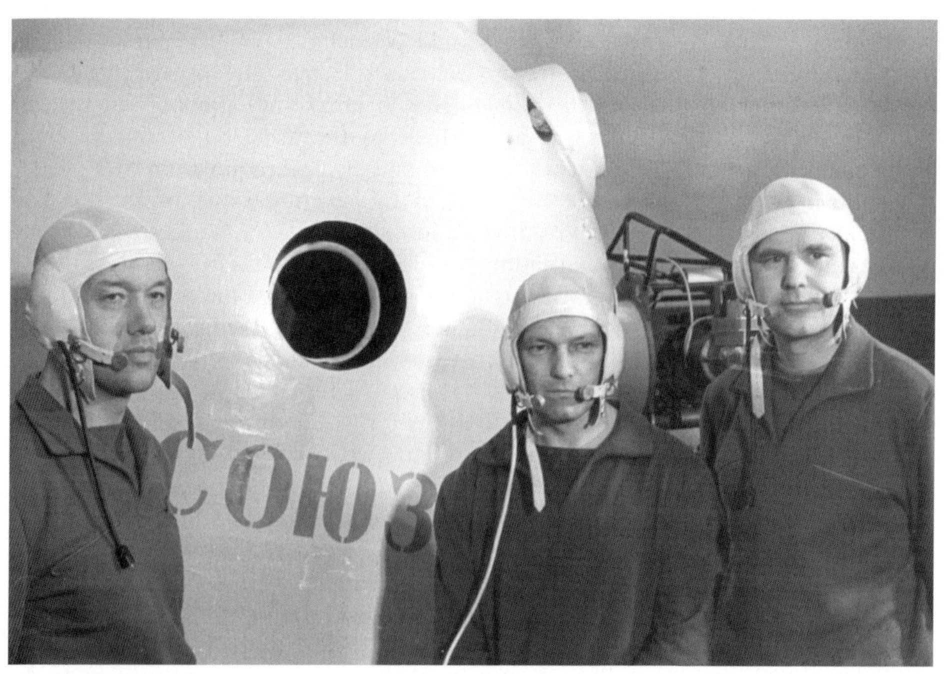

The Soyuz 10 crew in front of the Volga docking simulator. (from left) Energiya engineers Aleksey Yeliseyev and Nikolay Rukavishnikov with Air Force commander Vladimir Shatalov

Tension between the TsKBEM and the Air Force remained, however, with Kamanin only assigning Air Force personnel to the early Soyuz crews. He was sceptical about the health of some of the civilian candidates, calling them invalids. This tension had to be resolved and it was done at the highest level, with Air Force Chief of Staff Sergey Rudenko ordering Kamanin to solve the problem. Mishin requested that the eight testers be examined at the TsNIAG hospital by the Medical Commission and on 31 August, Anokhin presented the civilian team to Kamanin. On 5 September, Kubasov, Grechko and Dolgopolov were certified as being eligible for training, and Yeliseyev joined them when they started training on 1 October. Grechko broke his leg while parachute jumping and was replaced by Makarov, and additions to the group came at the beginning of 1967 with the selection of Nikolay Rukavishnikov in January and Vitaliy Sevastyanov in February.

Yeliseyev and Kubasov were assigned in mid-November to the Soyuz 1 and 2 crews to be launched in the spring of 1967. Yeliseyev, who had adopted his mother's name, had problems with KGB screening when they discovered that his father had been jailed in the 1930s for anti-Soviet agitation. He had also just been divorced and had remarried, which caused concern but, unlike Air Force pilots who had divorced, it did not lead to his disqualification.

Formal structure established at Energiya

On 27 March 1967, the Central Committee and the Council of Ministers approved a new policy for cosmonauts such as flight engineers and cosmonaut researchers. Authority over this new class of cosmonauts was given to the Ministry of General Machine Building (MOM) and medical control was moved from the Air Force to the Ministry of Health. From then on IMBP would be the organisation that medically cleared all future cosmonauts. A number of organisations wanted to use this new policy to create their own cosmonaut teams, including TsKBEM, TsKBM (Chelomey's bureau), IMBP, Paton Institute and other organisations. This structure took time to put together and it took almost a year before the first selections were made. In the mean time, Anokhin brought in three more engineer 'testers' on 18 August 1967. They were Vladimir Nikitskiy, Viktor Patsayev and Valeriy Yazdovskiy.

The Soviets started to operate the new system of putting potential cosmonauts before the Medical Commission (GMK) on 30 November 1967, when three men were put forward. The first meeting of the State Commission was on 27 May 1968, although it did not meet again until March 1972. The full list of candidates, their affiliation and dates appears as Appendix 4.

The structure of the organisation of cosmonaut selection and training within Energiya has always been very difficult to track. The controlling department, number 29, is called the Flight Test Department and has at least three groups under its management (currently departments 291, 292 and 293). The flight test service has as its staff both flown and unflown current cosmonauts, and a number of testers and engineers who have been identified in some cases as potential cosmonauts. The Flight Test Department is currently headed by ex-cosmonaut Aleksandr Aleksandrov. Department 291 is the Energiya cosmonaut team, which also has two other

The Soyuz 11 crew in front of a Soyuz simulator. from left) Energiya engineers Viktor Patsayev and Vladislav Volkov, with Air Force commander Georgiy Dobrovolskiy

departments that support cosmonaut training; flight documentation and EVA operations, each with its own group of engineers and specialists. A number of these men were and are potential cosmonaut candidates and have been described as such in the press, but a number who have died while undertaking training have also been identified in the Soviet press as cosmonauts or trainees. This is not an accurate description and has caused confusion for those trying to track the make up of the team. TsUP, the mission control centre, is also located at Korolyov. It is another department responsible to Aleksandrov and another source for potential cosmonauts.

In May 1968, the first official order from MOM, signed by Sergey Afanasyev, authorised the first civilian cosmonaut team. There were eleven in all, with nine from TsKBEM and one, Vladimir Fartushniy, from Paton Institute (see Other Cosmonaut Selections on page 176). The other person was Feoktistov, who was not a formal member of the TsKBEM team even though he was designated a cosmonaut. The nine from TsKBEM were Volkov, Grechko, Kubasov, Yeliseyev, Makarov, Patsayev, Rukavishnikov, Sevastyanov and Yazdovskiy. Four other candidates – Anokhin, Dolgopolov, Nikitskiy and Bugrov – had been stood down due to health problems.

The Soyuz T 7 crew in the Mir simulator. (from left) Air Force commander Vladimir Titov and Energiya engineers Aleksandr Serebrov and Gennadiy Strekalov

This group formed the core of engineer assignments within the Soyuz and Lunar programmes from 1967 to the mid 1970s. On 30 June 1971, Patsayev and Volkov were killed when the Soyuz 11 descent craft depressurised during its landing, after its record breaking mission to Salyut 1. By 1972, flight opportunities were reduced when the position of cosmonaut researcher was abolished following the Soyuz 11 disaster and the crews were reduced to two wearing full space suits. On 27 March 1972, however, three additional engineers, Boris Andreyev, Valentin Lebedev and Yuri Ponomaryov, reported for training. The following year, the fourth selection was made, with Vladimir Aksyonov, Aleksandr Ivanchenkov, Valeriy Ryumin and Gennadiy Strekalov starting on 27 March 1973. The engineers all worked for the bureau already and they had first hand knowledge of the equipment they would fly and the space stations they would work on. Yazdovskiy was removed from the Soyuz 13 mission in December 1973, due to being 'psychologically incapable of shifting to the role of cosmonaut operator.' He was a man who spoke his mind and had clearly made enemies. He was stood down (and replaced on the mission by Lebedev) and never got another assignment, leaving the team on 1 July 1982. Ponomaryov, who was married to Ponomaryova, a member of the 1962 female selection, was assigned to a number of back-up crews in the Salyut programme, but was stood down due to medical problems, resigning on 11 April 1983. Andreyev served as a back-up crew member on ASTP and then on a number of Salyut 6 and 7 crews before being paired with Zudov for a flight. However, he failed a medical and left on 5 September 1983.

A new chief designer

On 22 May 1974, TsKBEM was merged with KB EnergoMash and renamed again as the Scientific and Production Association (NPO) Energiya, with Valentin Glushko as its new head. The deputy director responsible for manned space flight was Yuri Semenov. On 1 January 1977, Konstantin Feoktistov was added to the team as he has been short-listed for a mission. He trained for a number of missions, but kept missing out due to failing medicals. One of these missions related to flying an older cosmonaut, a similar mission profile and objective to that flown by John Glenn on STS-95 in 1998. Feoktistov finally stood down on 28 October 1988 without making a second flight.

In 1977, the medical requirements to be a candidate were very strict. They were described by Sergey Bedziouk, a candidate for the team who passed the GMVK in

The back-up crew to a number of long duration crews posing in front of the Mir simulator. (from left) Air Force commander Vyacheslav Zudov and Energiya engineer Boris Andreyev

March 1978: 'Only a few individuals out of hundreds of candidates passed the medical evaluation. I remember someone told me that there was one successful candidate out of five hundred people undergoing medical evaluation. First there was a one-day outpatient check-up. If everything was in order then the candidate was sent for a comprehensive inpatient check-up at a medical facility, which lasted 30 to 40 days.' This was the same for all civilian selections.

On 1 December 1978, the GMVK considered fourteen more candidates from Energiya, of which seven were selected. They were Aleksandr Aleksandrov, Aleksandr Balandin, Aleksandr Laveykin, Musa Manarov, Viktor Savinykh, Aleksandr Serebrov and Vladimir Solovyov. Four of this group reported for training immediately and the rest were called up a couple of years later to start training for Salyut 7 assignments. The candidates who did not pass the GMVK were Valeriy Chervyakov, Gennadiy Isayev, Aleksandr Kulik, Viktor Pyotrenko, Aleksandr Khaustov, Sergey Bedziouk and Nikolay Pyotrov.[1]

The next selection related to the desire to fly another female before the Americans flew one on the Shuttle. In 1978, the United States had selected six women astronauts and as the launch of the first Shuttle got closer, there was pressure on the Soviet mission planners to upstage the Americans again. On 30 July 1980, the mandate commission selected Nataliya Kuleshova and Irina Pronina, as well as other women candidates from other bureaus, and on 16 May 1983, Svetlana Savitskaya transferred from her establishment to join NPO Energiya. Kuleshova was initially assigned to a crew, but had medical issues and was replaced by Pronina. Pronina served as back-up to Savitskaya on Soyuz T 7 and was then assigned to the prime expedition crew of T 8 with Titov and Strekalov. She was replaced very late on, due to the long standing opposition to flying women from senior figures within the programme. Savitskaya escaped these problems, perhaps because her father was a Marshal within the Soviet Army and she had a great deal of very high political support for her flights. She flew two short missions and would have commanded a third, the first such assignment for a woman, until it was cancelled when problems developed with Salyut 7. She stood down on 27 October 1993 to take up a position in Energiya and is now a leading member of the Communist Party in the Russian Duma.

On 15 February 1984, two more engineers were selected, Aleksandr Kaleri and Sergey Yemelyanov. This was the seventh selection of engineers. Kaleri went on to fly four missions to Mir and ISS and also served as Deputy Commander of the team to Gennadiy Strekalov, who had been selected as Head of the Energiya cosmonaut detachment in 1985.

On 2 September 1985, the eighth selection, Sergey Krikalev and Yuri Zaytsev, started training. Zaytsev went on to serve as a back-up cosmonaut on a couple of occasions, but was dismissed from the team on 14 March 1996, by the order of Yuri Koptev, the Director General of the Rossiyskoye Kosmicheskoye Agentstvo (RKA, the Russian Space Agency). He continues to work in the design bureau.[2]

The ninth selection contained only one candidate. Sergey Avdeyev was passed by the GMVK on 6 March 1987 and started his basic training programme in December 1987, which lasted 18 months. He went on to make three long-duration expedition flights to the Mir station, including one of over a year in duration.

Energiya engineer Aleksandr Kaleri during sea training

The tenth selection was passed by the GMVK on 25 January 1989. The four candidates were Nikolay Budarin, Yelena Kondakova, Aleksandr Poleshchuk and Yuri Usachev, who started their basic training in September 1989 and finished in January 1991. The eleventh selection reported on 3 March 1992, with three candidates starting training. They were Aleksandr Lazutkin, Pavel Vinogradov and Sergey Treshchev. All seven went on to fly missions to Mir and/or ISS.

Russian engineers and more changes

In 1992, four of the Energiya cosmonauts left – Musa Manarov, Nataliya Kuleshova and Irina Pronina on 27 July, and Sergey Yemelyanov on 9 July after failing the Medical Commission. Yemelyanov had been a back-up on a couple of occasions but had been removed due to a medical problem. He continued to work within the Energiya cosmonaut detachment, but died of a heart attack on 5 December 1992.[3] The next selection was made on 1 April 1994, with the arrival of Nadezdha Kuzhelnaya and Mikhail Tyurin. They underwent their basic training, completing it on 25 April 1996 with the rating of excellent.[2]

On 10 May 1994, it was reported that the cosmonaut teams of various departments had been reduced, leaving Energiya with twelve cosmonauts.[4] In a list published a few days earlier, the team had sixteen in its complement, with thirteen in the main team and three candidates.[5] In July 1994, NPO Energiya was renamed once more as RKK (S.P. Korolyov Rocket and Spacecraft Corporation) Energiya, under the control of Director Yuri Semenov, who had replaced Glushko on 22 March 1992.

A formal crew portrait of the unflown Soyuz T 5 crew. (from left) Air Force Commander Yuri Isaulov and Energiya engineer Valentin Lebedev

The thirteenth selection, of Energiya engineers Konstantin Kozeyev and Sergey Revin, was made on 9 February 1996. A third engineer, Oleg Kononenko, representing TsSKB in Samara, trained with his Energiya colleagues and on 5 January 1999, he transferred to the Energiya cosmonaut team. They started their training on 3 June 1996 and completed their basic training on 18 March 1998.

The fourteenth selection was made by the GMVK on 26 July 1997, when Oleg Skripochka and Fyodor Yurchikhin joined the group. This selection was supplemented when Mikhail Korniyenko joined on 24 February 1998, having passed the GMVK. Yurchikhin flew a mission on STS-112 to the ISS, as a Mission Specialist. After this, there were no more selections for nearly five years, but during this period, the civilian engineers completed a number of long missions. On 19 October 2002, Rukavishnikov died after a long illness.

By an order of the President of RKK Energiya, on 6 February 2003, Pavel Vinogradov was assigned to the post of Chief of Department 291 while remaining a test cosmonaut instructor. He succeeded Gennadiy Strekalov, who had been transferred to another job in the bureau. Strekalov died of cancer on 25 December 2004. Shortly after Vinogradov's appointment, cosmonaut Sergey Avdeyev stood down from the team, on 14 February 2003. He had completed three space flights, reaching the record breaking total duration of over 748 days in space.

In 2003, three more engineers were passed for entry to the team. They passed the Mandate Commission on 26 May and reported for their basic training on 29 May.

The State Commission formal meeting on 29 May 2003 to select new candidates. (seated, from left) Yuri Koptev (Rosaviakosmos), Yuri Semenov (Energiya), Grin (Rocket Forces). (standing) Candidate Mark Serov and Valeriy Ryumin (Energiya). (Courtesy Novosti Kosmonavtiki)

Meeting in the White Room of the Cosmonaut Training Centre confirming a crew. In the centre, standing, is TsPK commander Vasiliy Tsibliyev and next to him on his right is his deputy, Valeriy Korzun

The 2003 group of cosmonaut candidates. (from left) Sergey Zhukov, Yevgeniy Tarelkin, Aleksandr Samokutyayev, Anton Shkaplerov, Anatoliy Ivanishin, Sergey Ryazanskiy, Andrey Borisenko, Mark Serov and Oleg Artemyev. (Courtesy Novosti Kosmonavtiki)

They were Oleg Artemyev, Andrey Borisenko and Mark Serov. Another candidate's entry was denied. Anna Zavyalova had passed the Medical Commission in 2000, and her failure to be confirmed by the State Commission probably reflects the attitude towards women within the higher levels of Russian space management. Another loss to the team came on 28 May 2003, when Oleg Makarov died of a heart attack.

On 25 March 2004, Poleshchuk retired due to medical issues. He remains the head of Energiya department 293, responsible for coordinating EVA training in the civilian detachment. A month later, on 5 April, Yuri Usachev also retired. He continues to work in department 291 to support the detachment. A third Energiya cosmonaut retired on 27 May 2004 when Kuzhelnaya stood down. She had decided that she was unlikely to fly a mission and blamed Russian attitudes towards women undertaking such flights. She has joined Aeroflot as a trainee pilot on Tupolev 134s. The fourth retirement in this year was that of Nikolay Budarin, who stood down on 7 September 2004 after having made three flights. These latest departures left twelve cosmonauts, of whom eight have flown (as of 31 October 2004), plus three candidates who should complete their basic training in 2005. On completion of his fourth flight, Kaleri was made deputy to Vinogradov, who is himself due to make a long duration flight in 2005 or 2006

Krikalev, who was launched to ISS on TMA 6 in April 2005, became the first Russian to fly six missions. If he completes the full six-month mission, he will surpass Avdeyev's 748-day individual duration record in October 2005. So far, 54 engineers from the 'Korolyov' design bureau have trained to be cosmonauts, with 35 having flown and eight of those having spent over a year in space.

Table 8: Organisation and Structure of the Energiya Cosmonaut Team

RKK Energiya is the lead design bureau responsible for the construction of flown manned spacecraft, the Soyuz spacecraft and all its variants. Over the years, it has had a number of different names and was originally part of another bureau, NII-88. It is based in Podlipki, which was renamed Kaliningrad and finally renamed Korolyov.

1956–1966	OKB-1 The Korolyov design bureau
1966–1974	TsKBEM Central Construction Bureau of Experimental Machine Building
1974–1994	NPO Energiya (Energiya Science and Production Association Named after Academician S.P.Korolyov)
1994–date	OAO S.P.Korolyov Rocket and Space Corporation Energiya (RKK Energiya)

The Flight Test Service was created on 1 January 1982. This is the lead department within the design bureau for cosmonaut training. It is designated Group 29 and oversees departments 291, 292 and 293, as well as Mission Control (TsUP) and a Flight Documentation department.

Head of Group 29

1982 Jan 1–1987 Jul 12	Valeriy Nikolayevich Kubasov
1987 Jul 12–date	Aleksandr Pavlovich Aleksandrov

Deputy Director

1993–date	Aleksandr Sergeyevich Ivanchenkov

The Energia cosmonaut team was run under the following departments.

1964 Apr–1966	Department 90
1966–1973 Oct 11	Department 731
1973 Oct 11–1975 Mar 1	Department 71
1975 Mar 1–1977 Feb 1	Department 111
1977 Feb 1–1982 Mar 1	Department 110
1982 Mar 1–date	Department 291

There is a close relationship with Department 292 and 293.

Heads

1964 Apr	Sergey Nikolayevich Anokhin (Acting)
1966 May	L.M. Kubshinov (Deputy A.I. Lobanov)
1966 Jul 17	Sergey Nikolayevich Anokhin

Tracking these positions after this date is less clear, as many of the dates were not formally announced.

1973 Oct 11	Oleg Grigoryevich Makarov
1977 Feb 1–1979?	Vitaliy Ivanovich Sevastyanov
1979?–unknown	Valeriy Nikolayevich Kubasov
1984 Dec–1985 Mar 24	Vladimir Alekseyevich Solovyov
1985 Mar 25–2003 Feb 6	Gennadiy Mikhailovich Strekalov
2003 Feb 6–date	Pavel Vladimirovich Vinogradov

Deputy to Department commander

1978–1982	Sergey Nikolayevich Anokhin

1983–unknown	Svetlana Yevgenyevna Savitskaya
1994 Mar 16	Aleksandr Yuriyevich Kaleri. He was the deputy to Strekalov for some time and certainly covered his training. He must have been replaced himself while he was training. A number of other cosmonauts headed the department while Strekalov trained and flew missions.

Department 292 is the EVA training department. It is currently headed by Aleksandr Poleshchuk, who was acting head for some time but was made its permanent Head in March 2004.

REFERENCES

1 Thorny Road to the Stars, *Spaceflight*, 44, September 2002, p 381–383, Aleksandr Zheleznyakov
2 *Novosti Kosmonavtiki*, 20 May–2 June 1996
3 Letter to *Spaceflight*, January 1994, from Vadim Molchanov
4 TASS, 10 May 94
5 *Novosti Kosmonavtiki*, 23 April–6 May 1994

Other Soviet and Russian cosmonaut selections

The selection processes for the new groups of civilian cosmonauts were the same as those for the Air Force, but the qualifications were different, reflecting the jobs envisaged by the heads of the design bureaus and the State Commission.

Medical screening from 1960 to 1966 took place at the Central Military Scientific Aviation Hospital in Moscow. This responsibility was moved to the Institute of Medical and Biological Problems in 1967. Screening is conducted by a medical expert Commission (VEK) and the results are then passed on to the Chief Medical Commission (GMK), which is a board of leading doctors representing the Ministries of Health and Defence. All are members of the Russian Academy of Medical Sciences and they pass candidates solely on medical evaluations and tests. Successful candidates then become aspiring cosmonauts.

The next step is to go before the State Interdepartmental Commission, which is also known as the GMVK. This is the government committee responsible for cosmonaut selection and has also been called the Credential Committee, or Mandate Commission. It is made up of Chief Designers, Air Force commanders and officials from the Russian Space Agency or its previous entities. It is a political committee which bases selection on political reliability (particularly before the fall of the Soviet Union) and both moral and human qualities or frailties. This system is still in place and the Commission meets on a regular basis as new selections are required.

If a candidate successfully passes the Commission selection, they are admitted to their respective teams as a cosmonaut candidate. They then have to undergo basic training (OKP), which usually lasts around two years and is organised and provided by the Cosmonaut Training Centre at Star City. Having achieved the status of cosmonaut, candidates continue to be medically screened every year. They are then involved in rounds of generic training, and only when assigned to a mission do they undergo mission specific training within a training group and a crew.

Appendix 4 is a full list of those who have taken their medicals and passed the State Commission. This list first appeared in 1997 but has been updated to include candidates who have been brought before the Commission since then.[1]

THE COSMONAUT GROUP OF THE ACADEMY OF SCIENCES

The first scientist to be considered for a mission was Georgiy Katys, who was on the short-list for the 1964 Voskhod programme. His participation was championed by Mstislav Keldysh, the head of the Academy of Sciences, who felt that it was important to have research at the heart of the programme. Katys was one of eight candidates being looked at for the mission and was passed by the Mandate Commission on 29 May 1964. He had also been subject to an enquiry by the KGB, who discovered he had relatives in France and that his father had been sentenced to death by Stalin. Kamanin commented that, 'All this spoils the candidate for flight. More suitable candidates can be found.' Katys had been considered for selection in 1962 but was over the imposed age limit. He served as the back-up to Feoktistov on Voskhod 1 and he was a potential candidate for a number of missions, including Voskhod 3. He was with Volynov on the prime crew in the autumn of 1965, but the experiments were probably not ready and he was dropped in favour of an Air Force cosmonaut, although the mission and subsequent Voskhod missions were eventually cancelled. Katys was always passed over for selection in the fight for flight positions between the Air Force and OKB-1, but he remained in the lists for some years and formally enrolled in the Academy of Sciences team as part of a supplementary selection in May 1968.[2]

In late 1966, eighteen scientists were selected, including seven from IZMIRAN, The Institute for Terrestrial Magnetism, Ionosphere and Radio Wave Propagation of Akademiya Nauk (Academy of Sciences). They were sent to the Central Military Scientific Research Aviation Hospital for screening, but only four passed the tests.

Permission was given in May 1967 for civilian agencies to select their own teams of cosmonauts. The Academy of Sciences confirmed their first selection on 22 May 1967 and only a few days later, four men were selected. They were Rudolf Gulyayev, Ordinard Kolomitsev and Mars Fatkullin from IZMIRAN, and Valentin Yershov from the Institute of Mathematics. They underwent a version of OKP basic training in that era, but it soon became clear that there were really no opportunities for them to fly missions. Gulyayev and Kolomitsev left the team in 1968, followed by Fatkullin in 1970, when he decided he had no chance of flying. Yershov was a very strong candidate and he was included in the L-1 training group, as he had done some work on developing a navigational device for use on the lunar missions. He was considered again in 1973 for possible inclusion in a Salyut crew but he failed his medical. He left in 1974 due to progressive deafness, but he had also refused to join the Communist Party, which had been reason enough for the dismissal of other candidates. In May 1970, following the creation of the Salyut programme, there was another attempt to create a scientist cosmonaut team to fly as third-seat researchers. The new team would include Katys, Yershov and four new candidates. This was due to rivalry within the scientific community and the fact that there were no members of the Academy of Science team left in training. Over the years a number of scientists had been passed by the Medical Commission, but none were ever passed by the GMVK to start training.[3] Yershov died on 15 February 1998 and Fatkullin died on 16 April 2004.

In 1980, as part of the women's selection, Irina Latysheva was approved. She worked for the Space Research Institute (IKI) and was part of a large group of women selected for potential missions on space stations on 30 July 1980. At that time, there was no official Academy of Sciences group, so she was attached to the Energiya team. She stayed in the team until 25 February 1992 but never played an active part in crew training.

A second woman, Yekaterina Ivanova, was approved on 9 March 1983. She had passed the Medical Commission in 1980 and then took part in a hypokinesis experiment in 1982 at IMBP. She underwent OKP training (completing it on 29 June 1984) and was then assigned as back-up researcher to Savitskaya on Soyuz T 12, training with Vasyutin and Savinykh. In 1984 she was then included in the all-female crew due to fly in 1985/6 to Salyut 7 as the flight engineer, alongside commander Savitskaya and researcher Dobrokvashina. This mission was cancelled due to problems with Salyut 7 and the illness of Vasyutin. She stood down from the team in April 1987 and returned to the Leningrad Institute of Mechanics in St. Petersburg.[4]

In the 1980s and 1990s, a number of cosmonauts left their design bureaus or the Air Force and entered scientific research institutes. The Academy of Sciences entered them into their cosmonaut team but there is no evidence that they were ever seriously considered for missions. They included Grechko, who entered the RAN team in July 1986 and left in 1992. He had joined the Institute of Physics having left his job at Energiya. Lebedev entered the team in November 1989 and left in 1993. He worked for the GeoInfo Centre of the Academy of Sciences. Artsebarskiy entered the team in September 1993 and left a year later. He worked as a department head in IT in the Laboratory of Large Scale Constructions. Stepanov entered the RAN team in March 1995 and left in February 1996.

A lot of research has been conducted on board space stations, but it has been mainly engineering and medical. Air Force pilots have also undertaken many experiments on a contract basis for a variety of institutes. However, in retrospect, it should be recognised that the role of the scientist in space has not been fully achieved in Russian cosmonautics.

In 1996, the Russian Space Agency rationalised the number of cosmonauts in training and which agencies could have their own teams. No science astronauts were included in the new structure and the Academy was not on the list of those who could have their own cosmonauts.[5]

THE COSMONAUT GROUP OF THE INSTITUTE OF MEDICAL AND BIOLOGICAL PROBLEMS

Of the 99 Soviet and Russian cosmonauts who have flown in space at the time of writing this book, only five have been doctors. In the very early days of the planning for human space flight, the Soviets envisaged engineers, scientists and doctors flying regularly on space missions.

The role of doctors in the manned programme

The first opportunity for such selections came in 1964, when the Vostok craft was converted to hold three cosmonauts rather than one. It was envisaged that one of the crew would be a doctor and four doctors were considered for selection from different bureaus and enterprises. Captain Boris Yegorov was short-listed from the Military Medical Services. He had become involved in the space programme when he joined the Test Institute of Aviation and Space Medicine and when he joined the parachute rescue team to recover cosmonauts after landing in 1961, for which he underwent a rigorous parachute training course.[6] The second candidate was Aleksey Sorokin. He was a military doctor on the staff of the Cosmonaut Training Centre, who trained from 1 June 1964 until the mission. He continued to work at the training centre until his death on 11 January 1976 and was a Lt-Colonel at that time. He is buried near Star City. The third candidate was Dr. Boris Polyakov, who was on the staff of IMBP and trained from 1 June to 2 July 1964. The fourth and last candidate was Lt-Colonel Vasiliy Lazarev from the Air Force. He had been involved in the Volga high-altitude balloon programme and as well as being a doctor, he was also a test pilot.

They all went before the Mandate Commission on 26 May 1964. After many different combinations were considered, Yegorov eventually flew the mission, which lasted for one day in October 1964. Lazarev, who joined the Air Force cosmonaut team in 1966, was his back-up. After his flight, Yegorov went to work at IMBP, where he stayed until 1984. He then became the Director of the Scientific Industrial Centre for Medical Biotechnology under the Ministry of Health. He died of a heart attack on 12 September 1994.

The Institute of Medical and Biological Problems was established as an independent entity by a decree of the USSR Council of Ministers on 28 October 1963. It was formed from sub-divisions of the Science and Research Institute for Aviation and Space Medicine of the Air Force and was responsible for all aspects of medical and biological support for humans in space, from selection to in-flight support and rehabilitation. All civilian cosmonauts do their medical tests at the Institute.[7]

When the Voskhod programme was planned, they wanted a medical research flight lasting over a week, and four doctors from the newly formed IMBP were identified as potential candidates. None of them considered themselves as cosmonauts, even though they did do some training on the experiments. This would have involved operating on a rabbit in a special chamber in flight to study the behaviour of blood and body organs in weightlessness. The candidates included Dr Yevgeniy Ilyin, who went on to head the Bion satellite programme and is a deputy director of the Institute for Science. He did undertake a five-day simulation of the Voskhod mission with two staff from IMBP and was a Captain of medical services at that time. Aleksandr Kisilyov (who was a Major of medical sciences) and Yuri Senkevich were the other prime candidates. Senkevich was a crewman on the raft 'Ra 2' and a well-known Soviet TV journalist, as well as a skilled doctor. He was in charge of the work on the rehabilitation and convalescence of humans. The reserve was Sergey Nikolayev, who was a Lieutenant of medical services at that time. (He still works at the institute.) By 1966, the mission was no longer on the agenda following the death of Korolyov. Senkevich tried to join the team in 1972 when IMBP recruited their own team, but he was turned down having just been divorced.[8]

In late 1967, to aid testing, the Institute recruited a group of professional test subjects. The group consisted of young athletes in perfect health, all of whom had been through military training. They were given the name of Cosmonaut Number Zero, and like some of their actual cosmonaut colleagues, their identities and even their existence only became clear twenty years later. They undertook many of the tests that would be used on the future cosmonaut team and these test subjects created the parameters that allowed the medical experts to perfect the training regime. Among the equipment used were the catapult (which tested the stress of a hard landing after the freefall of the capsule), the pressure chamber, the isolation chamber and many other machines and situations. A number of these men suffered serious injury, including one subject called Bogdan Guk, who used the catapult from a great height during one test and broke several bones. He was sent home after receiving honours. The group did dream of space, but they also lived on the 'drug of experiment' and a number of them committed suicide, had drink problems and suffered from depression after being removed from the group.

When the demands of long duration flight were exposed by the Soyuz 9 mission (when the crew returned in some distress), the testers were put through a set of experiments which looked at these issues. These experiments went on for many years. Flotation, for example, was an effective way of simulating weightlessness. Actual cosmonauts might spend up to three days in this sort of experiment, but one test subject spent 56 days in a flotation tank. In another test in 1977, one candidate lay down for 186 days. The centrifuge was used to simulate take off and landing and one tester, Sergey Nefiodov, described his relationship with the machine as follows: 'I always sort of bowed as I approached the monster. I was not afraid of it.' However, along with the respect towards the metallic giant went a hint of fear. The test group's identities have often been confused with actual cosmonauts and many hoped that if they performed well they might have been considered for assignments. But as Soviet and then Russian space plans were dropped or amended, they became resigned to just being testers. They were described in one article as the 'raw material' of the programme.[9]

Ground simulators

Another example in which testers were used to advance medical knowledge into new areas was when three scientists spent a year in a closed environment on the IMBP site between 5 November 1967 and 5 November 1968. They were studying the problems of supporting human life in a prolonged orbital mission. The 'commander' was a doctor, German Manovtsev, and the other crew members were biologist Andrey Bozhko and technician Boris Ulybyshev. This showed the early commitment of Soviet planners towards understanding the medical issues related to long duration space flight.[10]

At IMBP, there is a mission control facility where doctors and psychologists monitor the crew 24 hours a day. They can communicate directly with the crew, or via MCC outside Moscow. In 1970, a special base for medical and technical testing of life support systems was constructed, called the Ground Experimental Complex (NEK). This was a mock-up of a heavy planetary spaceship and between 1970 and 1973, several three-person crews were tested in isolation for up to fifty days. The unit, a modular design, is still used today and housed the 1995 HUBES test with ESA. The record duration so far is a 240-day test with international crews under the SFINCSS 1999 test. There was one core crew – Vasiliy Lukyanyuk and Vladimir Karashtin – which stayed there the whole time, but visiting crews came with fresh supplies. Visitors included Valeriy Polyakov and Sergey Ryazanskiy. The Institute has a 9-metre centrifuge for testing the cosmonauts during ascent and descent simulations and there is a training hall with mock-ups of space craft and simulators. It also has a deep water chamber to simulate long duration diving.

Medical cosmonauts selected

In 1972, a group of civilian cosmonauts was formed, based at IMBP, with the aim of medically supporting any mission in orbit, including surgery. It was envisaged that as the Soviets flew long duration missions of up to a year, the need for in-orbit support would increase. It was also planned that the long-term space stations which were on the drawing board would have a module devoted to medical research. The first selection for this group was made on 22 March 1972. Three cosmonaut physicians were selected and Dr Valeriy Polyakov was appointed the group's first commander.[11] The other two were Georgiy Machinskiy, who left in 1974, and Lev Smirenniy, who left in 1986. Both were medically disqualified. Valeriy Polyakov went on to fly two long duration missions on the Mir space station. He is the holder of the world duration record for a single mission and also held the cumulative duration total until recently surpassed by Avdeyev.

The second selection occurred on 1 December 1978. The three candidates selected this time were German Arzamazov, Aleksandr Borodin and Mikhail Potapov. Arzamazov served as a back-up to Polyakov on two occasions and it was suggested that he might fly a US Shuttle flight, but his falling out with Russian planners over his second assignment led to him leaving the team in December 1995. Borodin did not receive an assignment and failed a medical, leaving the group on 10 March 1993. Potapov served as a back-up for a Soyuz T mission in 1980. He trained with Isaulov and Rukavishnikov and also did support work on the Salyut missions. He left the team on 27 May 1985 due to medical problems but continues to work as a department head at IMBP.

Polyakov was considered as a doctor for a Soyuz mission in 1980 and he trained for a mission with Lazarev and Makarov, but instead he acted as a back-up to Soyuz T 3 when it was decided that a repair mission was needed. Potapov acted as his support.

The third selection, made up entirely of women doctors, came on 30 July 1980. They were Galina Amelkina, Elena Dobrokvashina, Tamara Zakharova, Larisa Pozharskaya, and Olga Klyushnikova. This group included representatives from Energiya and the Academy of Sciences. The idea was to fly women cosmonauts before the recent American female selections flew on the Shuttle and they trained to serve as cosmonaut researchers on missions. Klyushnikova left soon after selection

The support crew of Soyuz T 3 with their trainers. (crew from left) Dr Mikhail Potapov, commander Yuri Isaulov and Energiya engineer Nikolay Rukavishnikov

because she was a children's doctor. This was the only (and unofficial) reason ever stated and she has never spoken about the experience. The rest underwent a full basic training programme, including isolation tests, parachute jumping and other basic training schedules. Amelkina left in May 1983 having failed a medical test. She disclosed that she did all her training at Energiya at Podlipki and had never been to the training centre at Star City, but the women did undertake duration testing in the test facility at IMBP. When interviewed, Dobrokvashina and Pozharskaya described tests in this facility lasting up to a month at a time. Zakharova did not seem to be involved in any direct mission training and she left on 1 September 1995. Dobrokvashina came the closest to a flight in 1985.[12]

In 1983, a doctor was recruited from the Institute of Clinical Cardiology of the USSR Academy of Medical Sciences. Oleg Atkov was selected to act as a cosmonaut researcher on a long duration mission to Salyut 7. He was to test in orbit a machine called 'Argument'. This instrument makes it possible to record the functioning of the heart (particularly its left section), to measure the thickness of the heart walls and to determine a number of other indicators for the condition of the cardiovascular system. It worked well and was put into mass production. Atkov firmly believed that all long duration missions should have a doctor on board and he stayed on board the station for over 240 days. Polyakov acted as his back-up, training with Vasyutin and Savinykh and with Potapov acting in a support role.[13]

The fourth selection, on 2 September 1985, was of Yuri Stepanov, who was a

Doctor Elena Dobrokvashina undergoing sea training

medical research engineer. He did his basic training from November 1985 to February 1987 with colleagues from other groups but left the team on 20 March 1995 and joined the Institute of General Physics. He was enrolled as a member of the Academy of Sciences team.

Dobrokvashina trained as a Research Cosmonaut on an all-female crew (commanded by Savitskaya) to fly to the Salyut 7 station on a visiting mission. The mission crew was assigned and trained for some months but it was cancelled when the difficulties experienced with the station led to an extra repair mission being flown and no further Soyuz being available for Salyut use. The mission would have been designated Soyuz T 14 or 15.

In February 1988, Polyakov was included with Arzamazov in a training group to train for a visiting mission to Mir. The intention was that the doctor would stay on board to work with the long duration crew. Polyakov made his first flight on 29 August 1988 on board the Soyuz TM 6 space craft, with Arzamazov acting as his back-up. He was a physician researcher to the Mir station and this flight came soon after a decision was made that a doctor would fly on every mission lasting more than six months. This followed the Laveykin heart problem which cut short his mission in 1987. Polyakov monitored the crew of Titov and Manarov as they finished their year long mission, and his programme was put together by IMBP. He was in orbit for 240 days 22 hours 34 minutes and 47 seconds. Polyakov is an excellent scientist and a qualified clinical physician and has knowledge of both conventional medicine and folk remedies and acupuncture. While on board Mir, he successfully filled a tooth for Titov.[14]

The fifth selection was made on 25 January 1989. It consisted of three physicians – Vladimir Karashtin, Vasiliy Lukyanyuk and Boris Morukov – who completed their basic training (OKP) on 7 February 1992.

Russia's medical cosmonauts

On 10 March 1993 Borodin, Dobrokvashina and Pozharskaya left the cosmonaut team, even though in the same year it was decided to fly a doctor on a mission lasting more than one year, after much pressure from the Russian medical establishment. This related in part to a simulation of a flight to Mars. Polyakov had spent a lot of time lobbying for the mission, and on 2 February 1993, three candidates were selected to train for it. They were Polyakov himself, plus Arzamazov and Morukov, the most experienced of the new group who had just passed their OKP. When the crew was finalised, there was a major row, as Arzamazov felt he was a better candidate than Polyakov, who had been selected for the mission. Arzamazov said that Polyakov had spent too much time behind a desk and walked out from his assignment as back-up. The formal announcement said he had been removed due 'psychological incompatibility with his colleagues'. Morukov took his place but was not formally assigned to the back-up crew. The crew was launched on 8 January 1994.[15]

Polyakov went on to fly a mission which remains the world's long duration record flight of 437 days 17 hours 21 minutes and 31 seconds. It was very important in terms of knowledge gained. Soon after this mission, Polyakov retired and took up the post of Deputy Director at IMBP, responsible for human space flight. He formally stood down from the cosmonaut team on 1 June 1995. In 1994, the organisation was renamed the 'State Scientific Centre of Russian Federation Institute of Bio-Medical Problems of the Russian Academy of Sciences'.

In late 1998, Morukov was assigned to a Shuttle mission. He was initially assigned to STS-101 with Malenchenko, but for a variety of reasons, the crews were reorganised. Morukov was reassigned to STS-106 and became the second member of the IMBP selection to fly in space when STS-106 was launched on 8 September 2000 for a mission of 11 days 19 hours and 10 minutes. Morukov remains an active cosmonaut based at IMBP.

The flight opportunities for doctors are very few and for this reason, both Karashtin and Lukyanyuk left the cosmonaut team, on 17 January 2001 and 18 February 2003 respectively. Both had actively taken part in a number of isolation tests at IMBP, including the SFINCSS-99 test that lasted 240 days. Such lengthy tests are conducted on a regular basis to examine issues of crew compatibility and the tension involved in long duration missions. But there were simply no opportunities for them to fly, so they gave up their status to pursue their careers in the medical profession.

Larisa Pozharskaya died after a long fight against cancer on 18 February 2002. She had continued to work at IMBP on human flight issues and had continued to work closely with Dobrokvashina in the same department. Senkevich died of a heart attack on 25 September 2003.

In 2003, IMBP negotiated for a new candidate to train as a cosmonaut to represent the bureau. His name was Sergey Ryazanskiy, a biochemist who is very experienced in testing medical space equipment. He enrolled in the team on 29 May 2003 and is currently undertaking his basic training at the Cosmonaut Training Centre.

The need for doctors to be at the heart of the programme is well proven. IMBP has gained extensive knowledge of the physical and mental effects of long-term exposure to space on humans, both from monitoring station missions and ground based simulations. They continue to test new equipment and work on the goal of sending humans to Mars.

THE COSMONAUT GROUP OF MASHINOSTROYENIYA

The OKB-52 bureau was set up in 1955 and began to work on space projects in 1959. It was also called TsBKEM in the 1960s and was then renamed Mashinostroyeniya in 1985. They started work on a top secret project in 1964; the manned space complex 'Almaz', which would have crew complement of six. The first flight models were constructed at Khrunichev in 1970 and the crews started training at Star City.

In the late 1960s, designer V.N. Chelomey decided to select a number of specialists from his company to train as cosmonaut engineers. His design bureau was building the Almaz military space station and was also working on its own manned craft as a rival to Soyuz, the TKS. He felt that one member of any crew should be involved in the design of the craft, which was the same argument presented by Korolyov and Energiya. The first of their engineers were sent to the Medical Commission in October 1968. They were Viktor N. Yeremich, Eduard D. Sukhanov and Valeriy G. Makrushin, who was also made head of the team in 1968. Oleg N. Berkovich was sent to the Medical Commission on 20 May 1969, but later in the year he was medically disqualified. The team within the bureau was formed in July 1970 and was known as the 'Special Contingent'.

When creating his team, Chelomey included testers to support those selected as cosmonauts. The department was number 42, headed by Yemelyan D. Kamen, whose deputies included Ye. Zhernov, who developed simulators, G.M. Noneshnikov, who developed in-flight documentation, and Ye. N. Myslin, who developed a laboratory where cosmonauts would acquire practical skills. Their parachute training was led by L.D. Smirichevskiy and their zero-g training by V.N. Kalemin and A.A. Nuzhnov. One of the main aspects of their training was using the Analogue for real-time simulation, both in manned and unmanned mode. They duplicated the work on board Salyut 3 and 5 using their own engineers. It was ready for use to support the first military Almaz in 1972 and all commands to Salyut 3 were duplicated to the Analogue as well. Volynov and Zholobov learned to repair Salyut 5 while working on this vehicle and Glazkov and Gorbatko worked on the computer repair and worked out how to replenish the air supply on it as well. There was direct contact between crews in orbit and those working in the Analogue simulator, which was based at Reutov.

Chelomey's deputy, G.A. Yefremov, referred a number of his engineers to the Institute of Medical and Biological Problems for medical testing. They also worked very closely with the Institute on medical and physical training. Their links were Yuri Senkevich and Anatoliy Ragulin, a professional ice hockey player who coached the bureau's ice hockey team.

The selected team

The cosmonauts were selected in three groups and completed their basic training with civilian colleagues from IMBP and Energiya. The first formal selection was on 22 March 1972, when Valeriy Makrushin was selected, but Sukhanov was not and was removed from the team. The second selection – Dimitriy Yuyukov – was made on 27 March 1973. Internally, the department was expanded in 1976 to take future expectations into account.

The last selection, of four more engineers, was made on 1 December 1978. They were Vladimir Gevorkyan, Aleksey Grechanik, Valeriy Romanov, and Valeriy Khatulev. Khatulev left the team very soon after because he seems to have not turned up for training. He did not wish to leave design work and concentrated instead on a small satellite programme. Further selections were planned and to that end, four candidates passed the Medical Commission between 1979 and 1981 and joined Department 42. They were Anatoliy Chekh, Boris Morozov, Sergey Chelomey and Sergey Chuchin. It was intended that two more would be put forward (Sergey Kondratyev and Lev Tararin), but delays in manned flights and the KGB taking six months to process their documents meant that there was no further expansion. Sergey Chelomey was the son of the chief designer. He passed the Medical Commission but was never called by the GMVK so was never a real cosmonaut candidate. He died on 6 March 1999.

The cosmonauts underwent a full course of basic training, including parachute jumping at air clubs in the Moscow region, learning to pilot light aircraft (including a YAK 18a), and sea training in the Black Sea. Their training also included a four-day survival exercise in 1982 in the Tyan-Shan Mountains at a height of 4052 metres. They were led by Senkevich and Rukavishnikov.

Flight opportunities for the Almaz engineers were always going to be restricted, as they were confined to those flights involving military Salyuts. When Almaz started flying (in 1974 as Salyut 3, and then two years later as Salyut 5), it was decided to use the Soyuz as the ferry craft. At this time they were flying the two-person configuration following the disaster of Soyuz 11, so each was crewed by an all-Air Force commander and engineer. Another factor was the political intrigue of the leading bureaus.

TKS crews

In 1979, three crews were formed within TsPK. These crews underwent training in the TKS-VA simulator at TsPK, studied the design of the Soyuz return capsule at NPO Energiya, and took part in the State Interdepartmental Tests of the TKS-VA flight programme aboard the Analogue in Chkalovskiy. The training lasted three years, as plans to launch a new generation military Salyut were prepared. Each crew

The welcome of a crew having completed an 8-day test in an Almaz simulator. (from left). Valeriy Romanov (OKB-52), Gennadiy Sarafanov, Maj.-General Aleksey Leonov (training manager) and Vladimir Preobrazhenskiy (Air Force engineer)

would include a flown Air Force officer from the Almaz programme and a number of crews were finalised and approved in late 1979:

1. Glazkov (FE) and Stepanov (CDR) with V.G. Makrushin
2. Sarafanov (CDR) and Preobrazhenskiy (FE) with V.A. Romanov
3. Artyukhin (FE) and Berezovoy (CDR) with D.A. Yuyukov.
4. Vasyutin (CDR) and Rozhdestvenskiy (FE) with A.A. Grechanik as the reserve crew.

Crew 2 completed an eight-day simulation on 20 November 1979, which constituted the State Interagency Test. Leonov was also present when they exited the station mock-up. The test was located at Chkalovskiy and no pictures of the mock-up have been released, reflecting the still-secret nature of this programme.

The flight of the TKS complex occurred under the designation of Kosmos 1267 on 25 April 1981. Sergey Chelomey did take part in the launch preparations of Cosmos 1267 in late April 1981, boarding the TKS vehicle to perform communications tests while it was being prepared for launch on top of a Proton. The mission was repeated on 2 March 1983 as Kosmos 1443.

By the early 1980s, it was clear that there were going to be no further dedicated manned military stations. Chekh left the team in 1983 to work on cruise missiles and Chelomey left soon after. In 1984, the remaining cosmonauts spent eight days in a simulation of hyperkinesis tests of weightlessness by lying down with their bodies slightly inclined, but by late 1984 training was being curtailed.

The Almaz craft that would have been designated 'Salyut 7' – with two docking ports which could have accommodated two TKS Proton-launched vehicles – remains at Reutov. The TKS, which by all accounts was a superb craft, was destined never to fly. Almaz T flew in unmanned mode on an Earth resources flight in 1986.

The group was disbanded on 8 April 1987. There was no formal moment, but it just became clear that training had ceased and that the assignment was over. Five of the group were still active at the point of cancellation and many of them remained as staff members of the design bureau, working on rocket and satellite technology.[16]

Means and measures to train crews and ensure their work during a space flight

The following report was prepared by Cosmonaut Valeriy A. Romanov and published in Russia. It lists the training expectations on the civilian cosmonauts selected by the Mashinostroyeniya team. It could equally apply to civilian space station training:

> *The training of crews for space flights on the Almaz space complex was carried out in the following main directions:*
>
> *Medical and physical training of the organism to make sure it can withstand space conditions. [This was done at] the Institute of Medical and Biological Problems of the Ministry of Public Health of the USSR (IMBP MZ) according to programmes approved by the Ministry of Public Health. These included*
>
> - *training of the vestibular apparatus on turning chairs and on swings*
> - *the adaptation of the body to various loads (up to 10-g), reproduced on centrifuges*
> - *the adaptation of the body to abrupt changes in atmospheric pressure and temperature in thermal altitude chambers*
> - *physical training in the mountains (at medium altitude)*
> - *adaptation of the body to zero-g in vacuum [chambers], water tanks and by maintaining the body in a negative angle (–8°) for a certain time (feet higher than the head)*
> - *experimental training in isolation chambers and [survival] training in the Taiga and the desert to determine psychological readiness and compatibility*
> - *learning how to use rescue devices and survival equipment in case of emergency landings in the water, in the mountains, the Taiga etc.*
> - *first aid training [on dummies and persons]*
>
> *Special technical training at and according to programmes of TsPK, NPO Mashinostroyeniya, NPO Zvezda and the enterprises of the Ministry of Defence mainly included:*
>
> *working out the crew's activities:*
>
> - *in conditions of zero-g on flying laboratories (TU-104, IL-76), in the water tank (control, moving gear and [injured persons], spacewalks, assembly work in space)*

- *after landing of the return capsule on water (undulation of 3)*
- *in case of the need to transport the return capsule by helicopter*
- *in simulators, mimicking the flight programme and non-standard situations, including docking and undocking*
- *on Analogues of the Almaz complex on the basis of flight documentation and flight programmes during all stages of a mission*
- *on flight-rated vehicles of the Almaz complex on the basis of flight documentation in simulation mode*
- *during entry and [emergency] evacuation at the launch pad*
- *during installation of cargos delivered to the station or to be returned to Earth*
- *with food rations, life support systems, thermal control systems, location devices and radio equipment*
- *with individual protective gear (pressure suits, EVA suits)*
- *with the Mission Control Centre*

aircraft and parachute training:

- *studying the night sky and performing astrophysical measurements, both from the ground and from a flying laboratory*
- *studying the theoretical foundations of space flight (astronomy, navigation, dynamics, ballistics, communications etc.)*
- *studying the systems and equipment of the Almaz complex, flight documentation and ways of controlling the systems and the equipment*

Flight control is carried out by the Chief Operations and Control Group (GOGU) from the Mission Control Centre (TsUP). The divisions of the GOGU carry out the following tasks to support the crew:

- *planning the crew's activities four days in advance, refining [the plan] day by day*
- *planning the crew's activities and the radio exchanges, ensuring that the crew carries out the planned work*
- *planning radio exchanges for the present orbit*
- *compiling radiograms*
- *maintaining radio communications with the crew*
- *analysing the crew's activities, their work capacity and also their medical and psychological condition*

On the basis of the conclusions of the operational analysis groups, the specialists of NPO Mashinostroyeniya and the back-up crew members, simulating standard and non-standard situations on the Analogue, the GOGU makes recommendations about the crew's activities aboard the Almaz complex.

Signed by Romanov, 13.08.1999

Training schedule of the group

This group's training highlights have been published as part of their official biographies and it can be related to the tasks outlined in the previous document written by Romanov: These items have been taken from the biographies of cosmonauts Makrushin, Yuyukov and Romanov.

1972: They underwent astronomical training at the Crimean Astrophysical Observatory.

During the next few years when missions to Salyut 3 and 5 flew, there were crews operating in the ground-based Analogue doing a real-time simulation of space-based crew activities. A number of this team's cosmonauts and engineers operated this training. There also is some evidence that the Analogue was used to support the Salyut 2 mission. Although this failed in orbit, the missions were run through based on plans.

1975: Between 17 June and 25 July, they undertook parachute training at the Air Club of DOSAAF in the town of Serkhupov

1976: In April they took part in VA No 0010/1 Aircraft drop tests. In September and October, the group took part in testing Sokol, Sokol KV and Orlan D spacesuits.

1978: From 19 January to 8 Feburary, there was training in a mock-up of the OPS Almaz in conditions of weightlessness in the flying laboratory, Tu 104LL. During April-June, they underwent water zero-g tests in the Black Sea, performing assembly operations in depths of 10 to 15 metres. From 2–20 October, they conducted theoretical training related to the TKS transport supply, in order to carry out ground-based tests at the Chkalovskiy Air Base.

1979: The group of cosmonaut testers continued its medical and physical training at IMBP. A lot of attention was paid to special technical training, which included:

working out ways of working in spacesuits
working out ways for the crew to deal with a splashdown
working out ways for the crew to deal with a non-standard landing of the VA
studying and perfecting flight documentation
EVA training at the underwater tank at TsPK

From 20–28 November, they undertook State Department tests of the TKS and the VA crew test aboard the Analogue of the TKS at the Chkalovskiy Air Base.

1979–1982: The crews performed direct flight training in the simulators of TsPK and studied the design of the Soyuz descent capsule at NPO Energiya. In August 1982, as part of their survival training, they carried out a four-day trek through the AK-SUU mountain pass in the Tyan-Shan mountain range at an altitude of 4052 m. This was under the leadership of Yuriy Senkevich of

IMBP, a former cosmonaut candidate who hosted the popular television series *Travellers' Club* and is world famous for having taken part in the expeditions of Thor Heyerdahl. Also part of the group was veteran cosmonaut Nikolay Rukavishnikov.

1984: The cosmonauts took part in eight-day training sessions at IMBP to simulate weightlessness, spending all that time lying down with their bodies slightly inclined (the feet eight degrees higher than the head).

However, special training for a manned flight on the TKS and VA began to be curtailed. In the autumn of 1984, Vladimir Chelomey convened the leaders of the firm and the cosmonauts and ordered them to continue their training, hoping that his spacecraft would eventually also fly in piloted mode. However, Chelomey died on 8 December 1984.

OTHER COSMONAUT SELECTIONS 1964–2005

One of the features of cosmonaut selection was the inclusion of candidates who were brought together or identified as potential 'flyers' due to mission requirements, political reasons or design bureau politics. They usually only trained for short periods and for a single mission. They were not formally attached to an organisation that had permission to assemble a team, but for administrative and training purposes, many were in the establishment of the training centre.

1964: The selection for Voskhod 1

In May 1964, when it was agreed to fly the Voskhod 1 craft with three cosmonauts, only one would be an Air Force pilot and a number of candidates from other disciplines were considered. Four were doctors from various bureaus and organisations (Yegorov, Sorokin, Lazarev and B. Polyakov – see section on IMBP cosmonaut selection page 163); one was a senior engineer from OKB-1 (Konstantin Feoktistov – see section on Energiya cosmonaut selection page 147); and one was a scientist from the Academy of Sciences (Katys – see section on Academy of Sciences selection page 162) The seventh person was Engineer-Colonel Vladimir Benderov, who was a test pilot for the Tupolev design bureau. He passed the Mandate Commission on 1 June 1964 but stood down on 2 July. Benderov was recommended to Kamanin by a senior Air Force colleague in February 1964. It is not clear why he was included and in what role he was being considered, but it is assumed he was a potential commander. This may have reflected concerns about the lack of experience within the first Air Force selection. Benderov was promoted to the rank of Major-General in 1971 and was later killed in the Tupolev 144 crash at the Paris Air Show on 3 June 1973. He is buried in central Moscow with other members of the crew. His involvement in the space programme only became known many years after his death.[17]

1965: The Journalists for Voskhod missions

Some senior officials, particularly Korolyov, felt that there would be a benefit in flying a journalist, who could write about the wonders of space travel and report first hand about what they saw and felt. Korolyov had been disappointed with the way cosmonauts and astronauts had been unable to describe their experiences. This did throw up one interesting dilemma, however, which was that early space travel and the identities of many of those involved were state secrets, and it is not clear what would have happened if a journalist had flown in that era. In 1964, Korolyov asked three journalists to undertake medical screening for a possible mission on the Voskhod space ship. In a book called 'Seven days in the life of Korolyov', Yaroslav Golovanov, one of the candidates, wrote: 'Once in early 1965, Korolyov offered me (a chance) to fly into space. Only then did I find out that he had wanted for a long time to send a journalist into space. I went for a physical along with Yuri Letunov.' To their amazement, they passed, and their acceptance was signed by Korolyov, but when he died in January 1966, the idea died with him. Given the ambitious plans for the Voskhod programme it cannot be discounted that, if Korolyov had lived, his idea might have led to a flight for one of the candidates.

Letunov was a radio journalist who had worked at Baykonur and later, was the chief editor of the '*Vremia*' news programme. He died in 1983. Golovanov, who was a famous writer on space for a number of newspapers, including the *Komsomolskaya Pravda*, and author of a number of books including a major biography on Korolyov, died on 21 May 2001.[18]

Another possible candidate at this time was Colonel Mikhail Rebrov, who was the journalist for the military newspaper *Red Star*. He had covered the space programme and continued to do so until his death on 24 April 1998. He was also the author of many books about the cosmonauts.

1968: Paton Institute candidates for the welding experiment

In 1968, it was decided to fly an experiment to use a welding device in the Orbital Module on a solo Soyuz flight because it was felt that welding would be a technique that would be used in the construction of large space stations. This experiment was developed by the Paton Institute in Kiev. One member of the bureau was entered into the team to work on this mission as a candidate to operate the system in orbit. He was Vladimir Fartushniy, an engineer, who passed GMVK on 24 May 1968. He completed his basic training with Yazdovskiy and Patsayev and was assigned to the mission with Shonin and Kubasov. It was scheduled for mid-1969 but slipped back and in the end, he did not fly, going on instead to work on a possible Soyuz Kontakt mission. He left the team in the middle of 1971 when he failed a medical due to a car crash. There have been other experiments using welding in space, all developed by the Paton Institute and used on Russian stations.

1983: Additional doctor assigned to a Salyut 7 mission

On 9 March 1983, Oleg Atkov passed the Medical Commission and was assigned to the team, in part to serve as a doctor in orbit for when they flew Feoktistov on a long duration flight to test the effects of long-term weightlessness on an elderly person.

Atkov was a cardiologist based at the Institute of Clinical Cardiology of the USSR Academy of Medical Sciences. Feoktistov failed a medical and his mission was cancelled, but Atkov was reassigned on 1 September 1983 as cosmonaut researcher on the Soyuz T 10 mission with Kizim and Vladimir Solovyov. His back-up was Valeriy Polyakov from the IMBP group. Atkov flew the mission, which was launched on 8 February 1984 and lasted 237 days, during which he operated the 'Argument' system that had been installed on the station at launch. He left the cosmonaut team in December 1984, returning to work at his institute. Atkov was the Soviet Union's first real payload researcher, being selected for a mission and leaving soon after completing his assignment.[19]

1989: The journalist in space

In 1989, GlavKosmos signed an agreement with a Japanese TV station to fly a reporter on a visiting mission to the Mir station. This produced a backlash from the Russian press, demanding to know why the Soviet Union was going to fly a foreign journalist before a Russian one. The campaign in the press was very high profile and the USSR Journalists Union created a space commission to offer a short-list of possible candidates for screening.[20] The first group of candidates undertook their medicals on 27 November 1989 and there were 37 candidates in all, including four women.[21] The scheme was highlighted in the press using the slogan 'Space to children', as the royalties from published videos and books would be donated to children's health causes.[22]

The names of successful candidates began to be published in Moscow newspapers on 28 February 1990 and the six short-listed candidates were selected on 11 May 1990. They were Colonels Aleksandr Andryushkov and Valeriy Baberdin from the military newspaper *Krasnaya Zvezda* (Red Star); Yuri Krikun, a film director from the Ukrtelfilm Studio in Kiev; Pavel Mukhortov, who wrote for *Sovetskaya Molodezh* in Riga; Svetlana Omelchenko of *Vozdushny Transport* based in Moscow; and Valeriy Sharov from *Literaturnaya Gazeta*, who was their Far East correspondent. The group started their training at the beginning of October 1990 for a flight scheduled for 1992, which was the United Nations' International Year of Space.[23]

A number of articles appeared about their training, written by the candidates. This gave western observers a real insight into their training and how OKP worked in the Russian space programme. They took their examinations at the end of January 1991 at the Cosmonaut Training Centre (with Baberdin and Omelchenko achieving straight 'As' in all the subjects) and completed their basic training on 7 February 1992, graduating as cosmonaut researchers.[24]

Due to the fall of the Soviet Union, it became obvious that there would be no finance to fly the mission. Krikun, who was a Ukrainian, did have a possibility of a flight on an all-Ukrainian crew to look at the effects of the Chernobyl accident and this did dominate the space news for a while after the break up of the Soviet Union. Kadenyuk was another possible candidate, but it would seem that it was not a serious venture and never got off the drawing board. They all went back to their journalist and TV careers and their mission never occurred. Ironically, the flight of the Japanese TV journalist occurred in 1991 on board Soyuz T 11, while they were still in training. Baberdin died on 2 October 2003.[25]

1990: The selection on behalf of Zvezda

The Research and Development and Production enterprise, 'Zvezda', was established in 1952. It is based in Moscow and is the design bureau responsible for the design and construction of space suits since the early 1960s. They have designed all the suits used in the programme, including the lunar suits, and they have also built the 'Orlan' design EVA suits (which are based on the lunar suit), the MMUs used on Mir, and the SAFER system which will be used on ISS in the future. They are also famous for their work on ejector seats and designed the ones which would have been used on Buran. This work has been performed under the supervision of Gay Severin, who is the General Director and designer.[30]

On 11 May 1990, the GMVK selected Vladimir Severin as a cosmonaut. He was the son of the Chief Designer and a representative of the bureau, as well as an engineer with a background as a test engineer within the bureau. He undertook his basic training from 1 October 1990 until 6 March 1992, although there is no indication that he trained for a specific mission. He did undertake some general training on Mir systems, but stood down in 1995 when the cosmonaut team was halved. After he left the team, he was made a Hero of the Russian Federation on 21 June 1996, which was awarded for his work on testing the ejector seat for the newly constructed Su 29KS sports plane. The SKS 94 system has been installed in the new generation of sports plane.[27]

1990: Kazakhstan space

The Baykonur launch site is located in the autonomous republic of Kazakhstan. Since the flight of Gagarin, though a number of cosmonauts launched into space had been born in Kazakhstan, all had flown as ethnic Russians, not ethnic Kazakhs. The Russian space authorities entered into an agreement with the President of Kazakhstan, Nursultan Abishevich Nazarbayev, to fly an ethnic Kazakh into space on a Soyuz to the Mir station, and two candidates were selected. On 11 May 1990, Talgat Musabayev passed the State Commission. He was a civilian airline pilot but was also a Master of Sport in aerobatic flying and a member of the Soviet team, and had won the USSR's aerobatics championship in 1983. In March 1991, he transferred to the Russian Air Force team. The second candidate arrived at Star City on 21 January 1991. Toktar Aubakirov was one of the most famous test pilots in the Soviet Union. For testing the MiG 29 and Su 27 supersonic fighters, he had been awarded the title of Hero of the Soviet Union on 31 October 1988, and in 1990, he was given the award of Meritorious Test Pilot of the Soviet Union.[28]

Aubakirov was assigned as a researcher on a visiting mission crew alongside Korzun and Aleksandrov, with Musabayev as his back-up, paired with Tsibliyev and Laveykin. On 15 July 1991, this mission was cancelled due to budget problems. The new crew alongside Aubakirov would be Alexandr Volkov and Austrian researcher Viehböck, who were confirmed on 10 August. The mission flew on 2 October 1991, lasting seven days, and on 11 October, Aubakirov was awarded the title of Pilot Cosmonaut of the USSR. He became Cosmonaut No. 72 and the last from the Soviet Union, which broke up just weeks later. Aubakirov also became Cosmonaut No. 1 for the new Kazakhstan Republic. He later became a deputy Minister of

Defence and then head of the Kazakh Space Research Agency and a deputy in the Parliament. In 1994, he became a Major-General and was made a Hero of the Republic of Kazakhstan. Currently, he is the advisor to the President for Aviation and Space Exploration.[29]

Musabayev remained part of the cosmonaut team of Russia and flew three missions to Mir and ISS. He currently has the rank of Major-General and was also awarded the title of Cosmonaut No. 2 of the Kazakhstan Republic.

1996: The cosmonaut corps of the space rocket military forces

The Rocket Forces (VKS) wanted to include their own representative in the cosmonaut team. Lt-Colonel Yuri Shargin passed the GMVK on 9 February 1996 and started his basic training on 3 June 1996. Shargin had been assigned, prior to his selection, as commander of the 1382nd military representative office group to Energiya RSC and had been based at the Baykonur cosmodrome for a number of years. He was the commander of this group of candidate cosmonauts, completing training on 20 March 1998, and was assigned as a cosmonaut tester of the Central Test and Control Office for the Space Rocket Military Forces.[30]

Soon after completing his basic training, Shargin was assigned as the back-up to Baturin for his flight on TM 28, but was replaced by Kotov in late May when he failed a medical. On 2 September 1998, he transferred from the Rocket Forces to the Air Force and was incorporated into the TsPK team. Then, on 28 December 2001, he transferred back to the Rocket Forces. It is not clear why these two transfers were made. In 2004, he was assigned to the third seat on board Soyuz TMA 5 with Sharipov and Chiao. The mission launched on 14 October 2004.

1996: TsSKB-Progress at Samara

Oleg Kononenko passed the GMVK on 9 February 1996, as a cosmonaut candidate for the State Scientific Industrial Rocket and Space Centre, TsSKB-Progress. He started basic training on 3 June 1996 and completed OKP on 20 March 1998. On 5 January 1999, he transferred formally to Energiya and joined their cosmonaut team. For more details see the selection of cosmonauts Energiya on page 147.

1997: Khrunichev State Scientific Production Centre

The Khrunichev State Scientific Production Centre has been responsible for the design and construction of the Proton booster and the modules used on Salyut, Mir and ISS. It has been at the forefront of the development of human space flight and for some years they have been trying to get one of their engineers aboard a mission to a station. On 16 June 1992, the main Medical Commission (GMK) passed Sergey Moshchenko as a prospective cosmonaut and in February 1997, he passed the GMVK. He was selected as a test cosmonaut candidate for the Khrunichev Centre Cosmonaut Office and started his basic training at the training centre on 16 January 1998, completing it on 1 December 1999. He was assigned to a long duration expedition mission to ISS and was paired with Malenchenko and Lu on ISS-7. He was removed from the crew in October 2002 because he was having problems with the training regime. Ironically, he would have lost his seat anyway due to the crewing changes post-Columbia (STS-107), when they moved to two-person crews. He remains an active cosmonaut and is available for future missions.[31]

1997: Representative of the President

Yuri Baturin was an advisor to President Boris Yeltsin. He passed OKP on 17 February 1998 and was attached to the Air Force group for administration purposes. He was assigned to a mission to the Mir station as a researcher and flew on TM 28 with Padalka and Avdeyev. On his return, he was awarded the title Pilot Cosmonaut of the Russian Federation on 25 December 1998. In 2001 he was assigned to his second flight, TM 32, and flew with Musabayev and Tito to ISS. He was awarded the title Hero of the Russian Federation on 28 September 2001 and remains an active cosmonaut.

1997–2000: the film project

In December 1997, came the first reports of film director Yuri Kara's desire to make a feature film using the Mir space station. The story line was that in 1999, the Russians were going to abandon Mir, but one cosmonaut refuses to leave and says he will orbit the Earth for the rest of his days. The controllers decide to send a female cosmonaut up to the station to lure him back. The cost of the project was estimated at around US$ 20 million and the film would be called '*The Mark of Cassandra*', as it was based on a novel of the same name by Chingiz Aitmatov. They had identified three actors to play the parts: Vladimir Steklov, Nataliya Gramushkina (who was 22 and a recent drama school graduate) and Olga Kabo.[32] By 1999 it was decided that only one actor would fly to Mir, to play the renegade cosmonaut. The scenes involving other actors would be shot on the ground and in transport aircraft in free-fall mode.[33]

In April 1999, an agreement was signed between the training centre and the State Committee on Cinematography to train actor Vladimir Steklov, starting on 1 July. He was introduced to his trainers on 7 June and was scheduled to undertake a six-month training programme. The female cosmonaut would now be played by Masha Shukshina and there are pictures of her undertaking sea training in the Black Sea. Steklov did his sea training with Sergey Volkov and US astronaut Joan Higginbotham.[34]

In January 2000, the issue of finance was still the major block to the mission. The main crew of Zalyotin and Kaleri, who would be the last crew on Mir, were only a couple of months away from their launch and a crew picture had been taken that included Steklov. It became clear that no state funding would be available to support the mission, but there were rumours that a Virgin Island venture capital company was interested. By now, the title of the film had also changed to '*The Final Journey*'.[35] Steklov passed his medical at Star City on 26 February, as mission planners wanted to take the project to the wire, but he had no back-up.[36] On 16 March, it was decided to remove Steklov from the crew as no money was forthcoming and the crew was due to leave for Baykonur for their pre-launch tests a couple of days later.

A spokesperson for the Russian Aerospace Agency said that Steklov, 'will not go to Mir as earlier planned because of a failure to meet the terms of the contract.' The producers of the movie claimed that the initial payment to get the project underway had been made, but Russian space officials complained that they had not even been paid for Steklov's training, let alone the multi-million dollar bill for the flight. This was the end of the project, because in early April, the Soyuz TM 30 craft blasted off for the last flight to Mir with only a two-person crew on board.[41]

Selections since 2000

Since the Steklov selection, a number of other individuals that were not attached to formal groups have passed the Medical Commission. One long-standing candidate is Yuri Loktionov, who has passed GMK but has been passed over on a number of occasions by the GMVK because he is having medical problems. Another candidate is Sergey Zhukov, who is General Director of the ZAO 'Centre for Technology Transfer', which is part of Rosaviakosmos. ZAO works on patent licence work and the commercialisation of the results of scientific and technical activities. Zhukov passed the Commission on 31 May 2002 and was included in the 2003 selection when he passed the GMVK on 29 May 2003. He is currently doing his basic training, which should be completed in 2005.

Sergey Polonskiy is the general director of the Moscow branch of OAO 'Stroymontazh', which is a very large building firm in Russia. He passed the commission on 12 September 2002 and was considered for the 2003 selection, but was not passed by the State Commission. When the Russians had a vacancy for a 'tourist' flight in October 2004, Polonskiy offered up to US$ 5 million to fly, but was turned down. He had trained in Moscow and in Houston and was the first millionaire Russian to be considered for a flight.

REFERENCES

1 The Selection of Cosmonauts, *JBIS,* **50**, 1997, pp 311 -316, Sergey Shamsutdinov

2 *Red Star*, 5 April 1988, Voskhod candidates, by Tyulin, a famous space administrator; Kamanin Diaries, 1 September 1965

3 Michael Cassutt, 7 February 1993 correspondence; Selection of Cosmonauts, already cited

4 Letter dated 12 December 1991 to Boris Yesin

6 Russian Science Cosmonauts, *Spaceflight*, **38**, November 1996, by Igor Marinin and Igor Lissov

6 Experiment RISK, Vitaliy Volovich, Progress Publishers 1986

7 *Novosti Kosmonavtiki* number 22, 1993; Directory of the Russian Space Industry, 1993

8 Interview with Ilyin by Hall and Vis, IMBP 2001

9 Cosmonauts Number Zero, Nina Chugunova, *The Bulletin of Atomic Scientists*, May-June 1994

10 A year aboard a spaceship, Bozhko and Gorodinskaya, 1969

11 Izvestia, 30 August 1988, Cosmonaut Physician Polyakov's roles during planned stay on Mir

12 Interview with Galina Amelkina, Hall and Vis, August 2004

13 Meditsinskaya Gazeta, January 1988 No 5, Atkov's work in space noted

14 Zemlya I Vselennaya, No 6 Nov-Dec 1990, pp 32–36; Izvestia, 30 August 1990; Presentation at BIS HQ in 2001; *Spaceflight News*, August 1990, interviews at the ASE in Holland
15 Moscow Radio, 6 Jan 1994, reported by the BBC
16 The difficult Road to Space, V. Romanov, Tribune Internal Publications Mashinostroyeniya; The History of the Group of Cosmonaut testers of TsKBM (NPO Mashinostroyeniya); *Novosti Kosmonavtiki* No 12, 2000, p 78, Almaz cosmonauts, Shamsutdinov; Chelomey's cosmonauts: Why there are no crews from NPO Mashinostroyeniya in Outer Space, Nina Chugunova, NASA translation TT 21771
17 Soviet and Russian Cosmonauts 1960 – 2000, *Novosti Kosmonavtiki 2001*; Kamanin Diaries 1964–1966, Bart Hendrickx, *JBIS*, **51**, 413–440
18 *Komsomolskaya Pravda*, 30 March 1989 by Golovanov; Letter to Bert Vis, 25 December 1990; Obituary of Golovanov, *Novosti Kosmonavtiki*, July 2003
19 *Novosti Kosmonavtiki*, Soviet and Russian Cosmonauts 1960–2000 in Russian; Interview with Oleg Atkov, Bert Vis, 9 June 1991 transcript; *Meditsinskaya Gazeta*, 15 January 1988
20 *Pravda*, 6 April 1989
21 TASS, 23 November 1989 and 30 November 1989
22 TASS, 27 Nov 1989
23 TASS, 2 October 1990
24 *Red Star*, 26 July 1991, 27 July 1991, 31 August 1991, 11 September 1991 Krikun diary, published in *Soviet Soldier*; Svetlana's story, Colin Burgess, *Spaceflight*, **44**, July 2002
25 Radio Kiev, 13 November 1991
26 Russian Space Suits, Abramov and Skoog, Springer-Praxis 2003
27 *Novosti Kosmonavtiki*, 12–13, 1996, p 4; *Aerospace Journal*, July-August 1996
28 *Kazakhstanskaye Pravda*, 19 January and 12 April 1991; *Rabochaya Tribuna*, 21 February 1991
29 *Nezavisimaya Gazeta*, 4 August 1992; *Red Star*, 13 January 1993; Moscow radio, 27 February 1993; Red Star, 4 March 1995; Moscow Interfax, 8 July 1999; Biography of Aubakirov, Heroes of the Soviet Union web site, www.warheros.ru
30 *Aerospace Journal*, 11/12 1996; Biography, www.gctc.ru
31 NASA biography, dated February 2001
32 Reuters, 22 December 1997; Associated Press, 22 December 1997
33 Associated Press, 26 June 1999
34 *Novosti Kosmonavtiki*, emails from Igor Lissov, dated 30 April and 18 June 1999; *Novosti Kosmonavtiki*, October 1999
35 *Florida Today*, 13 January 2000; 17 January 2000
36 RIA news agency, 26 February 2000
37 Associated Press, 16 March 2000

Cosmonauts selected to fly the Buran Shuttle

The construction of Buran was started in the mid 1970s. The Soviet authorities decided to put Buran testing in the hands of the pilots of LII, which consisted of a large group of civilian pilots and engineers with a long history of successful testing and development of all types of aircraft. The operational missions, however, would be given to Air Force test pilots based at the Air Force Test Pilot School at Akhtubinsk, although military pilots were not expected to fly the Buran before 1985. These decisions led to the creation of two more cosmonaut groups.

THE STATE SCIENTIFIC TEST RED BANNER INSTITUTE NAMED FOR CHKALOV (GKNII VVS)

The Council of Ministers approved the development of the Buran programme and the Soviet Air Force selected nine pilots in 1976 as potential candidates to fly Buran operationally. This group was assigned to the Chkalovskiy Test Pilot School at Akhtubinsk. The pool of pilots was very small but it also became clear that the Buran system needed test flying, so it was agreed to select a group of very experienced test pilots to work with colleagues from LII. In May 1977, the nine trainee cosmonauts graduated from test pilot school ranked as test pilot third class and were assigned to the Cosmonaut Training Centre for two years of basic training. In 1978 when the Air Force tried again, they found only two acceptable candidates to start test pilot training. They would also be assigned to the training centre when they graduated from the test pilot school. All these men were eventually transferred to space station operations.

In 1978, the Ministry of Defence decided to recruit their team to fly the Buran system and on 1 December 1978, six very experienced test pilots were selected. They were Colonel Ivan Bachurin, the group's commander, and Lt-Colonels Aleksey Boroday, Viktor Chirkin, Nail Sattarov, Vladimir Mosolov and Anatoliy Sokovykh. They started their basic training course, which was evenly divided between the test pilot school and the Cosmonaut Training Centre at Star City. Two other short-listed candidates were considered for this selection. One was Vladimir Gorbunov, who turned down the opportunity. He was involved in testing the MiG 29 and went on to have an illustrious career as a test pilot at the Mikoyan design bureau, including being awarded a Hero's star in 1992. The other candidate was a pilot called Oleynikov.

This selection meant that the 1976 group of pilots (who were designated test cosmonauts in 1979) were not all needed for Buran immediately. Two were assigned permanently to Star City to work on Almaz, while six others were assigned to flight test work relating to Buran. They were Nikolay Moskalenko, Anatoliy Solovyov, Yevgeniy Saley, Leonid Ivanov, Leonid Kadenyuk and Aleksandr Volkov. The ninth, Sergey Protchenko, was dismissed for medical reasons. He remains working as a test pilot.

On 23 October 1980, Ivanov was killed while test flying a MiG 23 and Sattarov left the team in May 1980 after performing an unauthorised barrel roll in a Tupolev 134. In 1981, Chirkin withdrew from the team, because he did not believe that the Buran would ever fly and wanted to pursue his test flying career. He was testing the SU 27 aircraft.

On 12 February 1982, the four remaining VVS test pilots, Bachurin, Boroday, Mosolov and Sokovykh, were registered as test cosmonauts. In 1982, Bachurin became a Merited Test Pilot of the USSR and in 1983, Solovyov, Volkov, Saley and Moskalenko reported full time to Star City as potential commanders of Soyuz T missions. Kadenyuk was removed from the cosmonaut team due to his divorce, but he remained at the flight test centre as a test pilot.

The next group of Air Force pilots was selected for the programme by the GMVK. They were Colonel Viktor Afanasyev and Lt-Colonels Anatoliy Artsebarskiy and Gennadiy Manakov, who started formal basic training at the Cosmonaut Training Centre. Due to an incident which resulted in a major accident, Sokovykh was removed from the team and later, in August 1987, Mosolov was removed from the team following his divorce. This coincided with the formation of the formal team and on 8 January 1988, Afanasyev, Artsebarskiy and Manakov transferred formally to the staff of the training centre at TsPK. This possibly reflected the low expectations of flight opportunities for Buran.

On 25 January 1989, the GMVK selected three more pilots to join the Air Force Buran programme: Colonel Valeriy Tokarev, and Lt-Colonels Anatoliy Polonskiy and Aleksandr Yablontsev. A year later, the GMVK selected a further three candidates; Colonels Valeriy Maksimenko, Aleksandr Puchkov, and Nikolay Pushenko. They started work on 11 May and trained with their colleagues from the previous selection. They all passed their basic training on 9 April 1991 and responsibility for the group now passed to the new structure which had been set up in 1987.

The GKNII VVS group was formed by an order of the Ministry of Defence and it became the formal team for the Air Force's involvement in the Buran programme. The first cosmonauts were assigned to the team on 7 August 1987. They were Colonel Ivan Bachurin, the group commander, and Colonel Aleksey Boroday. Between January and April 1988, Bachurin and Boroday flew six missions on the BTS-02 atmospheric analogue of Buran.

On 25 October 1988, Kadenyuk joined the group to work on the Buran landing system in a MiG 25. From November 1990 to March 1992, Bachurin, Boroday and Kadenyuk underwent a complete training programme for a flight on the Rescue Soyuz to dock with Buran. This work was conducted at Star City, with Bachurin as

Disconnected electrical wiring hanging from the Buran flight deck simulator in the KTOK is a solemn reminder of the end of the ambitious Soviet space shuttle programme

The Buran motion base simulator in the KTOK was meant for training crews for the atmospheric portion of the return to Earth. In this picture, taken in 1999, the simulator was still ready for use, but by 2003, it had been partly dismantled

prime commander, Boroday as back-up commander and Kadenyuk as reserve commander. When funding for the Buran programme was stopped, the mission was cancelled. Further additions to the team came on 8 April 1992 when Puchkov and Yablontsev joined them. Bachurin retired in December 1992 but was replaced by Tokarev on 30 January 1993. Boroday, who replaced Bachurin as group commander, was himself the next to leave, in December 1993, and Tokarev replaced him as commander. The last member of the group arrived in February 1995 when Pushenko joined the team. After the cancellation of the Soyuz mission, the team divided their time between Star City and Akhtubinsk. Polonskiy was offered the opportunity to transfer to Soyuz missions within the cosmonaut team but refused, preferring to fly and test aircraft. He became a Merited Test Pilot of the Russian Federation.

The institute commanders were unsure of the purpose of this group and did not give them appropriate tasks. The commander of the Test Pilot School suggested that the group should be disbanded, but the Air Staff were also looking at possible future flights of the MAKS system. Tokarev, as commander of the group, tried to get them more test flying, including cargo work. He arranged for them to test fly aircraft for the Moscow branch of the NII and also asked for the team to be transferred to the main cosmonaut team at Star City. They all lived at Zvyozdnyy Gorodok anyway, but he did not rate their chances as high.

The group was officially disbanded on 30 September 1996 by the order of the Air Force Chief of Staff (order number 123/3/0716) and the group members returned to their careers. Tokarev transferred to the cosmonaut group at Star City in 1997. The State Commission agreed this on 25 July 1997 and he formally joined a few months later. He was assigned to Houston as a Mission Specialist and flew on the Space Shuttle on STS-96. He remains an active Russian cosmonaut. Kadenyuk went to serve in the Ukrainian Air Force and then flew on Space Shuttle mission STS-87 in November 1987 as a Payload Specialist and the Ukraine's first astronaut. Boroday went to fly heavy lift aircraft but crashed in October 1996 near Turin, Italy, while flying an Antonov 124. He nearly died from his injuries and he did lose his legs, although he had experimental prosthetic limbs fitted in the USA. He still works in aviation and lives at Star City. Pushenko and Yablontsev went to work as test pilots, developing the new generation of civilian airliners.

FLIGHT RESEARCH INSTITUTE OF THE MINISTRY OF AVIATION INDUSTRY LII MAP

The formal establishment date of this group varies according to sources. In 1977, the Ministry of Aviation Industry decided to form a group of test pilots from its test centre based at Zhukovskiy, to work on the Buran programme. It was the role of the LII test centre to develop new types of aircraft but there was also a test pilot school for civilian test pilots at the centre. They had been involved in the development of the Soviet lifting bodies such as BOR 4 and Lapot. In 1981, the flight centre was named after M.M. Gromov, a very famous test pilot. Nine pilots – all civilian – were identified for this group and sent to IMBP for medical testing. The original group was selected by the State Interdepartmental Commission GMVK on 1 December 1978. They combined the role of cosmonaut with continuing their flight testing and conducted work on some of the most advanced fighter planes, as well as flying air displays all over the world. They were some of the best pilots in the Soviet Union.

The Group of the Branch Complex of the Test Cosmonaut Preparation (OKPKI)
The commander of the group was identified as Igor Volk and the team gained a nickname based on his name. 'Volk' means 'wolf', so the group became the 'Wolf Pack' and the trainees were 'cubs'. The five pilots selected were Oleg Kononenko, Anatoliy Levchenko, Aleksandr Shchukin, Rimantas Stankyavichus, and Volk. The four unsuccessful candidates included Vladimir Turovets, who failed the medical test. He was a candidate for both the 1978 and 1983 selections but was killed in a MiG 8 crash on 8 February 1982 just days after being considered by the Commission for the second selection. Also unsuccessful were Viktor Bukreyev, who was killed on 17 May 1977 while testing a MiG 25; Aleksandr Lysenko, who was killed on 3 June 1977 while testing a MiG 23 soon after taking the medical tests; and Nikolay Sadovnikov, who withdrew and joined the Sukhoi OKB as a test pilot.

Oleg Kononenko was included in the first Buran group, but was killed while undertaking his basic training when testing a Yak 38 on the aircraft carrier '*Minsk*' in the South China Sea on 8 September 1980. He failed to eject. He was a Merited Test Pilot of the USSR, which was awarded to him on 15 August 1980. The remaining four completed their basic training on 10 August 1981 and were approved by the Ministry of Aviation Industry, and this is viewed by some as the formal start date for the group. They were awarded their test cosmonaut qualification on 12 February 1982. Volk attained the status of Merited Test Pilot of the Soviet Union on 18 August 1983 while he was training for a mission to the Salyut 7 space station. He initially trained with Kizim and Solovyov but the mission was delayed due to problems with re-supplying the station during that period. He eventually flew as a Cosmonaut Researcher, on board Soyuz T 12 with Dzhanibekov and Savitskaya on 17 July 1984. He was in space for 11 days 19 hours and 14 minutes. On returning to Earth he flew to Moscow and back using the Tupolev 154 and the MiG 25 with the adapted Buran control systems to test the effects of zero-gravity on his flying skills. He was awarded the titles Pilot Cosmonaut of the Soviet Union and Hero of the USSR. His back-up was Savinykh, who was paired with Ivanova and Vasyutin. Interestingly, the experiments relating to the effects on zero-gravity had previously been conducted by Dzhanibekov and Popov on their return from orbit on Soyuz T 6 and 7 respectively.[1]

The search for additional pilots started in 1979, with four candidates being considered, but the selection was delayed until 1982 when a handful of other candidates were added. As well as Magomed Tolboyev and Ural Sultanov, the other candidates were Viktor Zabolotskiy, whose selection was delayed until 1984, and Pyotr Gladkov. Gladkov declined the assignment, having taken advice that the chances of Buran flying were very low. The second selection was chosen by the GMVK on 9 March 1983 and approved by MAP on 25 April 1983. It consisted of Sultanov and Tolboyev. In 1984, the third selection (consisting of just Zaboloskiy) was approved by the GMVK on 15 February and by the ministry on 12 April. The fourth selection, Yuri Sheffer and Sergey Tresvyatskiy, joined on 2 September 1985, with ministry approval on 21 November 1985. The five LII cosmonauts from these selections all did their basic training together, from 13 November 1985 to 22 May 1987. On 5 June 1987, they were awarded the certificate of Test Cosmonaut.

Test flying the BTS-02 Analogue

From the autumn of 1984 to April 1988, the test programme of the BTS-02 Buran Analogue was undertaken by some of the LII team. The first flight occurred on 10 November 1985 following a series of ground tests. The LII pilots provided two crews for this test programme: Volk and Stankyavichus as the number one crew, and Levchenko and Shchukin as the number two. They flew in different combinations as commander and pilot. Levchenko was a Merited Test Pilot of the Soviet Union, which was awarded to him on 15 August 1986. The last flight of BTS-02 occurred on 15 April 1988, flown by Volk and Stankyavichus. Two military Air Force pilots, Bachurin and Boroday, also took part in the programme. There were a number of Buran simulators built and operated and one of these still exists at the Cosmonaut Training Centre.

In March 1987, Levchenko and Shchukin started training for an assignment on a Soyuz TM flight. Levchenko was the prime pilot, with cosmonauts Vladimir Titov and Musa Manarov. He flew on 21 December 1987 on an eight-day mission, during which he conducted an experiment to test the feasibility of flying Buran during the early stages of weightlessness, using a special stand mounted in the Orbital Module of TM 4. When he landed, he carried out the same experiment to test the effects of zero-gravity as undertaken by Volk. He was the second Buran LII pilot to fly a space mission and became a Hero of the Soviet Union and a Pilot Cosmonaut of the USSR. Shchukin was Levchenko's back-up and trained in a crew with Aleksandr Volkov and Aleksandr Kaleri. Levchenko died of a brain tumour on 6 August 1988.

On 18 August 1988, Shchukin was killed during a training flight on a Su 26M sporting plane while preparing for an air show. When Buran made its return in November 1988 from its flight in automatic mode, Tolboyev flew the MiG 25 chase plane which sent the landing images back to mission control. From 1988 until the spring of 1989, Stankyavichus trained as a prime crew member with Viktorenko and Balandin, while Zabolotskiy trained as a member of a back-up crew with Manakov and Strekalov, to visit Mir. The mission was cancelled and the third seats on missions began to be assigned to foreign astronauts in return for money to ease the costs of what was becoming a cash strapped programme.

All the Buran cosmonauts assigned to space station flights had to be trained in Soyuz and station systems, as well as keeping their flight skills at a high level.

The identities of the LII group were disclosed in an article in the *Trud* newspaper on 4 January 1989, following the successful Buran automatic mission. The last member, Yuri Pridhodko, joined in 1989, being passed by the GMVK on 25 January and achieving approval from the ministry on 22 March. Most likely, he joined because of the deaths of Levchenko and Shchukin in 1988. He underwent the OKP training programme, finishing in 1990.

Stankyavichus was awarded the title Merited Test Pilot of the USSR on 4 October 1989, as was Zabolotskiy on 13 December 1989. Sheffer also achieved this status on 16 August 1990. On 9 September 1990, Stankyavichus was killed while performing advanced aerial aerobatics at an air show near Treviso in Italy. He is buried in his home town of Kaunas in Lithuania. On 28 July 1992, while flying a Yak 38 vertical take off plane at the Moscow Air Show, Zabolotskiy's ejection system was spontaneously activated but he was not injured. On 16 November 1992, Sultanov was named deputy Commander of the 'Fedotov' Test Pilot School (ShLI), a division of LII.

Training for the Buran programme tailed off and certainly by 1993, the feeling was that Buran was a dead programme. Tolboyev became a Hero of the Russian Federation on 17 November 1992 and he retired on 12 January 1994 when he was elected to the State Duma as a member for Dagestan. Meanwhile, Tresvyatskiy had joined a team of LII pilots in a display team flying Su 27s. At the Farnborough Air Show on 25 July 1993, he ejected from a MiG 29 after being hit in the tail by another MiG fighter. Pridhodko retired on 27 April 1994 and moved to the USA on a contract with an American company. Sheffer lost his status due to medical problems and by late 1994, only four pilots were still attached to the programme; Volk, Zabolotskiy, Sultanov and Tresvyatskiy, with Zabolotskiy as the group commander. In 1994, when the numbers of cosmonauts were halved, the fate of the Buran pilots was still undecided. By 1995, the funding for Buran was zero, but no one wanted to take responsibility for disbanding the group. Volk had been promoted to Commander of the Flight Research Centre (LITs) in 1995 and Sheffer also worked at LITs in an administrative role. Tolboyev became a Merited Test Pilot of the Russian Federation on 19 October 1996 and Sultanov received the same award on 18 December 1997, as did Tresvyatskiy on 17 June 1999. Sheffer was awarded the title of Hero of the Russian Federation on 7 December 1998.

Sheffer died of a heart attack on 5 June 2001. He is buried in the same cemetery as Levchenko, Shchukin and Kononenko. Pridhodko died of cancer in the USA on 25 July 2001. He is buried near Zhukovskiy.

The number of pilots in this selection who achieved the highest award for test flying reflects the skills of the team. Nine of the eleven achieved the Merited Test Pilot award and four achieved the highest award of a Hero's star. This selection has also had a considerable death rate, with three dying in air crashes, reflecting the dangerous nature of test flying. The group does not seem to have a formal date of disbanding.

REFERENCES

1 Pravda, 4 March 1988

'How do Buran Cosmonauts live?', Marinin, *Novosti Kosmonavtiki*, Nov 5–18 1995
Buran time line, Michael Cassutt, 31 August 1993
Nevavisimaya Gazeta, 13 April 1993

Moscow Radio, 10 May 1994, BBC transcript
Testers at LII, V. P Vasin and A.A.Simonov
Biographies of all pilots at LII

International training

The training facilities and procedures at TsPK evolved for the Soviet national manned space programme in the early 1960s, and for the next decade or so, the thought of hosting crew members from other countries was not seriously entertained. However, with the creation of the US/USSR Apollo-Soyuz Test Project (ASTP), or as the Soviets preferred to call it, the Soyuz-Apollo Experimental Flight (EPAS), the Russians had to allow some American astronaut training at the centre for familiarisation with the Soyuz spacecraft and to get to know the cosmonauts they would work with in orbit.

Following the decision to develop the US Space Shuttle in 1972, there were discussions with the European Space Agency (ESA) about developing a pressurised laboratory module to fly on scientifically orientated missions of between a week and ten days, with European astronauts flying as specialists to handle some of the experiments in the payload. This news was not lost on the Russians who, still looking for the 'space spectacular' to demonstrate their space prowess and rich space exploration heritage, created an international cooperative programme of their own. This became part of the politically based 'Interkosmos', and 'offered' one-week flights of Eastern bloc citizens to Soviet space stations, hopefully before the US Space Shuttle could fly in the late 1970s and early 1980s.

By the mid-1980s, in a changing world, the American Shuttle was flying but was struggling to generate enough commercial payloads to sustain its flight programme and reduce the huge launch costs. The Soviets were looking to expand their international cooperation by offering commercial agreements to international partners in order to generate much needed funds for their own struggling programme, in a country also undergoing significant change. By the early 1990s, the two great space-faring nations had joined together with Europe, Canada, Japan and Brazil to create the International Space Station (ISS). All this meant reviewing the training of Russian cosmonauts and allowing more foreign candidates into the Cosmonaut Training Centre to prepare for joint missions. It is expected that this international training capability will be expanded in the future to become a regular feature of activities at TsPK.

COSMONAUTS TRAIN WITH ASTRONAUTS 1973–1975

The agreement for a Russian spacecraft to dock with an American spacecraft was signed on 24 May 1972, after about three years of meetings and negotiations that had included the development of a suitable docking and transfer system between the Soviet and American hardware.

Development of a joint training programme

The nature of the ASTP/EPAS programme required close collaboration between the Americans and the Russians. This included visits to each other's facilities for meetings, reviews of facilities and inspection of hardware, all prior to the crews working together. The first American delegation arrived in Moscow on 24 October 1970 and visited TsPK the following day, where they were shown the simulator facility, one of the highlights of the trip.[1] They were not only shown the Soyuz simulator, but were also allowed to climb inside for instruction in its systems and features. After a long period of studying the Soviet programme from international meetings and American aerospace publications, they could at last listen to first hand experiences from those who had worked on and flown the spacecraft, and could get a hands-on feel for the hardware, offering a clearer insight into the Soviet approach to human space flight at that time. The tour also included familiarisation with the active-passive Soyuz docking simulator.

Over the next two years, visits by Soviet officials to the US and by Americans to Russia allowed the profile of the mission to be established and the political, technical, and cultural differences to be smoothed out, leading to the signing of the joint mission document in the summer of 1972. The question of crew training was factored into planning activities from 1971, which recognised that the two crews would be launched, flown and returned in their own national spacecraft. However, they would still require an understanding of each other's spacecraft, an integrated programme of rendezvous, docking, undocking and station-keeping training, a coordinated programme of crew transfers, and lessons in each other's language. This would take around two years, from selection of the crews in 1973 to the actual mission in 1975. Part of the 4–6 April 1972 negotiations in Moscow was the '17 Points of Agreement' that included the statement: 'As a minimum, flight crews should be trained in the other country's language well enough to understand it and act in response as appropriate to establish voice communication regarding normal and contingency courses of action.' This statement required foreign language studies to be added to the Soviet (and American) crew training programme for the first time, a requirement that lapsed after ASTP flew but was reintroduced for the Shuttle-Mir and ISS programmes. The July 1972 meeting in Houston confirmed that mandatory joint crew training would include familiarisation with both spacecraft, preparation of a Crew Activities Plan and Detailed Operational Procedures document, development of training procedures for both teams of flight controllers, and a programme of joint simulations between control centres. Once again, agreeing to such detailed time lines meant that the Soviets would have to provide access to Soyuz systems and cosmonaut training procedures for the first time, to satisfy their American partners.[2]

In October 1972, another American delegation visited Moscow and, after discussions about training the crews with astronaut Tom Stafford and cosmonauts Andriyan Nikolayev (then Deputy Commander of the training centre) and Aleksey Yeliseyev, the initial training session was scheduled for the middle of 1973 in Houston, with the second meeting to be held in Moscow that autumn. The Soviets announced that they were considering the selection of two prime crews and two back-up crews, with the second set of cosmonauts being trained in the event that a second Soyuz needed to be launched.[3] During the October 1973 mid-term review tour of Soviet space facilities, NASA Deputy Administrator George M. Low reported that he had seen more of Star City than in any previous visit. 'Of major significance is the amount of new construction underway. A new training building is being put up especially for ASTP.' This new four-storey building would include classrooms, lecture halls and display rooms for spacecraft subsystems. A new hotel and dispensary would be provided for the American team during ASTP, as well as space for simulators and 'a new and very large centrifuge.'[4] Though some Americans attributed the expansion of facilities at TsPK to the forthcoming joint flight, in reality the Soviets were implementing plans already established to accommodate the long-term space station programme and the prospects for a future space shuttle programme. Certainly, the ASTP project helped support the expansion of the centre, but it was the decision to develop long-term space stations to replace the disappointment and failure in the Soviet manned lunar landing programme that generated most of the new work at TsPK.

Crews for ASTP/EPAS

The American crew had been identified on 30 January 1973, a month after the return of the final Apollo crew from the Moon. In command of the mission was veteran astronaut Thomas Stafford (Gemini 6, Gemini 9, Apollo 10). Command Module Pilot was rookie Vance Brand and Docking Module Pilot was rookie Mercury astronaut Donald 'Deke' Slayton, who had recently been returned to flight status after a decade of being grounded due to a heart murmur. Along with the three crews assigned to the Skylab space station, this would be the final assignment in the Apollo era and the last US astronauts in space before the Shuttle, which would not fly until 1981. The back-up crew was former Apollo 12 astronaut Alan Bean (then in training for the second Skylab mission), Apollo 17 astronaut Ronald Evans, and Jack Lousma, who was also in training for the second Skylab mission. A support team was also formed, consisting of rookie astronauts Richard Truly, Robert Overmyer, Robert Crippen and Karol Bobko, all formerly of the USAF Manned Orbiting Laboratory Program, who had transferred to the NASA astronaut team in August 1969.

The Soviets officially named their crews on 24 May 1973. Crew One was Commander Aleksey Leonov and Flight Engineer Valeriy Kubasov, with back-up rookie cosmonauts Vladimir Dzhanibekov and Boris Andreyev. The second Soyuz crew would be Commander Anatoliy Filipchenko and Flight Engineer Nikolay Rukavishnikov, backed up by rookies Yuri Romanenko and Aleksandr Ivanchenkov. This was the first time that the Soviets had announced the names of cosmonauts assigned to a mission in advance of the launch, and the first time unflown rookie cosmonauts had been identified prior to making a space flight, reflecting the relaxation of some of the secret restrictions of the Soviet programme required by ASTP.

Training for ASTP

A formal document outlining the study and training sessions for the crew, as well as flight controllers and associated staff (designated a 'Crew and Ground Personnel Training plan' – ASTP 40 700), was worked out at the March 1973 Working Group meetings held in Houston, where Vladimir Shatalov, at this point holding the post of Director of Cosmonaut Training, worked with ASTP support astronaut Bob Overmyer to develop the overall plan of approach. There would be three joint training sessions in Russia and three in America. The first, in Houston in July, would be followed by a trip to Moscow, and the rest would alternate accordingly. As the details of the second or third training session would be dependent upon the pace of training and matters requiring attention, the host country would advise its guest team of the training agenda a month or so in advance, and updated material would be notified by formal document change notices. Allowing for orientation tours and

ASTP astronauts and cosmonauts participate in a joint crew training session in Building 35 at JSC in February 1975. (l to r) Shatalov (Director of Cosmonaut Training, TsPK), with crewmen (in flight suits) Stafford, Slayton, Kubasov, Brand (in CM hatch) Leonov and Dzhanibekov. (Courtesy NASA)

flexibility, each meeting would be open-ended, but not long enough to waste time.[5] Each crew would continue a daily training routine in preparation for their participation in the mission either in a prime, back-up or support role. Though the cosmonauts concentrated on their own hardware, they had to become familiar with the Apollo spacecraft for the rendezvous, docking and joint operations, undocking and, in case of a contingency situation while docked, a return to Earth in the US spacecraft.

8–21 July 1973 (Houston) The first team of Soviet cosmonauts to visit dedicated training facilities outside the Soviet Union while preparing for a mission arrived at the newly renamed Lyndon B. Johnson Space Center (JSC), Houston, on 8 July 1973. (Nikolayev and Sevastyanov had visited the the Manned Spacecraft Center in Houston in 1970 as official guests.) In all, ten cosmonauts (the four two-man teams plus Shatalov and Soyuz Flight Director Aleksey Yeliseyev) completed the orientation and familiarisation session. To give the Soviet team a clearer understanding of what they were likely to encounter during the mission with regard to voice communication from Apollo, they listened to taped recordings from earlier Apollo missions and, in a step towards standardising the 'mission language', they reviewed the 'Glossary of Conversational Expressions between Cosmonauts and Astronauts during ASTP' document. They also viewed a series of video tapes, narrated in Russian (by Russian speaking engineers at Rockwell International), which described the systems, components and operation of the Apollo Command and Service Modules and the Docking Module that would be used to join the two spacecraft in orbit. Allowing the cosmonauts to view the tapes in Russian saved a lot of time and these tapes were taken back to Russia with supportive documents and illustrations, allowing the cosmonauts to continue familiarisation with the American spacecraft at their leisure. Following a review of the planned joint crew activities, each of the cosmonauts participated in a session in the CM simulator, for a real-time interpretation of the workings of the spacecraft, and also reviewed the Docking Module mock-up and sub systems. They later toured the Rockwell facility in Downey, California, where they viewed a higher fidelity mock-up of the DM and several flown CMs stored there. A demonstration of the Apollo docking and entry simulation was followed by a return to Houston for another week, before returning to Moscow on 21 July.[6]

19–30 Nov 1973 (Soviet Union) The first visit to Russia by American astronauts for specific mission training (Scott, Armstrong and Borman had done earlier PR visits) was an orientation tour of the TsPK facility, with a series of Soyuz briefing videos in English (flight profile, systems, operations) and a tour of training facilities, including spacecraft simulators, the docking simulator and a mock-up of both Soyuz and the Salyut space station. The American team included the three prime and three back-up crew members, plus Overmyer and Bobko from the support crew and Gene Cernan as a Special Assistant. Refinements in joint procedures, participation in a cosmonaut physical training programme and a number of social events were completed. The American delegation stayed at the Intourist Hotel in Moscow, some 40 km from the training centre and, driving to and from the centre in a bus escorted by police, they

ASTP crewmembers Leonov, Slayton and Stafford sample an Apollo meal in the CM trainer, Bldg 35, JSC, February 1975. (Courtesy NASA)

had time for a small snowball fight 'along the road from Star City to Moscow!' Relaxation meant a visit to the Marine Bar in the American Embassy, because it would be some years before international facilities would be available inside the grounds of Star City.[7] The Americans returned with copies of the videos, transcripts of Soyuz air-to-ground commentary and line drawings, to help the astronauts familiarise themselves with the Soviet spacecraft when time allowed in their own training programme for Apollo.

8 Apr–3 May 1974 (America) Veteran cosmonaut Valeriy Bykovskiy, who was described as the Soviet Technical Director for ASTP, served as instructor to the Soyuz prime crew of Leonov and Kubasov, assisting with the development of Soyuz flight procedures they were to practice during the joint training sessions in Moscow and Houston later that year. On 15 April, Overmyer and the prime crews discussed the remaining issues in the training programme for over an hour. During the first week, the cosmonauts worked with their American colleagues in a variety of sessions. They worked with the Apollo back-up crew in the Apollo simulator, reviewed the Russian-English glossary, conducted Public Affairs Office (PAO) and press conferences, received briefings on crew equipment (including the 16 mm onboard movie camera), worked out in the astronauts' gym, reviewed the flight plan, and attended briefings on the Docking Module and trainer. The second week saw over twelve hours familiarisation training with the Docking Module equipment to be used during transfers, using the high fidelity mock-up to slowly work though the multitude of procedural steps required to operate the equipment safely and efficiently. The cosmonaut team also worked on the developing flight plan, practiced on the communications equipment, received a briefing on the Apollo TV camera system, and spent three hours with Brand and Bean in the Command Module simulator focusing on the final stages of the planned Apollo approach and rendezvous with the Soyuz. They also continued English language exercises.

24 Jun–11 Jul 1974 (Soviet Union) In June 1974, the astronauts stayed at the new three-storey Hotel Kosmonavt, which proved to be very hot and was also 'bugged'. This proved useful, however, as 'talking to the walls' usually ensured that a supply of international beer was provided at the end of a hard day of training.[8] During the joint training session held at TsPK, each crewman conducted ten hours of communication practice in the Soyuz spacecraft following a script simulating the conversation between the two spacecraft in orbit, although for this exercise, the participants spoke over an intercom to a fellow crew member seated in a different office. As the Americans were in Russia for the 4 July celebrations, they caused some concern for the security staff at TsPK when they celebrated in the astronauts' hotel at Star City by launching celebration rockets from empty mineral water bottles. After explaining to the guards that it was their day of revolution, the guards left their guests to their party, once again amazed at the antics of American astronauts.[9]

Interior view of the Soyuz Orbital Module mock-up located in Bldg 35 at JSC in Houston. Stafford and Leonov review a checklist and photographic equipment during the February 1975 joint training session. (Courtesy NASA)

9–27 Sep 1974 (America) Communications training was again the focus of this joint training programme, held at JSC in Houston. The training occurred in the Flight Crew Training Facility, with the cosmonauts seated on one side of a glass partitioned laboratory and the astronauts on the other side, to develop their inter-communication skills and language experience further. The cosmonauts also gained additional experience with the mock-ups at JSC. In December 1974, the Soviets mounted the Soyuz 16 ASTP dress rehearsal of Soyuz operations for the forthcoming mission. This gave the cosmonauts a real-time run through of their operations in orbit for the joint mission seven months later. While in Houston, the cosmonauts completed medical examinations related to medical experiments on the joint mission.

7–28 Feb 1975 (America) This training session started in Washington DC before moving to the Kennedy Space Center (KSC) in Florida for three days, where they toured the facilities and received briefings on the Apollo-Saturn launch system. The cosmonauts were able to view the American flight hardware (CSM 111, Docking Module DM-2, and Saturn 1B SA-210). The trip to Florida also included a visit to the Disney World facility before flying over to Houston to continue with joint training at JSC. During their time at JSC, the cosmonauts received briefings on the five joint experiments planned for the mission and updates on the rules and procedures in the event of contingency or emergency situations. Over several days the cosmonauts participated in joint sessions with their American colleagues, using

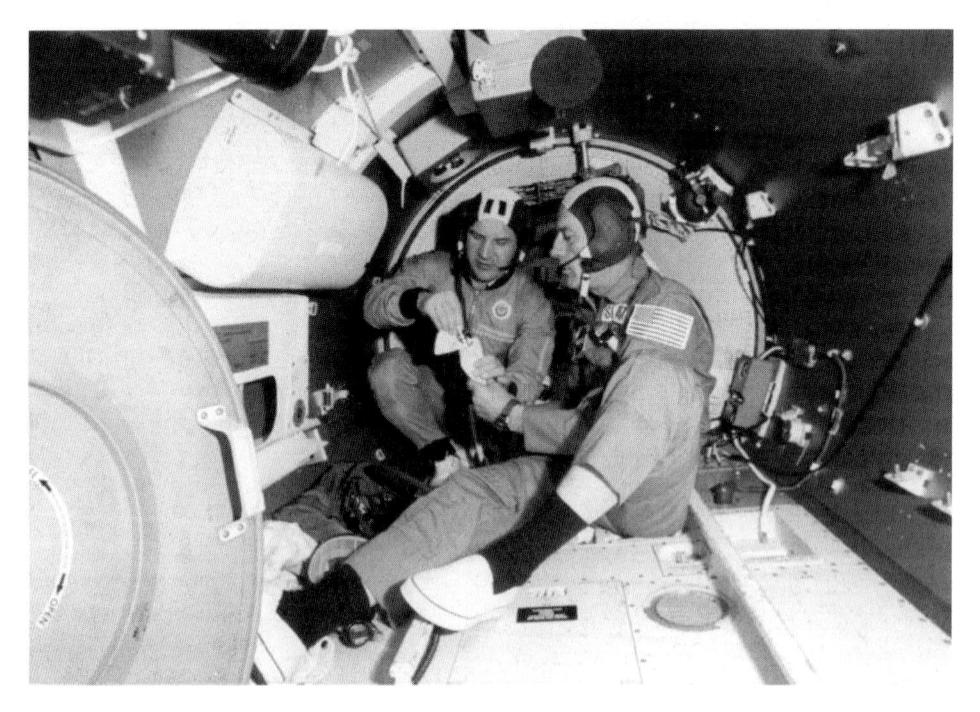

Kubasov and Slayton in the Docking Module mock-up at Bldg 35 at JSC during the February 1975 joint training session. (Courtesy NASA)

the latest updated flight documentation and running through the complete joint activity programme using the mock-ups of the Apollo CM and DM, and Soyuz OM and DM. As the prime crews went through their 1-g walk through, the back-up crews were in the simulators, until the prime crews moved over to communication techniques practice and the back-up crews were able to conduct their own 1-g walk through activities. In all, the crews each had two run throughs of the crew transfer sequence. This joint training session also included over sixty hours of language training, press and PAO events, and further gymnasium work. Leonov and Kubasov tasted the food they would eat in the Apollo spacecraft during the docked phase of the mission and the cosmonauts also flew the Apollo CSM simulator to experience the rendezvous and docking sequence from the American point of view.

14–30 Apr 1975 (Soviet Union) The final joint training session followed a similar schedule to the previous one in Houston, with the crews practicing transfer and joint activities in the various Soyuz mock-ups and conducting contingency situations using Soyuz simulators. They also practiced radio communication skills while using the rendezvous and docking simulators. During this visit, the American astronauts visited the mission control centre at Kaliningrad on 19 April and completed a trip to the launch facilities at Baykonur, viewing the Russian flight hardware undergoing processing (Soyuz spacecraft, the R-7 and Sokol pressure garments), and taking a trip to the launch pad. Even after all the cooperation and friendship, security in the Soviet Union demanded that they flew around the country at night.

ASTP review of flight plans during a joint training session at JSC in February 1975. (from left) Slayton, Kubasov, Dzhanibekov, Brand and Leonov. (Courtesy NASA)

13–20 May 1975 (Inter-Control Centre Simulations) In addition to the joint crew training sessions, the different mission control teams held joint training sessions in March, May and June 1975. The crews also participated in these sessions. For example, during the second of the three inter-centre training sessions which began on 13 May, the cosmonauts in the Soyuz simulator at TsPK began the countdown at 1 hour before the projected launch of Soyuz and ran the first 25 hours of the joint mission. With their American colleagues in their own Apollo simulator in Houston, they practiced both launches and a pre-planned programme of manoeuvres, while linked to the NASA and Soviet mission controls. On 15 May, the crews began a 56-hour simulation that would cover the flight plan from 47 hours 10 minutes GET to 103 hours GET. This included the final rendezvous, docking and crew transfer, undocking, second docking, and the final separation manoeuvre. The crews were able to evaluate plans covering contingencies and emergencies that might occur during the actual flight. Four days later, on 20 May, a re-run of the rendezvous and docking phase was conducted, lasting 9 hours. A Soviet Union team of controllers was based in Houston, including another Soviet rookie cosmonaut, Valeriy Illarionov (who never got to fly into space himself), as Capcom (Capsule communicator – an old American term for one who talks to the crew on orbit).

Do you speak American?

When Aleksey Leonov was informed of his selection to command the Soyuz crew for ASTP, he was thrilled, but stated that he did not know a word of English. Vladimir Shatalov told him he had two years and two months to learn. 'Learning English did prove the most difficult part of our training. The first time I entered a classroom, I saw that our teacher was some twelve years younger than I. He introduced himself in English, though he was Russian like me, and from then on, I could not draw a word of Russian out of him. He just could not be persuaded to speak Russian. At best he'd point to an object and name it in English.'[10] Leonov wrote that retaining the English words proved difficult, but he persevered, often staying up after midnight studying the language. Eventually he grasped the basics of the language and drew satisfaction from the fact that the Americans also had difficulty in mastering Russian. The mixture of Russian/English was termed 'Russton' (from RUSSian and HousTON) and the pronunciation of syllables by both the Americans into Russian and the Russians into English was often the source of great humour between the crews. For example, the English word 'manoeuvre' sounded like 'manure' to the cosmonauts and raised a few laughs when talking about forthcoming Apollo 'manure'.[11] During the November 1973 training sessions in Moscow, the Americans learned from the Russians that they had made significant progress in learning English due to each member of the prime crew having his own individual instructor. The Americans had begun Russian language lessons during 1972, but this was only in short stretches and affected their progression in such a difficult language. The news that Russian cosmonauts had full-time instructors prompted Stafford to request two full-time Russian language instructors in order for the Americans to be as efficient in Russian as the Russians were becoming in English.

ASTP astronauts Bob Overmyer (l) and Vance Brand (c) are instructed on the Russian language by Anatovoy Forestanko at JSC in June 1974. (Courtesy NASA)

Reviewing and flying the mission

Though formal training would continue right up to the moment of launch, these were the final joint activities simulated prior to launch day. A joint Flight Readiness Review (FRR) was held on 22 May 1975 at the Presidium of the Soviet Academy of Sciences and took the form of a review of preparations, based on a long established FRR programme at NASA. A report by Vladimir A. Timchenko outlined the progress of Soviet flight crew training, stating that the training in mock-ups and simulators had revealed no difficulties in completing the mission as planned, and that the crews had successfully completed their training and appeared ready for the flight. The prime crew spent eighteen weeks in total in joint training sessions with their American colleagues, though the exact number of training hours they undertook is unclear. In the NASA official history of the ASTP programme (*The Partnership*, NASA SP-4209, 1978), it is stated that the American prime crew spent between 2600–3100 hours each in preparation for the mission – but that was for flying the more complex Apollo CSM and included the complex rendezvous and docking training.

Soyuz was launched on 15 July 1975, followed a few hours later by Apollo. The docking occurred on 17 July and joint activities continued until undocking and separation on 19 July. The Soyuz landed safely on 21 July, after a highly successful flight, demonstrating the Soviets' ability to amend – sometimes grudgingly – their cosmonaut training programme to incorporate foreign participation in a joint mission goal. It would be twenty years before American and Russian spacecraft joined in orbit again. During the joint transfers, Leonov had spent 5 hours 43 minutes aboard the American Apollo CM, while Kubasov logged 4 hours 47 minutes. Stafford, by contrast, spent 7 hours 10 minutes in the Soyuz, Brand 6 hours 30 minutes and Slayton 1 hour 35 minutes.[12] However, only Soviet cosmonauts had launched and landed aboard the Soviet spacecraft, though this would change the next year with the introduction of the Interkosmos programme of joint space flights.

Lessons learned from Apollo-Soyuz

The project and flight had been highly successful and gave rise to continued cooperation, for some years, in various fields of space sciences. This led to initial discussions in the late 1970s about the possibility of a second rendezvous and docking mission, but cooperation was not easy and there was still much to be learned on both sides, including trust.

For example, in early planning for the mission, the Soviets had to release details of the Soyuz 11 fatality in 1971, to convince the Americans that such an incident was not likely to occur during the joint mission phase and put the safety of the Americans at risk while onboard the Soyuz. Then, in 1973, two space stations were lost prior to a crew boarding them, giving rise to concerns over the reliability of Soviet hardware. A year later, Soyuz 15 failed to dock with Salyut 3, which again prompted a full report to NASA, explaining that the docking system was different and the failure

ASTP cosmonaut Yuri Romanenko tries his hand at an early US Shuttle landing simulator at JSC during the July 1973 joint training session. Standing behind Romanenko is Apollo astronaut Dave Scott and looking on Vladimir Dzhanibekov. (Courtesy NASA)

was unrelated to ASTP operating procedures. In 1975, during final preparations for the joint flight, the Soviets were also operating a Salyut mission (Salyut 4). They had to convince the Americans that they were able to control two separate manned missions safely at the same time. This was complicated when the second mission launched to the station in April 1975 had to be aborted due to the faulty separation of rocket stages. Yet again, the Soviets had to give full details to US engineers on the mishaps, and when a new crew remained on the Salyut during the whole of the ASTP mission, the Soviets again had to convince the Americans that they could handle the simultaneous missions.

The successful implementation of the flight and conclusion of the Salyut 4 mission helped to generate a growing awareness of the capabilities of the Soviet programme in the West, and instilled confidence in the Soviet Union after some years of setbacks. The response to these incidents and acceptance of the explanations can be seen as significant milestones in the relationships between the Cold War rivals. This mission was the first time that the two pioneering space powers had worked to a common goal in each other's spacecraft and as a result, enduring friendships were formed. Apollo-Soyuz might also have planted the seeds of the idea that evolved into the Interkosmos programme of manned space flights.

INTERKOSMOS TRAINING

From the selection of the first cosmonauts in 1960 until the mid 1970s, all potential crew members on Soviet manned spacecraft were citizens of the USSR. This ran parallel to the selection and training of American citizens for NASA and US Department of Defense manned space programmes conducted up to 1975. However, with the advent of regular flights of the American Space Shuttle from the 1980s, opportunities existed to fly non-career 'payload specialist' astronauts on dedicated science missions, as well as certain categories of 'passengers' (commercial, political, educational and social) for one-off missions to promote the wider access to space. As part of this marketing of Shuttle capabilities and services, cooperation was sought from international partners in space, such as Canada, Europe, and Asia. By 1976, NASA was calling for Space Shuttle astronauts who would not necessarily be qualified pilots, nor be required to attain a jet pilot rating prior to assignment to a space flight crew. In 1977, the European Space Agency began the selection of 'payload specialists' for consideration as crew members on the first flights of the European Space Laboratory (Spacelab) research missions aboard the Shuttle. This led to news reports of the potential for the first space explorers other than American astronauts or Russian cosmonauts to finally leave Earth. However, the Soviets had already announced a broadening of cosmonaut training to include representatives from members of communist states under the Interkosmos programme, in a bid to place 'foreign' citizens in orbit on a Soviet spacecraft before the American Shuttle.

The Interkosmos programme

The Council for International Co-operation in the Exploration and Use of Outer Space (Soviet acronym 'Interkosmos') was created in 1966 and formed part of the

Soviet Academy of Sciences. Its aim was to focus and coordinate all the work of countless Soviet ministries and departments involved in international cooperative space research programmes with members of east European and other communist countries. The formal adaptation of the Interkosmos programme came in April 1967 and five major areas of cooperation were pursued, under the space fields of physics, meteorology, communications, biology and medicine, and remote sensing. Over the next decade, an expanding programme of cooperation focused on unmanned and ground-based activities, but in 1976, this was expanded to include participation by Eastern Bloc countries in Soviet manned space flights.

Interkosmos cosmonaut programme

On 8 July 1976, NASA issued the call for the first Space Shuttle candidates, to be selected late in 1977 and, after a two-year training programme, to be ready to fly on Shuttle missions from 1980. In the same month during talks held in Moscow, delegates from the Soviet Union and the Interkosmos countries of Bulgaria, Cuba, Czechoslovakia, East Germany, Hungary, Mongolia, Poland and Romania discussed the possibilities of flying guest cosmonauts aboard Soviet spacecraft and space stations. They discussed the selection criteria for suitable candidates, their training programme in the Soviet Union, and possible fields of science to be incorporated in future flights. The resulting agreement of 13 July, signed in Moscow, noted continued cooperative development of the peaceful exploration of space. This was expanded on 14 September with flights for Interkosmos cosmonauts on Soviet spacecraft during the period 1978 to 1983. During 1979, three further counties were invited to participate in joint space flights, these being France (the first 'international' selection offered in April 1979), Vietnam (who joined the Interkosmos programme in May 1979) and India (whose flight was offered in June 1979).[13] The opportunity of further seats on Soviet (and subsequently Russian) flights would be offered under international and commercial agreements from 1985 to the present day.

Interkosmonaut selection and training

Comments by Lt-General Vladimir Shatalov, a former cosmonaut then in charge of cosmonaut training, were released by TASS on 14 September 1976. These indicated that a cadre of cosmonauts from eight socialist countries would fly in space from 1978. The commander of the crew would always be a Soviet cosmonaut, while 'the flight engineer and research engineer would represent other participating countries'.[14] The scheduling of these flights would depend on the availability of flight hardware and the preparedness of flight crews. Shatalov also indicated that the training programme would last about eighteen months, depending largely on adequate mastery of the Russian language. Selection criteria included prior qualification as a jet pilot, with an engineering education. As many of the Communist Bloc countries already sent their prospective pilots to the Soviet Union for pilot or graduate academy training, a basic knowledge of Russian was already assured, as well as the required security screening. Though restricted initially to military jet pilots, further Interkosmos and international selections of non-military pilots would not be ruled out. The field of scientific research proposed by each country would also be a factor in selecting potential cosmonauts to operate the experiments.

The first Interkosmonaut Selection – December 1976

During November 1976, four candidates from each of the three initial Interkosmos countries arrived for final screening at Star City. They were accompanied by a support team from each of their country's armed forces. They were reduced to a final two by a commission and this pattern was followed by all subsequent Interkosmos cosmonaut selections. The first group of trainees arrived at TsPK in December 1976[15] and consisted of six cosmonauts (two each from Czechoslovakia, Poland and the GDR), three of whom would fly to the Salyut 6 space station in 1978. All were military officers and most had already studied in the Soviet Union.[16]

Czechoslovakia

The Czech authorities examined the records of all serving Czech Air Force pilots early in the discussions for potential flights on Soviet spacecraft, and the top 100 were selected for national medical examinations held in Prague. In July 1976, twenty were short-listed for more extensive tests, completing a series of medical and psychological tests at the Czech Military Aviation Institute. Late in October, four finalists were chosen for further tests at TsPK, arriving there for two weeks of tests (along with four from Poland and four from the GDR) on 10 November. The day after returning to Czechoslovakia, the two finalists were named as Captain Vladimir Remek (who had to follow a strict diet to lose 20 kg and meet the strict weight requirements in order to qualify for Soyuz mass calculations) and Major Oldrzhikh Pelchak. The other candidates who were not selected were Michal Vondrousek and Ladislav Klima. Remek and Pelchak arrived at TsPK for training on 6 December 1976 and were paired with their Soviet commanders in August 1977. They sat their final exams in February 1978, with Remek taking the coveted seat on Soyuz 28 and becoming the first non-Soviet, non-American citizen to fly in space in March 1978.

Poland

For the selection of their national cosmonauts, the minimum criteria were established by the Polish Military Institute for Air Force Medicine and Psychology, who looked at university education and qualifications, as well as flying experience and a knowledge of the research fields. A search of military records identified a few thousand pilots for further tests, and the tests reduced the group to several dozen potential candidates. A short-list of eighteen was finalised in August 1976 and after a further period of selection and training activities, a group of five candidates was identified: Majors Andrzej Bugala, Henryk Halka, Mirsolaw Hermaszewski and Zenon Jankowski, and pilot-engineer Captain Tadeusz Kuziora. A two-week series of tests held at TsPK began on 10 November, with all five being declared fit for space flight training by the Soviet Medical Commission. Based on the results the two finalists, Hermaszewski and Jankowski, were sent back to TsPK for cosmonaut training in December 1976. In July 1978, Hermaszewski became the first Polish citizen in space, making the flight aboard Soyuz 30 with Jankowski as his back-up.

German Democratic Republic (GDR or East Germany)

Several hundred suitably experienced Air Force pilots were personally interviewed by Lt-General Wolfgang Reinhold, the serving Vice Minister of Defence and Commander-in-Chief of the GDR Air Force, in late July 1976. Each man was told that volunteers were being sought for cosmonaut training and that they had 48 hours to decide if they wished to continue in the selection process. The selection criteria required the candidates to be between the ages of 35 and 45, with over 1,000 hours pilot time in high performance jet aircraft, an exemplary military career record and notable academic qualifications, preferably as a graduate of a military academy (the Yu. Gagarin Air Force Academy in Moscow being preferred). By the end of September, thirty of the top candidates were selected for a further programme of detailed medical and psychological tests, and further instruction in physics, mathematics, various sports and Russian language studies at the Medical Institute for Pilots. At the end of October four finalists, all with the rank of Lt-Colonel, were selected for further tests at TsPK, arriving on 10 November for the two-week programme. They were Rolf Berger, Eberhard Golbs, Sigmund Jähn, and Eberhard Köllner. At the end of November, Jähn, and Köllner were selected as the two finalists to undergo cosmonaut training at TsPK, arriving on 6 December. Jähn eventually flew the mission aboard Soyuz 31 in August/September 1978, with Köllner serving as back-up.

The second Interkosmonaut Selection – March 1978

On 2 March 1978, Vladimir Remek was launched with Russian cosmonaut Aleksey Gubarev aboard Soyuz 28, on a one-week mission to Salyut 6. This allowed Czechoslovakia to claim to have become the third nation to place a citizen in space. The same month saw ten new candidates (two each from Bulgaria, Cuba, Hungary, Mongolia and Romania) arrive at TsPK to begin the cosmonaut training course, from which five would qualify for flights to a Salyut station. Their delayed arrival was probably due to the need to establish the training course pioneered by the Czech, German and Polish cosmonauts, and to advance these initial candidates through the training cycle to the final stages of mission training, freeing up the resources for the new arrivals. The latest candidates were again selected from a short-list drawn up by their respective military and medical authorities.

Bulgaria

Six candidates were identified as finalists to be the first Bulgarian cosmonaut. They were: Senior Lieutenant Aleksandr Aleksandrov, Captains Ivan Nakov and Chavdar Djurov, Majors Georgiy Lovchev and Georgiy Ivanov (who had to change his name from Kakalov, an obscene word in Russian!), and Lt-Colonel Kyril Radev. Ivanov was assigned to the prime crew of Soyuz 33 (which failed to dock with Salyut 6 in April 1979), backed-up by Aleksandrov. Following the failure to dock, Bulgaria was offered a second flight in 1988 to the Mir space station. Back-up

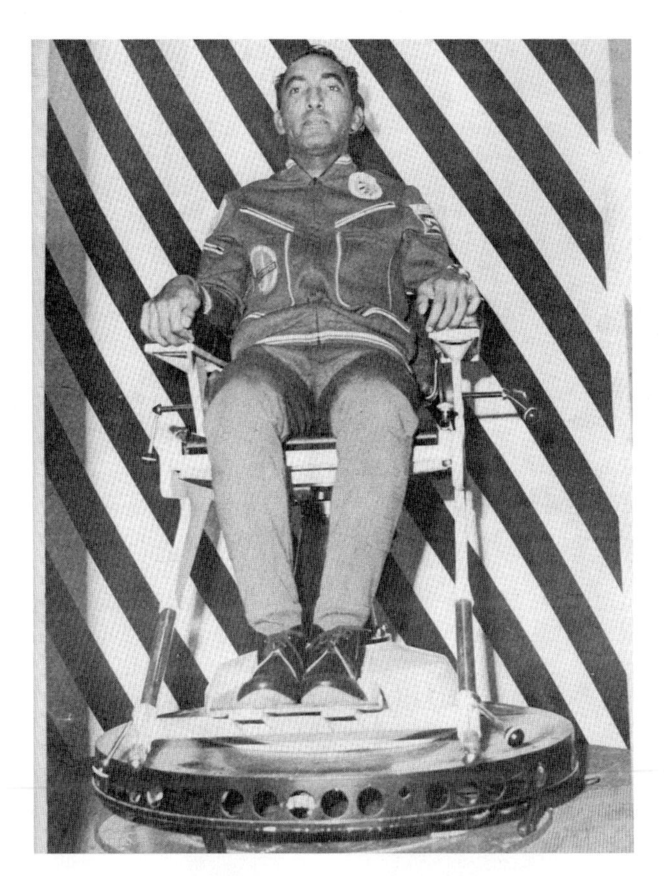

Cuban cosmonaut Tamayo Mendez on one of the spinning chairs based in the medical department of the training centre. The diagonal striping is meant to disorientate the trainee

cosmonaut Aleksandrov was selected for the new mission and subsequently became the only Interkosmos back-up cosmonaut to fly his own mission.

Cuba

The results of aptitude tests, medical examinations, and reviews of flying careers reduced a large group of Cuban Air Force pilots to nine finalists, from which four were sent to TsPK in mid-January 1978, spending six weeks completing further evaluations and tests before retuning to Cuba. On 1 March, Arnaldo Tamayo Méndez and José Lopez Falcon were notified of their success and they began training at TsPK on 22 March. Tamayo Méndez flew aboard Soyuz 38 in 1980, becoming the first Cuban (and first black person) in space, with Lopez Falcon serving as his back-up.

Hungary

Selection criteria for the Hungarian candidates included a maximum age of 35 and a maximum height of 180 cm (5 ft 10 in) – to fit inside the Soyuz Descent Module.

They also had to be a proven pilot and navigator, with a proven ability to learn the Russian language. After initial screening of all serving Hungarian Air Force pilots, over 1,000 met the basic requirements and were invited to take part in further selection processes. One candidate by the name of Bertalan Farkas replied that he was very keen and thought that this new way of 'flying' had to be the ultimate, even though his mother and wife did not seem so enthusiastic about his potential new assignment. They were not appeased when he explained that he had very little chance of selection, let alone flying! By September 1976 and a further review of each application, this group had been narrowed to a top twenty, who then completed an extensive two-week selection programme of over 100 individual tests at the Scientific Institute for Aviation at Kecskeme. The range of tests included will power, intelligence, temperament, character, self-control and physical endurance. Two of the leading Hungarian aviation medicine experts were involved in the selection progress; the medical tests were supervised by physician Colonel Dr. János Hideg, and the psychological tests were led by Colonel Dr. Jozef Szabó. As a result of these tests, four finalists, all serving Captains in the Hungarian Air Force, were selected at the end of 1977. They were: Imre Buczkó, László Elek, Bertalan Farkas and Bela Magyari. Following scrutiny of each of the finalists by Russian officials, the four arrived at TsPK on 21 January 1978 for a five-week evaluation process, before returning to Hungary at the end of February. Farkas (who thought having to stop smoking was one of the toughest criteria to match up to) and Magyari were named as the two finalists on 13 March, returning to TsPK on 20 March to begin training

Survival training on ground using the sea capsule 'Ocean'. The 'crew' is Hungarian cosmonaut Bertalan Farkas (top left) and Russian cosmonaut Valeriy Kubasov (top right)

for the Soyuz 36 mission. Now used to the idea of her husband's new 'career', Farkas's wife had become used to the thought of him flying in space. His mother, on the other hand, never got used to the idea and remained very scared throughout his cosmonaut career. Farkas eventually became the first Hungarian in space in 1980, with Magyari serving as his back-up.[17]

Mongolia

There has been very little released on the selection of the two Mongolian cosmonauts in 1978, but it is known that there were forty candidates.[18] There has been no detail of candidates or tests conducted, but the two finalists were both Air Force engineers: Captain Zhugderderdemidiyyn Guragchaa, who flew as prime Cosmonaut Researcher on Soyuz 39 in 1981, and Captain Maidarzhavyn Ganzorig, who served as his back-up.

Romania

Details of the Romanian selection are also sparse. It is known that 100 candidates were nominated, from which three military engineers underwent tests at TsPK in December 1977. They were Majors Dumitru Dediu, Cristian Guran, and Major-Engineer Dumitru Prunariu. Prunariu and Dediu were selected, with Prunariu being commissioned in the Romanian Air Force upon selection. He flew as primary Cosmonaut Researcher on Soyuz 40, the final Interkosmos flight of the series, in 1981.

With the first group of Interkosmonaut candidates undergoing training in the Soviet Union, the second group had to return to their former military units, resume normal duties and wait for about a year before finally learning if they were one of the final two. It was not until 21 January 1978 that the finalists from the second selection reported to TsPK for a final five-week selection process, ending on 28 February. The results were announced the following day, with a reporting date at Star City of 20 March 1978.

Vietnam joins Interkosmos – April 1979

In May 1979, it was announced that Vietnam had officially joined the Interkosmos programme, making them eligible for inclusion in the manned space programme. It was also revealed that four candidates were already at TsPK, and had been undergoing the selection process since April. Three of the four are known: Captain Bui Thanh Liem, Colonel Nguyen Van Quoc and Lt-Colonel Pham Tuan. The finalists were Tuan and Liem, who began training for Soyuz 37, with Pham Tuan flying the mission in 1980 and Bui Thanh Liem serving as his back-up. Following the mission, Liem returned to operational flying with the Vietnamese Air Force, but was killed in a plane crash on 30 September 1981.

Overview of Interkosmos training

All flights of the Interkosmos cosmonauts took place between March 1978 and May 1981 to the Salyut 6 space station. They used the older version of Soyuz, adapted for two-person operation, but with more responsibility placed on the Soviet commander than the Interkosmos cosmonaut, who flew as a 'Researcher' rather than the normal 'Flight Engineer' designation. As a result, their training included courses on the basic cosmonaut training programme (academic and wilderness/survival training) and on the systems and procedures of the Soyuz ferry spacecraft, the facilities at Baykonur, the R-7 launch vehicle, and the Salyut 6 space station.

Most of the candidates had no intention of flying in space at the start of their careers, since such an option was not open to them, and though several were military pilots, many did not have extensive experience or logged flying hours at selection. Going from obscurity to national fame was a daunting challenge – especially for a citizen from a communist state.

Mongolian cosmonaut Guragchaa revealed: 'At the beginning, I was not sure that I'd be able to cope with such sophisticated equipment, but Zvyozdnyy has excellent facilities for training and there were specialists who helped us a great deal in mastering space technology. When preparing for scientific experiments, we met scientists from various institutions and attended special lectures. We had to work rather hard at Zvyozdnyy.'[19]

Hungarian cosmonaut Bertalan Farkas reported that the early training was '... extremely hard. We had to get used to sitting behind a desk again to study very difficult subjects. We were taught cosmology, took part in flights simulating the feeling of weightlessness, and parachuted.' A working week for training lasted twelve hours a day and seven days a week.[20] On weekends, the day ended at 1600 hours, but was always followed by home study.

Remek prepared for his space flight by often sleeping with the foot of his bed propped up by a pile of books, to keep his head lower than his feet and help him prepare for the increased blood flow to the head. During his first training session in a Soyuz simulator, he lost 2 kg in weight during a three-hour session.

Interkosmos training schedule

Each Interkosmonaut candidate arrived at Star City with their immediate family and was accommodated in the apartments in Hotel Kosmonavt, where all the Soviet career cosmonauts lived. In order to help acclimatise the new arrivals, they were billeted on different floors. The following day, the group of candidates reported for their first session of instruction. The training course was divided into four periods, lasting about 18 months.[21]

First Training Period (6–7 months)

Most candidates found this to be the hardest part of the training programme, with lectures in basic theoretical education covering astronomy, navigation, mathematics and the Russian language. During these months, the candidates visited several factories responsible for the production of space flight hardware and systems and, occasionally, sites of national cosmonautics importance, such as the house of Konstantin Tsiolkovskiy and the museum of space achievements. They also completed a parachute jumping course and participated in fit tests for Sokol pressure garments. Training advisors included former cosmonauts Leonov, Volynov,

Yeliseyev and Lazarev and towards the end of the first training period, a special training group of veteran cosmonauts was formed to prepare in parallel for the role of Soviet commander on the joint flights.

Second Training Period (over 2 months)
For the Interkosmos candidates, this period consisted of lessons on the basics of spacecraft technology, and it proved difficult even to the ardent students of technology. Used to the 'aeronautics' of aircraft, the candidates found that the 'astronautics' of spacecraft posed a completely different methodology of flying. Meanwhile, the experienced cadre of Soviet commanders worked on their role of taking a Soyuz to and from a Salyut station almost single-handed. Towards the end of this period, the Interkosmos candidates were 'paired' with their Soviet commanders and identified as primary or back-up candidate. Chief of the training centre, former cosmonaut Lt-General Georgiy Beregovoy, announced the crew assignments, selecting from training performance and in sequence to mark significant events in the respective communist country's history, as well as for requirements in the Salyut manifest. However, poor performances, illness or better performance from the assigned back-up crew could result in a change of assignments, so all candidates had to be sure to put in 100% effort if they wanted to keep their flight.

Third Training Period (about 4 months)
The training programme became even busier during the third period. Teamed with an experienced Soviet commander, this was when they put theory into practice with numerous sessions (lasting up to six hours a day) in both the Soyuz and Salyut simulators, coupled with a series of endurance tests in the decompression chamber and centrifuges, while learning to operate as a crew. To achieve this meant long working days, extending an hour or two after the 'official' TsPK working day finished at 2100 hours. The paired crew also socialised together, getting to know each other's families and accompanying each other on off-duty social events and visits. Though becoming closer as a crew, the Interkosmos cosmonauts were still restricted in what they could or could not handle inside the spacecraft. Czech cosmonaut Vladimir Remek indicated that tasks were strictly divided and observed between them. If he tried to touch something he was not supposed to, his hand would quickly be smacked by his commander (Aleksey Gubarev), creating what was termed 'the red hands joke', something the first Western guest cosmonauts (the French) also experienced a few years later.

Fourth Training Period (about 4–6 months)
The final training period included specific mission training for the crews and familiarisation with the scientific instruments and experiments supplied by the relevant Interkosmos nation. This period covered repeated simulations of key mission events, such as launch, rendezvous, docking, work aboard the Salyut, undocking and landing. In addition, the crews completed water landing and water egress survival training in the Black Sea, and other wilderness survival training in the

event of emergency landing situations. The physical endurance tests were stepped up, and simulator runs became daily as launch day approached. About one month prior to the planned mission, final exams were taken to confirm the primary and back-up positions.

Down to Baykonur

Two weeks prior to launch, the crews flew down to the Baykonur cosmodrome to begin familiarisation with their spacecraft, and the launch process and sequence of events they would follow on the day of launch. A few days before launch, the crews were confirmed and authorised for flight. Several of the Interkosmonaut candidates who were inspired by stories of Gagarin and the first cosmonauts were now about to follow them and make the journey from Earth to space themselves.

First civilian cosmonaut commanders

Between 1961 and 1978, all commanders of Soviet spacecraft were, by tradition, military pilots, with previous space flight experience where possible. This enabled them to 'fly' the spacecraft during critical phases of the mission, including rendezvous, docking and undocking with other spacecraft, manual orientation of the spacecraft, manual alignment for re-entry, and landing. With the advent of Soyuz in 1967, the opportunity arose to regularly fly up to three cosmonauts on one crew. The second crew member was termed Flight Engineer, which was normally a former 'test engineer' from the leading design bureau, Energiya (formerly OKB-1). They were responsible for monitoring onboard systems, and their training allowed them to take control over those systems from their control console if required. They also assisted the commander in phases of the flight. If a third member was aboard, they were termed 'Research Engineer' and these, too, were normally from Energiya. Their tasks focused on spacecraft control during autonomous flight and upon docking to an orbital station, where they would work with the scientific research experiments. However, following the loss of the three Soyuz 11 cosmonauts due to sudden spacecraft decompression after a 23-day flight to Salyut 1 in 1971 (the result of not being issued with pressure suits for critical phases of the flight, including re-entry), the three-person Soyuz was redesigned to accommodate two people wearing pressure suits, the support equipment for which replaced the 'research cosmonaut' seat. A three-person Soyuz was not reintroduced until 1980, so with the introduction of Soyuz Interkosmos flights from 1978, a change in the crew training was required.

Commanders would now be trained to combine the responsibilities of both commander and flight engineer, which raised the question about which would be the most suitable to perform this function – a military pilot or a civilian engineer? When the first training group for Interkosmos Soyuz commanders was formed from six experienced cosmonauts, two were civilian engineers from Energiya – Nikolay Rukavishnikov and Valeriy Kubasov. In the Soviet crew selection system, competition between prime and back-up crews can often be beneficial and raise training standards. In some cases, however, it can be detrimental, raising issues that are often not resolved until the point of final selection. This could also be said of members of the different 'groups' of cosmonauts (pilots, scientists, engineers, doctors, etc.) in securing selection to training groups in the first place. There was also the long standing 'argument' over assigning commanders from the military over civilian engineers. The American astronaut corps experienced a similar rift between the original pilot groups and the scientist astronauts recruited in the mid-1960s. As space flights in the 1960s and 1970s were essentially test flights, the argument became heated at times between selecting experienced test pilots who were used to evaluating high performance vehicles, and experienced engineers who had designed and tested the vehicles on the ground.

The civilian cosmonauts apparently wanted to be given the chance to prove that, with the same training as a pilot cosmonaut, a 'competent and physically strong engineer' had the required skills to control a Soyuz. The argument that pilots of aircraft within the atmosphere make natural cosmonauts was countered by the fact that flying a ballistic spacecraft like Soyuz bore little resemblance to flying such aircraft. By the time the first Interkosmos missions were selected, however, the opportunity for civilian commanders was available, as there were insufficient flight experienced cosmonauts available to fulfil all the crew positions, including back-up and support roles. Kubasov and Rukavishnikov were probably selected because of their recent 'international' experience with Apollo-Soyuz (Kubasov was on the prime crew and Rukavishnikov on the crew of the reserve Soyuz). Both were also extremely capable engineers and veteran cosmonauts, with over a decade of training experience and two flights each behind them. Their assignment was perhaps secured, in part, as a result of an incident that happened on 13 February 1978.[22]

Gubarev and Remek were completing their final simulation 'exam' prior to determination of their status as a prime crew. This exam was monitored by fifty specialists in different departments across TsPK, listening in to the 'Earth-to-orbit' communications simulating the air-to-ground commentary during the mission. The simulation had progressed well for most of the day when an outside communication was heard, alerting the crew to a change in the forthcoming commands from the regular sequence and, in effect, helping them to prepare for an unplanned activity and affecting their potential reaction time. The Energiya members of the staff examination panel were very upset about this leak, which affected the impartial assessment of the crew's ability to fly the mission. However, wishing to avoid an embarrassing and diplomatically difficult situation so close to launch, the crews were passed and allowed to fly the mission. Energiya argued that such a crew of pilot commander and a non-Soviet pilot researcher was not sufficiently trained or experienced in dealing with diversions from scheduled commands without help.

Though not specifically citing this incident, Energiya apparently argued that a civilian engineer wouldn't need such prompting, being familiar with spacecraft systems and procedures from years in the design bureau and in ground testing. A case was put before the Central Committee of the Communist Party (responsible for the final decision of who or would not would fly on each mission) that civilians had the same right to command a mission as a military pilot. But it was an argument that was short-lived, because only two civilian cosmonauts commanded a national crew; one in 1979 (Rukavishnikov, who was not cycled to a second command and retired due to ill health) and one in 1980 (Kubasov, who also never flew again). It was not until ISS operations that civilian Energiya engineers were 'allowed' to 'command' a resident crew, but not until the first (Usachev) was taken to and from the station by the American Shuttle. The loss of Columbia and the lack of available experienced cosmonaut pilot commanders meant a number of 'civilian engineers' had to be trained as station commanders (and qualify as Soyuz commanders) for long duration resident crew training for ISS (Budarin, Kaleri, Krikalev).

Lessons learned from Interkosmos

The Interkosmos programme was a political and propaganda success. It also enabled Soviet space engineers to gain access to a lot of unique experiments prepared by key institutions and scientists from a number of countries. This was a cheap option and it did foster resentment in many Soviet cosmonauts, who would otherwise have occupied the second seat on these nine Soyuz missions. Having been removed from the flight schedule and passed over, some had to wait many more years to fly, while others were never called again.

INTERNATIONAL GUESTS AND COMMERICAL AGREEMENTS

In 1979, the Soviet space authorities began to expand their guest cosmonaut programme by offering flights to citizens of nations outside of Interkosmos. This was expanded in 1985 with a programme of international agreements based on commercial foundations, to attract much needed finances to the struggling Soviet programme. Though not totally successful, this programme was amended after the collapse of the Soviet Union to try to sell seats on Soyuz for flights to the Mir space station, until 2001 when the ISS station became the primary source of human endeavour in space. As well as the much needed injection of cash into the Soviet space programme, this programme created and delivered a range of new experiments to Salyut 6, Salyut 7 and Mir, to be worked on by various crews.

International guests

The offer of seats on Soyuz spacecraft to non-communist countries has its origins in April 1979, when the prospect of flying a French citizen was discussed during a visit to the Soviet Union by French President Valery Giscard d'Estaing. This led to joint discussions between the French and Soviet governments that summer. The success of the one-week flight of a French cosmonaut to the Salyut 7 station in 1982 prompted French scientists and officials to pursue longer flights in the future and led to a series of joint missions over the next fifteen years, making France one of the most successful countries to place its citizens in space on the spacecraft of other nations (Russian Soyuz or American Shuttle).

In 1981, the Indian government reached an agreement with the Soviet Union to send two of its citizens to train at TsPK for a one-week flight (in 1984 to Salyut 7) and this was followed by similar agreements with Syria in 1985 and Afghanistan in 1989, resulting in flights to the Mir space complex.

Commercial agreements

By 1985, the Soviets had recognised the political and financial benefits of offering 'space seats for sale'. This resulted in a programme of commercial agreements with other nations for the sale of Soviet launch services, including the possibility of flying a foreign national to the Mir space station. This was initially promoted under the GlavKosmos organisation, created in 1985 and planned as a 'Soviet NASA' to become a central organisation to coordinate all research, launches and cooperation on Soviet programmes, including that of development, management and marketing. Prior to this, a variety of governmental departments, institutions and facilities were involved in the decision making process of the Soviet programme. From the start, the operations of GlavKosmos were fraught with difficulties and restrictions, particularly financial, but though it would never really fulfil its promise of a commercial Soviet space service, it did generate a broad international cooperation in the marketing and sale of launch services. This brought in much needed investment to a struggling space programme, at a time of great economic, domestic and cultural difficulties after the fall of the Soviet Union and its evolution into the Commonwealth of Independent States (CIS) and today's Russia.[23]

During the period 1985–1990, GlavKosmos negotiated and agreed manned flights by citizens of Afghanistan, Austria, Japan, and the United Kingdom, as well as further flights by French cosmonauts. In 1990, GlavKosmos was replaced as the prime contractor by NPO Energiya and between 1990 and 2000, several more commercial agreements were negotiated with France, Germany, and organisations and individuals in the United States, independent of NASA. With mixed success, a more productive cooperative agreement also developed with ESA during the 1990s.

By 2000, Mir was nearing the end of its operational life and the first resident crew was aboard the ISS. Russia's emphasis shifted to support the construction and expansion of the larger complex, but they also continued to pursue agreements with foreign countries for flights on Soyuz to the ISS (which are detailed separately in this book). In the late 1980s, several other countries were approached, or agreed in principle, to support commercial flights to the Mir station. However, the demise of the Soviet Union, the restriction of funds and resources in the 1990s (hardware, propellants, ground support finances), the agreement with the Americans to fly a series of NASA astronauts on the station in the mid to late 1990s, the in-flight difficulties of a fire and collision in 1997, and the growing emphasis on ISS, all meant that most of these 'international Soyuz' flights (listed below) never materialised.

Soyuz seats for sale

The offer of joint manned flights in the 1980s and 1990s was extended to many countries around the world. Most were not taken up, but some led to other developments and cooperation in later programmes.

An invitation was extended to Argentina in 1987 and reports from German sources in 1989 suggested that Australia had been approached and contracts signed, and that negotiations were underway with Brazil. Preliminary offers were made to Canada in 1987 to fly a cosmonaut to Mir, and formal invitations were passed between the two governments in 1989/1990, to which the Canadian National

Research Council responded favourably. No firm agreement was reached, but Canadian Chris Hadfield did fly to Mir as part of a NASA Shuttle crew (STS-74) in November 1995. China was offered the opportunity to fly a cosmonaut on a Russian space flight in August 1986, and this was re-emphasised in 1988. This led to renewed cooperation between Russia and China and a series of visits to TsPK in the mid- to late 1990s, as China was developing its own manned space programme (see below). Finland was offered the opportunity of a flight during an official presidential visit to Baykonur in October 1987, but although the possibility was discussed for over a decade, the large fee of US$ 10 million was rejected by the Finnish government. In 1988, it was reported that Hungary was interested in a second flight, this time to Mir.

During the mid-1980s, an invitation to Indonesia gained some interest there, and during 1989, several discussions with officials from Iran pointed towards a joint flight, which Iran announced its intention to accept in March 1990, but which went no further. In October 1988, the newly formed Italian Space Agency (ASI) expressed a desire for a national space team of astronauts to crew future Shuttle missions and a possible flight to Mir was one option discussed at the time. But nothing would materialise for over a decade, until an Italian flew aboard a Soyuz to ISS. Malaysia was first approached to provide a national cosmonaut in 1987, but nothing developed beyond that until 2003, when discussions took place for a new selection on a Soyuz taxi flight to ISS. According to several reports in 1989, Spain was apparently approached for a flight to Mir, planned for December 1992. The mission never took place, although Spanish astronaut Pedro Duque, who was selected under the ESA astronaut programme, has completed cosmonaut training for a role as back-up to an ESA Mir mission and flew as part of a Soyuz Taxi crew to ISS in 2004. Sweden was offered a joint mission, reportedly in 1986, as was The Yemen and the former Yugoslavia sometime before 1987.[24] Sometimes these offers were mentioned by leading politicians on state visits, or as part of a scientific cooperation project. It would be appropriate to say that offering such opportunities to Third World countries was part of the Soviet/Russian foreign policy drive.

West European cosmonauts

For over 25 years, the Cosmonaut Training Centre has played host to a number of representatives from west European countries and space agencies, as a result of commercial and cooperative agreements, the demise of the Soviet Union and recognition of new states and borders.[25]

Austria

In a purely commercial agreement, Austria was offered the opportunity to fly a cosmonaut to Mir in 1987. The agreement was negotiated between July and October 1988, resulting in the AUSTROMIR project. The Austrians would provide fourteen scientific experiments, which clearly raised the interest from the Soviets, who only charged US$ 7 million for the flight instead of the nominal US$ 15 million fee. The call for candidates was issued in February 1989, resulting in 198 applicants (19 women) and, for once, pilots were excluded, emphasising the scientific nature of the mission. After a series of tests, the group was reduced to about fifty applicants, who were then screened in a series of medical and psychological tests to produce thirty suitable applicants. On 10 July 1989, seven finalists were chosen (two women – Elke Greidel and Gertraud Vieh, and five men – Manfred Eitler, Peter Friedrich, Lt-Colonel Robert Haas, Clemens Lothaller and Franz Viehböck). On 6 October 1989, Viehböck and Lothaller were selected for cosmonaut training, which they began on 8 January 1990. During their training, the two Austrians became friends with the Japanese and British pairings who were at TsPK at the same time. Viehböck was named to the prime crew of Soyuz TM 13 (with Lothaller as back-up), and was launched on 2 October 1991 for a standard one-week visiting mission to the Mir space complex. The training programme the Austrians and other 'commercial' programmes followed mirrored that of the British astronauts (see page 231).

Bulgaria

The first flight of a Bulgarian cosmonaut in April 1979 was the aborted docking attempt with Salyut 6 due to a faulty engine system on the Soyuz 33 spacecraft. As a result of that shortened mission, Bulgaria was offered the chance of a second mission, this time to Mir. This was not part of the Interkosmos programme, but was subject to a new agreement being signed between GlavKosmos and the Bulgarian Academy of Sciences in August 1986. As the Bulgarians were supplying a larger package of experiments and due to their significant involvement in the Interkosmos programme, this was treated more as a partnership mission than one of the newly created commercial ventures with other countries. A new selection process was started early in November 1986 and after a review of the candidates, four military pilots were short-listed. They were Aleksandr Aleksandrov (the former back-up CR for Soyuz 33), Plamen Aleksandrov (the younger brother of Aleksandr, making them the first brothers to be short-listed for space flight training), Georgiy Ivanov (the unlucky CR aboard Soyuz 33) and Krasimir Stoyanov. Unfortunately, Ivanov would not get a second chance to fly in space, as Aleksandr Aleksandrov and Krasimir Stoyanov were selected for further training in January 1987. In December, Aleksandrov was named as prime candidate and flew the Soyuz TM 5 mission to Mir in June 1988. The mission was designated 'Skipka 88.'

European Space Agency (ESA)

During a conference of ESA Member States, held in Granada, Spain, during November 1992, ESA decided to pursue stronger links with the Russians for further manned space flights, with the aim of gaining experience in long duration space flight, something that the American Shuttle could not provide. In the decade since Spacelab 1, only four representatives of ESA had flown on Shuttle flights, and with plans for the Columbus research laboratory to be attached to the International Space Station, a pair of precursor flights was planned as EUROMIR 94 (30 days) and EUROMIR 95 (135 days). On 7 May 1993, four ESA astronauts were named as a team to train for the missions. Ulf Merbold from Germany and Pedro Duque from Spain would train for EUROMIR 94, while Thomas Reiter from Germany and Christer Fuglesang from Sweden would train for EUROMIR 95. Of the four, only Merbold had flown in space (twice, in 1983 and 1992), aboard the US Shuttle and he had been a member of the original Spacelab selection in 1977. The four completed an ESA basic training course at the European Astronaut Centre in Cologne, Germany, before moving to TsPK in August 1993 for a second basic training course, this time focused on Russian cosmonaut training. In early 1994, the team began more specific mission training, but remained as a group until 30 May 1994, when it was announced that Merbold would fly the EUROMIR 94 mission (launched 4 October 1994 and landed 4 November 1994), with Duque as his back-up. With that assignment the EUROMIR 95 pair continued training for their own mission, until Reiter was named as prime and Fuglesang as back-up on 17 March 1995. A significant difference for this assignment was that both men had continued training at Star City and had qualified as Soyuz Flight Engineers over their previous qualification of Cosmonaut Researchers. The mission, flown between 3 September 1995 and 29 February 1996, included two EVAs by Reiter, and a highly successful experiment programme. ESA was so delighted with the results of EUROMIR 94 and 95 that they tried to gain support for a 45-day EUROMIR 97 mission, with Fuglesang as primary cosmonaut. But member states were not forthcoming with financial support for the plan and attention shifted to ISS, which would offer a whole new arena of discussions, opportunities, bartering, frustrations and disappointments (see Joint Programmes page 251).

France

Between 1979 and 1999, the French operated a highly successful cooperative programme of manned space flights with the Soviets/Russians. A total of seven missions were flown to Salyut 7 and Mir during this period, totalling over 292 days by five spationauts (four men and one woman). The CNES space agency was able to use this extensive series of missions to train a small cadre of spationauts to a very high standard and by assigning both a prime and a back-up, they were able to capitalise on the training opportunities and not waste the investment of training a back-up who would not fly.

Salyut 7 (1982) In September 1979, the French Space Agency, Centre National d'Etudes Spatiales (CNES), began a programme to select a cosmonaut to fly on a Soviet spacecraft, the first such offer to a non-communist country. By 15 November 1979, a total of 413 applications had been received, from which only 193 (including 26 women) were nominated for further selection. In the first phase, a programme of medical examinations and an overall assessment was emphasised, including general competence, familiarity with scientific equipment, and linguistic and sporting abilities. Only 72 remained for further consideration after this. During December 1979 and January 1980, an extensive programme of medical examinations and tests (lasting a week), psychological examinations (three days) and special physiological tests (three days), resulted in a short-list of seven candidates (including two women). In January and February 1980, tests of general competence and scientific aptitude, including linguistic skills, reduced the group to five finalists: Major Patrick Baudry, Lt-Colonel Jean-Loup Chrétien, Gerard Juin, Major Jean-Pierre Joban, and Françoise Varnier. Unfortunately the female candidate, Varnier, was eliminated

Claudie Haigneré, the French and ESA astronaut, training in the TM simulator as part of her training to gain Flight Engineer status

from the programme due to a parachute accident, but the other four went to TsPK for three months of intensive Russian language training in March 1980. On 12 June, Chrétien and Baudry were named for cosmonaut training, which began on 7 September 1980. A year later, Chrétien was named as prime for the mission (Soyuz T 6) to Salyut 7, which was launched on 24 June 1982.

Aragatz (1988) Following the Salyut 7 mission, French authorities lobbied for a second, longer flight of up to two months, which was agreed in October 1985. By then, a new cadre of French spationauts had been selected for future flights on European, American or Soviet missions. Selected from over 600 applicants in September 1985, they were Jean-François Clervoy; Claudie André-Deshays; Jean-Jacques Favier; Jean-Pierre Haigneré, Frédéric Patat, Michel Tognini and Michel Viso. In July 1986, four military pilot candidates (Chrétien; Tognini, Haigneré and Antoine Covette – a finalist of the spationaut group) went to TsPK for medical examinations, with Chrétien and Tognini being named for the mission training programme (that included EVA training) in August 1986. Chrétien was again named as prime and Tognini as back-up, due to the former's past experience and his competence both during and after his first mission. With an EVA planned, the Soviets decided to go with proven experience for this first long duration 'guest' mission. Training started on 15 November 1986 and the mission was launched on 26 November 1988, landing on 21 December after 24 days 18 hours. This was shorter than the planned two months, but far longer than any guest cosmonaut flight had logged to date. During his second mission, Chrétien completed a 4-hour 20-minute EVA, the first non-career cosmonaut to perform an EVA from a Soviet spacecraft. The mission had cost an estimated US$ 30 million, but was arranged as a scientific cooperative flight, rather than a purely commercial one.

Antares (1992) Another cooperative agreement was reached between the French and Soviets on 25 November 1988 (the day before Chrétien was launched on his second mission), allowing a French cosmonaut to visit Mir once every two years. After discussions on the amount France would pay the Soviets for each flight (originally the Soviets wanted US$ 15.4 million for a third mission and the French countered with US$ 5 million, but the agreement reached was US$ 12.3 million), the contracts were signed on 22 December 1989 for a fourteen-day flight in 1992, with twelve days aboard Mir. France continued to emphasise that the cooperative flights were scientifically based and not purely commercial, and that the fee included both the services of Soviet visiting experts during flight preparations and the supply of French-developed research experiments and hardware that would remain aboard the station. Six candidates were nominated for this mission in July 1990; Majors Leopold Eyharts, Jean-Marc Gasparini, Philippe Perrin and Benoit Silve, and Lt-Colonels Jean-Pierre Haigneré and Michel Tognini. Eyharts, Gasparini, Perrin and Silve had been selected by CNES in a second group of career spationauts earlier in 1990. In August, Tognini was named as prime candidate, with Haigneré as back-up, and both men arrived at TsPK on 5 January 1991 to begin their cosmonaut training programme for the mission. Tognini had been selected because of his previous knowledge of Russian as back-up to the Aragatz mission. The crews were confirmed on 7 July 1992 and Tognini flew the mission from 27 July to 9 August 1992.

Altair (1993) At the end of 1991, three spationauts were identified for preliminary weightless training at TsPK, as part of their preparations for possible flights to Mir; Jean-François Clervoy, Claudie André-Deshays and Major Leopold Eyharts. On 28 July 1992, the day after the launch of Tognini to Mir, a contract was signed for two more missions. A pattern was now emerging for these missions in which the back-up for one flight normally utilised this training to fly the next mission, so Haigneré was named to a mission that would last three weeks, including nineteen days aboard the station. His back-up was named as Claudie André-Deshays and their training began in November 1992 at TsPK. The mission was flown between 1 and 22 July 1993.

Cassiopee (1996) This mission was, as previously, signed off the day after the launch of the previous mission, on 2 July 1993, with André-Deshays assigned to make the sixteen-day flight and Leopold Eyharts as her back-up. The mission was flown between 17 August and 2 September 1996.

Pegasus (1998) The penultimate French flight to Mir was planned for early August 1997, but was delayed by events on the station (the collision of the Progress M-34 re-supply craft on 25 June) and medical issues concerning the back-up spationaut. Eyharts was scheduled to fly the mission with Haigneré as his back-up (replacing the originally assigned Viso), but Haigneré injured his leg during a badminton match on

Energiya cosmonaut Sergey Treschchev working on the ISS simulator at the training centre

14 July 1997, resulting in his return to France with a torn ligament. On 21 July it was decided to move the French flight back one mission, as the injury left Eyharts without a back-up. He was eventually launched to Mir on 29 January 1998, landing 21 days later on 19 February.

Perseus (1999) By the time the seventh Russian/French mission (sixth to Mir) took off on 20 February 1999, Mir was close to the end of its operational programme. The primary candidate was Haigneré and his back-up was André-Deshays. This was due in part to saving time on training, as both had experienced Mir residency and benefited from having gone through the training cycle before (in fact Haigneré completed four crew assignments within the Mir programme, and Deshays, three). In December 1996, CNES had negotiated an agreement with RKK Energiya for a 99-day mission. When it was decided that Haigneré would remain on Mir with the Russian resident crew, the mission was extended to 168 days. Launched on 20 February 1999, the mission actually lasted 188 days 20 hours (setting a new non-Russian endurance record), landing on 28 August and ending the highly successful French Salyut/Mir programme.

Germany

In 1989, discussions were underway with East Germany to fly a cosmonaut on a second mission, this time to Mir for a month in either 1992 or 1993. Both Interkosmos cosmonauts Jähn and Köllner were in line, with Köllner as favourite,

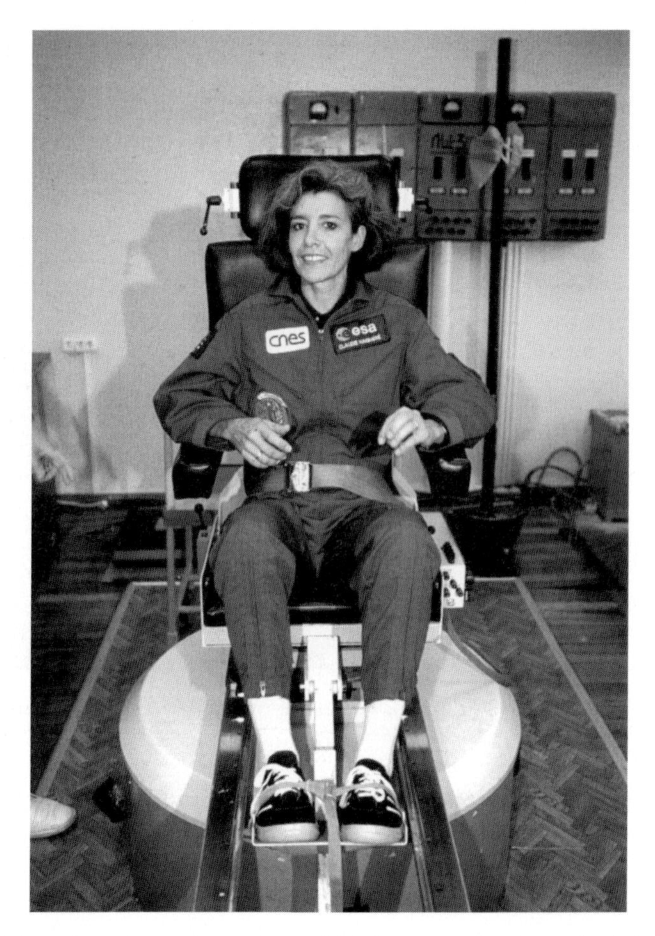

Claudie Haigneré on the cycle machine which is located in the medical department of the training centre

but historic events in both Germany and the former Soviet Union at this time effectively cancelled any plans to fly a cosmonaut from *East* Germany. However interest continued from *West* Germany, and subsequently from the unified country.

Mir 92 (1992) The preliminary agreements were signed on 25 October 1988 for a mission to Mir, but difficult negotiations, particularly financial, delayed the flight, which was planned as an eight-day visit in the first half of 1992. Unfortunately, this was also the scheduled launch time of the second German Spacelab D mission on the American Shuttle and it was announced that candidates for the Mir mission would be chosen from the team assigned to Spacelab D2, already in basic training at the German Space Agency (DLR) Crew Training Complex in Cologne. These candidates were Renate Brümmer, Ulrich Walter, Gerhard Thiele, Heike Walpot and Hans Schlegel, all selected in August 1987. With two to be assigned as prime on Spacelab D2 and three as back-ups, if two back-ups were reassigned from D1 to the

Mir mission, it was unclear whether a second back-up would be assigned to D2 from the former D1 astronauts (in the event this was not pursued). In May 1990, DLR named Reinhold Ewald and Klaus Flade (unselected finalists from the German August 1987 selection) to train for the proposed Mir 92 mission at TsPK from November 1990, with military pilot Flade assigned as prime candidate and Ewald as his back-up. Flade was launched to Mir on 17 March 1992, returning to Earth after a successful mission on 24 March.

Mir 97 (1997) After to the success of Mir 92, the Germans were eager to start discussions with the Russians in 1995 for a second flight. Originally designated Mir 96 for a late 1996 launch, the mission slipped into the following year and as a result, was redesignated Mir 97. The DLR German astronaut team had been disbanded in November 1993 (with ESA taking a lead in crew training for European nations), but as negotiations continued for the second German Mir mission, Walpot and Walter started Russian language lessons in the spring of 1995. With the Shuttle-Mir programme operating at this time, early plans suggested the return of the German on the Shuttle, which would have extended the training programme. But when it became clear that the Shuttle flight schedule would not be timed for the German mission, a launch and landing by Soyuz meant switching the training to a cosmonaut programme. Drawing on his past training with Mir 92, Ewald was assigned to the training programme, along with former Spacelab D2 astronaut Schlegel, who had to take a basic Russian language course before starting training for the mission in late 1995. In May 1996, Ewald was confirmed as prime candidate for the mission and was launched on 10 February 1997. After a largely uneventful mission, apart from the fire in one of the oxygen candles in the Kvant module on Mir on 23 February, the mission ended with a successful landing on 2 March. In the summer of 1997, Schlegel resumed Soyuz-Mir training, aimed at a possible third German mission in 1999 which was not forthcoming.

Kazakhstan

The Kazakhstan government was putting pressure on the Soviet authorities to include ethnic Kazakhs on a flight crew, due to the Republic's long association with the Soviet space programme. All manned launches have started from the Baykonur cosmodrome in Kazakhstan, and the region also hosts the primary recovery area. A number of Soviet cosmonauts had been born in Kazakhstan, but were of Russian descent, mainly from military families serving there. There was mounting political pressure to send a Kazakh national cosmonaut into space and as a result, in 1989, a selection process for new cosmonauts was initiated that included Kazakh applicants. They were to be confirmed in August 1991, but the abortive right-wing coup in the Soviet Union that month postponed the meeting of the commission, and the subsequent political disintegration of the Soviet Union over the next few months cancelled this selection process, although four candidates including one Kazakh – test pilot Toktar O. Aubakirov – were 'selected' unofficially. Final approval came in May 1992, after Aubakirov had already flown in space. He was one of the top pilots in the Soviet Union, working for the MiG design bureau. Originally to be flown as a separate mission, his flight was changed as a result of budget cuts and combined with the AustroMir mission flown between 2 and 10 October 1991. Aubakirov stated that he was flying for 'Kazakhstan in the USSR'. He completed only a four-month Cosmonaut Researcher training programme and, having been enrolled in the cosmonaut team at the personal request of the Kazakh president in January 1991, he never flew in space again. His back-up was former commercial airline pilot, Talgat Musabayev, who was selected for cosmonaut training in May 1990 and assigned to the Kazakh mission. He was selected when a search of the Soviet Air Force failed to find a suitable military candidate and arrived at TsPK in October 1990. After serving as back-up on the Kazakh flight, Musabayev transferred to the Russian Air Force and joined the cosmonaut team full time, helping to deflect criticism of the lack of ethnic Kazakhs in the team. He went on to have an impressive cosmonaut career, with two long duration missions to Mir (totalling over 341 days), a visiting mission to ISS and seven EVAs. Musabayev became one of the most successful 'international cosmonauts', who worked with a number of other international cosmonauts and astronauts.

Slovakia

The Slovak Republic began a cosmonaut selection programme during the summer of 1997 for a 1999 flight to Mir. The flight was a way of easing the debt owed to the Slovaks by the former Soviet Union and the selection programme was directed by General Stefan Gombika, a finalist in the 1976 Czech Interkosmos selection. From the thirty military pilots screened, four finalists were identified in February 1998: Colonel Martin Babyak, Lt-Col Michal Fulier, Major Ivan Bella and Captain Miroslaw Groshaft. On 23 March 1998, though the formal agreement remained to be signed, Fulier and Bella were selected to begin cosmonaut training immediately and as they were both fluent in Russian, this alleviated the need for language training. The mission was flown by Bella (with Fulier as back-up) during 20–28 February 1999, making Bella one of the last (with Haigneré) international cosmonauts to visit Mir. The mission followed the profile of the Interkosmos flights two decades earlier. Slovakia was a new nation and Remek, the first Czechoslovakian cosmonaut, was a Czech Republic citizen. National pride still plays a part in space exploration, and access to it.

United Kingdom

The first offer to fly a Briton to a Soviet space station was made during May 1986, but nothing came of it. GlavKosmos became involved in 1988 and, after much rumour and discussion, the agreement was signed on 29 June 1989. In July 1989 an advert was placed in UK national newspapers for astronaut applicants, with 'no experience necessary', for what was to be the first flight of a Briton in space, under the Project Juno mission to Mir in 1991. Around 13,000 applications were received and over the next five months, that number was reduced in stages, first to 3,000 that filled in application forms, then to 150 who were called for medical screenings and evaluations of scientific ability, and then to 36 put forward for extensive medical tests. Sixteen of these were nominated for decompression chamber and centrifuge tests at Farnborough Aeronautical Research Centre, and the short-list of six from these tests was further reduced to four by the Soviets on 5 November 1989. They were Lt-Commander Gordon J. Brooks, Major Timothy K. Mace, Ms. Helen P. Sharman and Mr. Clive P.G. Smith. On 25 November 1989, Sharman and Mace were named as finalists and began their training at TsPK five days later. Financial problems threatened the mission for some time, and it took almost a year to resolve the difficulties. Sharman was named primary candidate on 22 February 1991 and eventually flew the mission between 18 and 26 May 1991.

Near, Middle and Far Eastern Cosmonauts

In addition to cosmonauts from more 'local' nations, the Soviet Union, and more recently Russia, has pursued agreements to fly cosmonauts from a variety of eastern nations for over twenty years.[25]

Afghanistan

The agreement to fly an Afghan on a Soviet space mission was reached on 20 July 1987, with the formal agreement signed on 30 September 1987. The flight would take place in the first half of 1989, which was typical for this type of mission, allowing eighteen months of basic Cosmonaut Researcher training and mission training by the Afghan cosmonauts. During November and December 1987, the selection process was completed, resulting in 457 applicants. Of these, 52 were qualified for further consideration and then this group was reduced first to eighteen, then to eight, who went to Moscow in January 1988 for further selection, scrutiny and examination by the chief Medical Commission. The agreement for the flight was signed on 11 February and two days later, Captain Abdul Ahad Mohmand and Colonel Mohammad Dauran were selected. They began cosmonaut training at TsPK on 25 February. Both were experienced jet pilots, fluent in Russian, and familiar with aerospace terminology, which proved useful, as in April 1988, it was announced that the flight would be pulled forward to August 1988, shortening their training to just six months. No official explanation was given for this move, but at the time of negotiations for the flight, Russian forces were being pulled out of Afghanistan where they had been engaged in combat since they invaded in 1979. The Soviets wanted the flight to be completed before all troops were withdrawn, so the training programme was shortened and some parts cancelled altogether (such as the cold weather recovery training, considered unnecessary for a mission scheduled for August). The launch took place on 28 August 1988 and was planned for a 6 September landing, but difficulties with the Soyuz descent engine delayed the landing until 8 September, giving the cosmonauts an uncomfortable two days in the Soyuz Decent Module. Apparently Dauran was the original candidate for the mission, but only six weeks before the flight, he developed appendicitis and required an operation, effectively grounding him and giving Mohmand the flight. Mohmand is now a German citizen, having fled there when the Taliban came to power in Afghanistan. The status of Dauran is not known.

China

The Chinese were offered a joint space flight to the Mir space station by the Soviets in August 1986, but this proposal did not progress very far. China had long dreamed of flying citizens in space, ideally within their own programme, and after several feasibility studies in the 1980s, the official programme began as Project 921 in 1992.[26] Thirty years after severing relations with the Soviet Union, China re-established contact and communications in the early 1990s. Chinese officials soon visited Russia, including the facilities at TsPK, and expressed interest in purchasing space equipment that could be adapted for their own programme. Items obtained reportedly included a stripped down Soyuz (boilerplate) and designs of the proposed Zarya six-person Soyuz replacement. They also purchased a life support system, docking apparatus and a Sokol pressure garment. Between November 1996 and November 1997, two Chinese 'instructor cosmonauts' (Wu Tse and Li Tsinlung) completed the basic cosmonaut training programme as part of their preparation for assisting in the instruction of Chinese 'yuhangyuans' selected in 1998, though they did not complete any specific flight training. This experience helped the Chinese develop a training course for their own manned space programme. They were accompanied by a support team of twenty specialists, including translators. Since then, several Chinese 'yuhangyuans' have visited TsPK to complete partial or special training in preparation for further national space flights, though probably not on Russian spacecraft. Details of their visits are difficult to obtain but one of the more recent (in early 2005) was a six-week visit to use the Hydrolab to perfect EVA techniques, as part of the preparations for the first Chinese EVA expected in 2006 or 2007. These seem to be commercial agreements between the Chinese and the Russian space authorities.

India

In the late 1970s, offers were made to the Indian government for a joint space flight, with the agreement finally being signed in March 1981. This was a cooperative agreement, like the initial flight with France, rather than a commercial agreement. The selection process reduced the original 240 applicants to twenty and, by October 1981, to ten, who were all Indian Air Force officers. Each had been examined and conducted preliminary training at a variety of Air Force facilities before a final four were sent to TsPK for the final selection process. From this, the two finalists, Major Ravish Malhotra and Major Rakesh Sharma were selected in September 1982 and began cosmonaut training on 19 September. Sharma flew the mission between 3 and 11 April 1984 to the Salyut 7 space station. One of the more interesting and unusual, perhaps unique, experiments conducted on the flight was called 'Yoga', designed to reproduce five static positions (asanas) of the ancient art of physical exercise in space. The Indians were of the opinion that the use of Yoga might assist in the reduction of cardiovascular disorders caused by prolonged weightlessness and that such exercise could prove useful as an effective means of preventing muscular dystrophy during extended duration space flights. This meant additional special training for the Indian cosmonauts, which was not attempted by the Russians. During their training, the Indians completed a programme of daily exercises, which were reviewed by a team of doctors. During the mission, Sharma repeated these exercises, monitored by on board physician Dr. Oleg Atkov. Measurements and data collected from the groups of muscles in the legs, thighs and back of the Indian cosmonaut and the Soviet cosmonauts could then be compared and reviewed post-flight, with the aim of determining whether Yoga could help on long-term space flights. No results of this research appear to have been made known.

Japan

GlavKosmos signed a commercial agreement with the Japanese television network TBS to fly a Japanese journalist aboard Mir for a reported US$ 11 million. The selection was restricted to TBS staff and associated companies and resulted in 162 applications (including eighteen women), aged between 23 and 55. Of these, 98 candidates (including nine women) were examined in the first round of tests held in Japan, from which 45 passed. Then, after more detailed examinations, the group was narrowed to 21 and then to seven (including two women). The second round, also held in Japan, studied the remaining 64 applicants (including nine women), from which ten went forward for more detailed tests. From these two groups, the Japanese and Soviet specialists approved four candidates, although the Soviets indicated that at least six would be required to find one flight candidate. As a result, records were re-examined and two finalists from the first round were included. The candidates were Toshio Koiki (the TBS Moscow correspondent), Toyohiro Akiyama, Ms Ryoko Kikuchi, Atsuyoshi Murakami (all from TBS) and, from affiliated companies, Nobuhiro Yamamori, Kouichi Okada and Ms Naoko Goto. Additional tests began in Moscow from 25 August, finally approving Akiyama and Kikuchi. They were formally selected on 18 September for a mission originally scheduled for 1991, though this was later moved forward to 1990. On 1 October 1989, the two Japanese cosmonauts arrived at TsPK for cosmonaut training, with Akiyama flying the mission between 2 and 10 December 1990. A long time and dedicated smoker of four packs a day, Akiyama agreed to give up for the sake of a space flight – a personal challenge he had difficulty keeping to. He only abstained during the flight, and suffered from space sickness.

Malaysia

The initial invitation to fly a Malaysian cosmonaut was made in the summer of 1987, but nothing materialised for the next fifteen years, until a new round of negotiations began in 2003. In April 2005, the selection programme began for the planned flight in October 2007, as part of a Soyuz Taxi exchange mission to ISS. This seems to be related to commercial contracts relating to the purchase of fighter aircraft from the Sukhoi plant.

Syria

In 1985, an invitation was offered to Syria to train two candidates for a flight to a space station (initially thought to be Salyut 7, but when the Mir core was launched in February 1986, it was clear it would be to this new station). There is very little known about the selection process, but it resulted in four finalists who were sent to Moscow for final selection They were Kamal Arabi, Muhammad Fares, Munir Habib and Ahmad Rateb, all military pilots previously trained in the Soviet Union. Habib and Fares were selected and began cosmonaut training in October 1985 and in December

Inside the Mir simulator is the back up crew for the Soviet Syrian mission. (from left) Air Force commander Anatoliy Soloyvov, Energiya engineer Viktor Savinykh and Syrian cosmonaut Munir Habib

1986, Fares was selected as prime crew member with Habib serving as back-up. The mission was flown between 22 and 30 July 1987.

Millionaire cosmonauts

It soon became clear that with the dramatic changes resulting from the demise of the Soviet Union, the commercial cosmonaut seat programme would not generate the much needed funds to support the struggling Russian manned space programme. With a growing commitment to the International Space Station programme, the last few years of Mir saw the marketing of commercial seats to organisations or individuals who were able to pay around US$ 11–20 million per seat. In the 1990s, several schemes were proposed for fare-paying passengers or winners of lottery or reality TV shows that might (but never did) result in a flight to Mir as the top prize. The criteria for such flights, in addition to the funding, was to agree to learn Russian and spend a year in Russia in training, the prospect of which daunted many of those who expressed interest. Despite the prospect of at least one American millionaire flying to Mir, at the end of its operational life, the degradation of the veteran station and pressure to divert full attention (and precious resources) to the pending ISS programme led to all flights to Mir by non-Russian cosmonauts being abandoned, with only one short domestic mission flown in 2000 to assess the condition of the station and, effectively, seal its fate.

INTERNATIONAL COSMONAUT TRAINING

The training of guest cosmonauts for missions aboard the Soyuz spacecraft resembled that of the Interkosmonaut programme, but with the added task of mastering the Russian language, which most of the cosmonauts from Eastern Bloc countries had generally developed as a second language. The other challenge for western guest cosmonauts was the cultural shock of having to live and work in Russia for up to eighteen months and the difficulties, hardships and challenges this brought over and above dealing with the cosmonaut training programme and the mission itself.

Seizing the training

One of the most informative accounts of life as a 'part-time cosmonaut' is that of British astronaut Helen Sharman who, in her 1993 autobiography, *Seize the Moment*,[27] devoted a whole chapter to her training at TsPK. It is highly likely that her experiences mirrored that of all 'foreign' candidates training during the same period.

Sharman described the eighteen-month period between November 1989 and May 1991 as 'probably the most significant period of my life, in its own way even more influential on my outlook and ideas than the time I spent in space.' She explained that the theoretical and practical training was in itself a considerable challenge, and reflected the hurdle that most guest cosmonauts had before them, many of whom had never dreamed of flying into space before their selection. She also found herself not only thrust into the media spotlight in the UK, but also in a cultural void in what was then still the Soviet Union. She had no in-depth knowledge of the country she would live in for over a year, could not speak the language, and knew no one there who knew that much about England or the English language. There was also no certainty that she would fly the mission, or even if the funding for the training would be sustained.

Learning Russian was the first priority. She had listened to audio tapes in her car when she could during the selection process and had visited Russia for part of the medical tests, but she soon found out that the normal English method of 'getting by' in a foreign country with a few choice phrases or words, sign language and pointing would not suffice in Russia, and certainly not at TsPK, where English speaking was rare and German was more common. In addition, all the training, exams and air-to-ground commentary would be conducted in Russian, and coming to terms with the Cyrillic alphabet was a challenge. The language barrier has been one of the most challenging phases of any joint flight with the Russians since the days of ASTP. In 1984, the Indian cosmonauts explained that not only do new international cosmonauts have to overcome the Russian language, but they also have the additional challenge of learning 'space-Russian' – the abbreviations, terminology and meanings of the language of astronautics.

Living in Star City was, by most Russian standards, quite comfortable, with spacious if sparse apartments in one of the 'Cosmonaut Houses' tower blocks. And as she was in a 'privileged position', Sharman only had to order a phone call home one hour in advance, instead of the nominal two or three days most Russians had to endure, although even this request had to be made in Russian. The Russian language was intensive for the first three months, interspersed with sessions of physical training or sporting activities. Sharman found that Star City was equipped with excellent sporting facilities and with full access, most guest cosmonauts made full use of them to increase their physical condition and prepare their bodies for the stresses of g-forces and the onset of weightlessness. Mastery of the language finally seems to come from personal meetings with residents and staff at Star City, as the concentration and repetitive linguistic techniques in the classroom do not reproduce actually hearing and using the language in everyday life. Trips into Moscow on the local commuting train were possible, but not encouraged for safety reasons. The country was going through a monumental change and officials were not enthusiastic about risking the lives of expensive guest cosmonauts to the down side of modern life in Moscow. Inside the protected confines of Star City, it may have been restricting, but at least it was safe with everything on site, even if these facilities were basic at best. At least the Juno cosmonauts were not isolated for all of their eighteen months at TsPK. They managed six vacations, including the festive season and summer holidays, which were not always taken back in the UK. Though training for a space mission is intensive, it has to be remembered that even cosmonauts are allowed vacations, otherwise the intensity builds up to a point where training becomes ineffective and can become a safety or health risk.

Sharman explained that her training at TsPK could be divided into three general areas:

Suitability Training This included the physical and medical fitness programme and the Russian language, which Sharman reasoned was actually started with selection to the programme. As the training developed, this suitability training evolved from three months of intensive language study into a vast amount of academic study, where she learned the technical and scientific theory behind space operations.

Courses included the theory of (space) flight, orbital dynamics, cryogenics, rocket engine technology and communications.

Practical Training Essentially, the hands-on skills of becoming a cosmonaut. This phase started about a month or two into the training programme and initially she completed a course of parachute training. Then she experienced a programme of weightless training aboard padded Ilyushin 76 transport aircraft flying in parabolic curves, followed by survival training environments and repetitive emergency drills.

Parachute training started on the ground, then progressed to descents from helicopters and was restricted to just two jumps (at least for the Juno candidates), as safety was more important than enjoying the thrill of more parachute jumps. Though this type of training was a requirement for early Vostok training, where cosmonauts ejected from the descent capsule for landing, it has not been used since 1963 on a space mission, but is used as part of preparations for aircraft flights, such as in the Ilyushin 76 transport.

Emergency situations during a mission can be defined as those you can do something about (such as a Soyuz 18–1, Soyuz 23, T 10A or TM 5 situation), and those you cannot (Soyuz 1, Soyuz 11). Catastrophic situations are not covered in training, but everyone is aware of them. The use of the launch escape tower is also not 'trained for' as such, but is covered in briefings of what the sequence should be and what could be experienced by the crew (based upon actual experiences from the 1975 and 1983 Soyuz launch emergencies). Wilderness survival training is used to familiarise the guest cosmonauts with off-nominal landing situations and takes place in bush areas, and at the Black Sea for unplanned water landing situations in both hurried (in watertight Sokol pressure suits) and unhurried (in emergency survival gear) situations. Landing in the open ocean in a Soyuz is not something to look forward to, as the craft is cramped, pitches around, and is prone to filling with water. A crew therefore has to remove the pressure garment (Sokol suit), put on the survival gear, gather emergency equipment and supplies, and then get out of the spacecraft. When the Juno candidates did this, it was during June with temperatures of 30°C. After being placed in the cramped capsule and bobbing around in the water, trying to put on a multitude of survival garments (over the long underwear, they put on a red polo-neck sweater, sleeveless jumpsuit, a jacket, padded duvet-style trousers and a second jacket, a pair of padded boots, a thermal hat, the 'Trout' water proof suit with an integral flotation collar, and a rubber hat) and then struggling out took Sharman over three hours and her body temperature increased 2°C as a result. 'Because of the wave motion, there was a strong smell of vomit by the time we'd finished. This is an experience in which your relationship with the other people in the crew is tested to the limit,' she recalled. Once in the sea, the crew attached themselves to each other to prevent floating off in case of storm conditions, and prepared to stay like this for up to three days. 'Even though you know you can survive, you are left with the hope that the automatic landing systems will get you down where you wish to be – on the dry Kazakh plain with a rescue helicopter only minutes away,' Sharman reflected.

Crew Training This is where the candidates work in the ground-based simulators at TsPK as part of a crew, becoming accustomed to the equipment and displays of the

Soyuz spacecraft and the facilities aboard the space station. In this phase, the launch and re-entry procedures are practiced and the mission's experiment programme is focused upon. Sharman and Mace started crew training with their respective Russian colleagues in December 1990 and as they progressed through the training cycle, simulator time increased as they became the next crew in line for launch. Repeated simulations allow phases of the mission (launch, docking, undocking, descent) to become second nature. According to Sharman, 'This kind of training is brilliant. During the flight itself, almost nothing happened that surprised or scared me, so thorough were the preparations.' Both of the Indian cosmonauts, for example, had completed over fifty-four hours of 'flights' in the simulators during the course of their preparations. In addition to simulator time, the crew also completed habitability training with the crew facilities onboard the Soyuz and station (food, hygiene, sleeping and waste management issues) and also started scientific training for their experiment programme about two months into the crew training cycle. This required the experiment set up, operation and, where necessary, the collection or retrieval of several data items, mastering the operations and processes as close to perfection as possible to ensure that maximum data was collected during the flight. About a month before the planned launch, crew training wound down, exams were taken and the crews finally confirmed. Prior to the trip to Baykonur, ceremonial duties were performed, as part of the next Russian crew to leave Earth and fly into history. One thing that is unclear is whether any work is done on post-recovery activities and adjustment back to life on Earth. It seems that each cosmonaut is left to their own devices to blend back into life after space flight.

Recollections from a Spationaut

The first Westerner to experience a complete cosmonaut training cycle from selection to post-flight was French spationaut Jean-Loup Chrétien, who became the first Western citizen to fly on a Russian space mission in 1982. He flew again with the Russians in 1988 and has also completed an American Shuttle flight (to Mir) in 1997. This unique experience was recorded as part of the NASA Oral History Project in 2002 and during the interview, Chrétien recalled his experiences in training for a flight on a Russian vehicle.[28]

Chrétien also recalled great cultural shock, but was impressed with the efforts the Soviets showed in making them welcome, which helped when he was told he was staying for up to two years. For social events, they were driven to the Marine Bar in the American Embassy (as were the ASTP astronauts several years before). The training programme was being developed, and two years was a long time just to be 'a passenger' (which is one reason they shortened it to eighteen months by the time Sharman flew a decade later). Flying officially as a Cosmonaut Researcher, he was essentially an experiment operator, not a pilot, and despite his test pilot credentials, Chrétien stated that he had to endure a long, laborious training course of instruction on what the other members of the crew did on the mission, even though he would not participate. This was not required on this first international mission and could have been deleted, shortening the training time. The French cosmonauts also found themselves being seated in the centre of the cafeteria, screened off from other diners so they didn't learn too much about the Soviet programme over lunch!

Chrétien explained, 'I remember spending fifty hours on a very detailed course about the theory of space navigation and the lessons were more like school, with an exam every three weeks and a teacher coming in, teaching you at the blackboard the theory of everything. You had to take notes and write in your notebook, so we learned a lot.' All the training was, of course, in Russian and Chrétien explained that the first year was essentially theoretical courses, with survival training coming at the end of that year and simulator work beginnning only in the second year. For the first French mission, they completed an intensive Russian language training course of six to eight months in France before moving to Star City. 'I remember well, it was twelve hours a day including Saturdays – and it was a nightmare.' But it paid off and set the precedent that a good knowledge of Russian helps prior to moving to Star City – something the Americans found out a decade later in training for Mir.

Chrétien also explained the crew training once they entered their second year. As they were rookies to space flight, they still had a lot of classroom work and were assigned as much as forty per cent of the time on their own with teachers supporting theoretical studies, and sixty per cent with the experienced Soviet crew in simulators. Being interviewed twenty years after he completed the training programme, Chrétien was unclear of exactly how many hours he and Baudry spent in the simulators, but he did recall participating in three long simulations a week in the Soyuz simulator and two more in the Salyut 7 simulator, and this was repeated for his 1988 Mir mission. He also spent hours with the mission flight director, developing the checklists for ascent, orbital operations and entry. Experiment training with scientists occurred perhaps once a month for a week.

Chrétien also indicated the competition between him and Baudry during the first stages of training, because neither knew who would fly and who would not until shortly before the mission, a standard Soviet/Russian practice for many years. 'We knew that during that year we had to do our best so that we had the chance to get selected, and it's not a very pleasant situation, because we are two good friends and working hard together.' Chrétien and Baudry decided to do their best by trying to work together, knowing only one would be selected and trying to not think how much final selection depended not on their own performance, but on the opinion of others. As Chrétien recalled, 'That was probably the most challenging, most unpleasant and difficult part of the trip. We strongly recommended, when we came back, to never do that again, and to tell whoever is being sent which one is flying and which is not.' That was what the French decided to do with later selections, so they knew who would be prime and who would be back-up before they received detailed flight training.

Then there were the personality difficulties, which were not covered in the NASA oral history. When first assigned, Chrétien was teamed with Yuri Malyshev as his commander, but by early 1982 they were not getting along too well, mainly due to Malyshev's insistence on treating the experienced French test pilot as nothing more than a passenger and rebuking him for asking to participate in some of the spacecraft procedures, instead of just sitting still in his seat. It came to a head when Chrétien brought a pillow and duly went to sleep in the craft during one simulation run. The training was duly changed and so was the crew, with Malyshev replaced by veteran cosmonaut Vladimir Dzhanibekov, who had had experience with Western space explorers before (on ASTP). Malyshev was reassigned to command the Indian flight two years later, apparently with no problem.

For his second training cycle with the Russians a few years later (and with a stint of American Shuttle Payload Specialist training behind him as well), Chrétien noted the changes in the Soviet Union under the new leadership of President Mikhail Gorbachev. Though the social and political life of the country had begun to change, the space training had not changed that much, although for his second flight, he was the first foreign cosmonaut to undergo EVA training, which was an added bonus and, he felt, similar to the EVA training in the US.

As veteran Georgiy Grechko explained during his training as back-up FE on the Soyuz T 11 Indian mission during 1984, 'Months of preparation are filled with such incredible difficulties and trials that the space flight itself seems, in comparison, restful.'

REFERENCES

1 The Partnership, A history of ASTP, NASA SP4209, p 105–110
2 Ref 1, p 201
3 Ref 1, p 208
4 Ref 1, pp 232–234
5 Ref 1, p 252
6 Ref 1, p 253
7 We Have Capture, Thomas P. Stafford and Michael Cassutt, Smithsonian Institute Press, 2002, p 173
8 Ref 7, p 171
9 Ref 1, p 264
10 The Sun's Wind, Aleksey Leonov, Progress Publishers, Moscow, 1977, p 18
11 Ref 7, p 175
12 Ref 1, p 339
13 The Soviet Cosmonaut Team, **1**, Background Status, Gordon R. Hooper, GRH Publications, 1990, pp 97–127
14 TASS report in English, 15 September 1976, Foreign Broadcast Information Service (FBIS) – Russian section
15 FBIS TASS report in English, 24 December 1976
16 Additional details of Interkosmos selection and crew assignments can be found in: The Soviet Cosmonaut Team by Gordon R. Hooper (see Ref. 13)
17 Hungary and Space Research, Edited by Tibor Zádor, Hungarian Press Agency, 1980
18 Mongolia: Off to a good start in space, p 4, Novosti Press Agency Publishing House, 1981
19 Ref 18, pp 4–5
20 Ref 17, p 21
21 Vladimir Remek, biography sheet, Gordon R. Hooper, dated 16 March 1985; Bertalan Farkas, biography sheet, Gordon R. Hooper, 1980
22 Document throws light on Interkosmos politics, Aleksandr Zheleznyakov and Andrey

Koryakoiv, translated (into English) by Alex Greenberg, Spaceflight, **44** May 2002, pp 208–209

23 Who's Who In Space: The ISS Edition, Michael Cassutt, MacMillan, 1999, pp 470–473

24 Hooper, Soviet Cosmonaut Team, **2**, already cited, pp 125–127

25 Russian with a Foreign Accent: Non-Russian cosmonauts on Mir, Bert Vis, in The History of Mir 1986–2000, British Interplanetary Society, 2000, pp 86–99; The Soviet Cosmonaut Team, Hooper – previously cited, pp 94–127; and Who's Who in Space: The ISS Edition, Michael Cassutt, previously cited, pp 469–479

26 China's Space Programme, Brian Harvey, Springer-Praxis 2004, pp 239–255

27 Seize the Moment, Helen Sharman with Christopher Priest, Victor Gallancz publishers, London, 1993, pp 151–168

28 NASA JSC Oral History Project, Jean-Loup J.M. Chrétien, 2 and 8 May 2002

Surrounding the Salyut 6 simulator are several Soyuz descent modules. Almost obscured is the OKEAN ('Ocean') module used for water and winter survival training, alongside a second one (with the dolphin painted on the side) used for water survival training. In the foreground is the Soyuz 2 descent module, with two more descent modules to its right

This photo of Korpus 1 was released by NASA in 1974. In front is the Soyuz simulator, with the APAS docking mechanism as used during the ASTP programme. Behind it is the Salyut 4 simulator. In the background is another Soyuz simulator with APAS, as well as a full scale mock-up of the Soyuz spacecraft

Transition from the Salyut programme to Mir is evident in this photo, made around 1986. In the background, the Salyut 7 simulator is still in its place in Korpus 1, but the Mir and Kvant modules are already being used to train upcoming expedition crews

The TDK-7ST simulator

The Memorial of Flight, which stands outside the House of Cosmonauts

The simulators of the Priroda (left) and Spektr modules of the Mir complex

The Hydrolaboratory Annex

The trout suit (Forel) used for sea recovery training, being worn by German ESA astronaut Thomas Reiter

Desert survival training

During desert survival training, cosmonauts learn how to make use of the Soyuz parachute to make a shelter

The gate between Star City and the original TsPK grounds. The road leads to the Headquarters and Administration Building

The sign at the beginning of the access road from the town of Chkalovskiy to Star City. From here, it is another three kilometres to the main gate

The building next to the main gates to Star City and TsPK

The main hall in the House of Cosmonauts (Dom Kosmonavtov) is used for all sorts of official and unofficial functions, such as dance lessons

... festivities like jubilees and weddings ...

. . . and funerals of leading residents of Star City, such as cosmonauts. This is cosmonaut Aleksandr Kramarenko lying in state on 16 April 2002

Room dedicated to Yuri Gagarin in the museum in the House of Cosmonauts

The lake with the Profilactorium, in the old days, before the three NASA houses were constructed

When NASA took up permanent residency in Star City, three wooden houses were built for trainers and astronauts to stay in during training visits

The Star City sports hall. On the right, the abandoned Buran training facility is visible

This tennis hall in Star City burnt down in 2001, but was replaced by a new one in 2003

This view from Star City, looking north, shows a number of apartment blocks that are outside the Star City perimeter but are still considered part of the town. Many engineers and most of the young cosmonauts live in these buildings. The town cemetery is located in the woods behind these blocks

The view from the other side shows, from left, Blocks 49, 48, 47 and 46 and the Orbita hotel

Entrance to the building in Star City that houses the Cosmonaut Post Office and local Militsiya (police) unit

A unique photograph showing the plans for extending Star City in the mid-1980s. On the right, the Engineering and Simulator Building and Korpus 3A are easily recognisable. At bottom left are Blocks 2 and 4, where most of the veteran cosmonauts live. At the very top, is the Profilactorium and in the middle is a proposed new museum that was never built

Joint programmes

Following the highly successful Apollo-Soyuz Test Project (ASTP) in 1975, both the Soviet Union and the United States agreed to develop follow-on rendezvous and docking activity that would include an American Space Shuttle (which had yet to fly) and a Soviet Salyut space station, probably during 1981.[1] The intergovernmental agreement was signed in 1977 and extensive cooperative work was conducted during 1978, but the project faltered due to concerns over Soviet human rights issues, the international actions of the Soviet Union, and American fears over the transfer of technology. The increasingly strained relationship between the two countries over the 1978–1982 period saw the civil space agreement lapse in May 1982. Over the next decade, there was little cooperation in manned space operations, although cooperative unmanned and life sciences programmes were ongoing and a group of US astronauts visited Russia in February 1990.

SHUTTLE-MIR PROGRAMME – A NEW LEARNING CURVE

Just days before President Ronald Reagan's State of the Union address in January 1984, in which he invited US friends and allies to participate in the construction of a long-term space station (Freedom), private offers from the US to Russia suggested the idea of a simulated space rescue between the now flight-proven Space Shuttle and Russia's Salyut 7 space station. Nothing came of this, but it did begin to repair the differences between the two nations. However, it was not until 17 June 1992 that a new civil space agreement was concluded, which would see American astronauts and Russian cosmonauts flying on joint Shuttle or Mir space station missions. The agreement also included a suggestion to consider specific exchanges of astronaut and cosmonaut flight opportunities, and a Shuttle-Mir rendezvous and docking mission. This also led to discussions and studies about the potential of using the Soyuz TM spacecraft as a crew rescue vehicle for Space Station Freedom and, eventually, to look at elements of Soviet space station hardware and launch facilities that could support the early construction of the struggling space station.

In 1993, new president Bill Clinton ordered a substantial redesign of the over-budget space station Freedom, and the International Space Station programme took shape. By late 1993, Russia was a formal member of the ISS programme, which was now divided into three phases. Phase 1 (1994–1997) would see a programme of between seven and ten Shuttle missions to the Mir space station, including five medium to long duration flights of US astronauts and at least two flights of Russian cosmonauts on independent Shuttle missions (expanding the original 1992 agreement). Phase Two (1997–1998) would see the launch and construction of ISS elements from the US, Russia and Canada, capable of supporting three resident crew members in 1998 by means of a Soyuz TM spacecraft. Phase Three (1998–2002) would see the completion of ISS assembly, including elements from Europe and Japan. Though this has of course been delayed somewhat due to financial restraints, partner disagreements, launch difficulties and the loss of Columbia in 2003, the completion of the station remains a priority over the next few years.[2]

As this cooperative programme developed, so did the requirement for both astronauts and cosmonauts to be updated on each other's hardware and training techniques, and for an official joint training programme to be developed that could be transferred from the Shuttle-Mir programme to the ISS programme. It had been almost twenty years since the heyday of ASTP, and a new generation of astronauts and cosmonauts were at the forefront of space exploration. The learning curve had to start right back at the beginning.

Creating a joint training programme

For the joint Shuttle-Mir programme, a working group (number 5) was established for crew exchange and training. This was a small group of four individuals (two from each side), including cosmonaut Aleksandr Aleksandrov, now a leading figure in the Energiya OKB.[3] The objectives of the group were 'to determine the duties and responsibilities of cosmonauts and astronauts when completing flights on the Shuttle and Soyuz vehicles and the Mir space station; the content of crew training in Russia and in the US; and to develop training schedules and programmes.' To achieve this, regular meetings were scheduled, alternating between Russia and the US, with teleconferences in between. To help expand this development of a training protocol for long duration missions, a Johnson Space Center (JSC) NASA office was created at TsPK, where a NASA Astronaut Office (CB) representative would work permanently. Called the Director of Operations Russia (DOR), he or she had daily contact with Russian space officials, resolving issues relating to cosmonaut and/or astronaut training for joint flights. This position continued in the transition to ISS operations after the completion of Shuttle-Mir operations. A similar office and position was created by the Russians in JSC Houston, to oversee cosmonaut training in America.

In addition, a Crew Exchange and Training Working Group series of documents was created for defining the duties and responsibilities for each flight assignment, either on the Shuttle, Soyuz or Mir. These documents covered topics ranging from training plans for the spacecraft, science programmes, terminology, EVA systems, emergency evacuation systems and procedures, down to a Russian-English or English-Russian dictionary, crew equipment and personal items, including emblems. A second working group (number 6) handled the content of science training for US crews on Mir.

Table 9: NASA Director of Operations – Russia (Shuttle-Mir) TsPK 1994–1998

From	*To*	*Astronaut*
1994 Feb	1994 Jul	Ken Cameron
1994 Jul	1994 Nov	Bill Readdy
1994 Nov	1995 Mar	Ron Sega
1995 Mar	1995 Oct	Mike Baker
1995 Oct	1996 Mar	Charles Precourt
1996 Mar	1996 Oct	Wendy Lawrence
1996 Oct	1997 Jun	Mike Lopez-Alegria
1997 Jun	1998 Feb	Brent Jett
1998 Feb	1998 Aug	James Halsell

Table 10: Russian Director of Operations – Houston (Shuttle-Mir) JSC 1996–1998

From	*To*	*Cosmonaut*
1996 May	1996 Aug	Solovyov, A.
1996 Aug	1996 Dec	Malenchenko
1996 Aug	1996 Dec	Tsibliyev
1997 Jan	1997 Jun	Dezhurov
1998 Oct	1999 Mar	Zalyotin
1999 Mar	1999 Oct	Kotov

Summary of Shuttle-Mir training

During the Shuttle-Mir programme, a total of eleven NASA astronauts were selected to train at TsPK for participation in long duration space flights aboard the Mir complex, seven of whom completed such a mission. Four of these astronauts completed Russian EVA training, three of whom completed EVAs from Mir. The long duration Mir systems training was supposed to last fourteen months, but changes within the flight programme and delays in assignments meant this was not possible for some of the Americans.

Two training sessions each were completed, at NASA JSC in Houston and at TsPK, for the operation of joint Russian-American science programmes by the prime and back-up members of six Mir residency crews (Mir 18, 21, 22, 23, 24, and 25). In addition, five Russian cosmonauts (Krikalev, V. Titov, Kondakova, Sharipov and Ryumin) completed JSC training for six Shuttle flights (STS-60, 63, 84, 86, 89 and 91), with Titov completing two cycles of Shuttle training. Nine other Shuttle crews (STS-71, 74, 76, 79, 81, 84, 86, 89 and 91) completed a week of training at TsPK for joint activity at the Mir space complex, and the prime and back-up crews of Mir 20 through 25 completed a week of training each (six periods) for Shuttle crew compartment familiarisation and joint activity with visiting Shuttle crews.

COSMONAUTS ON THE SHUTTLE (SHUTTLE-MIR)

From 1960 until 1992, the majority of the Soviet/Russian cosmonaut team had trained at TsPK, with the exception of the Buran Shuttle pilots and a short period of familiarisation training by the eight ASTP cosmonauts in the mid-1970s. All the foreign cosmonauts came to TsPK for their training, and only ten American astronauts had performed familiarisation training at the centre for ASTP (three prime, three back-up and four support crew members). With the advent of Shuttle-Mir, all this would change. Though the main focus was on the American long duration missions aboard Mir and the series of Shuttle docking missions, the first opportunity came with the flight of the first cosmonaut on a Shuttle prior to the joint activities at Mir. For the first time in their programme, some cosmonauts would participate in crew mission training for a mission controlled from outside their own country.

Cosmonaut Shuttle training was according to the level of their assigned responsibility on each given mission, ranging from full Mission Specialist training to passengers only, or as part of a visiting Shuttle crew to Mir. The Mission Specialist-assigned cosmonauts did not all complete the same MS Ascan training programme as new NASA astronauts do, due to their previous cosmonaut training and flight experience:

Table 11: Cosmonaut Basic Shuttle Training – New Astronaut and Refresher Hours

Cosmonaut	*New*	*Refresher*	*Total* *
Krikalev**	80	128	208
Titov, V.**	11	74	85
Kondakova	57	21	78
Sharipov	7	16	23
Ryumin	43	23	66
Dezhurov	11	26	37
Strekalov	13	25	38
Onufriyenko	19	29	48
Usachev	5	0	5
Budarin	15	25	40
Solovyov, A.	17	3	20
Totals	278	370	648

* Totals only include formal training, not the time spent by individuals in initial preparation using workbooks in Russia.
** Includes back-up and prime training on STS-60 and STS-63

Mission Specialist training also included instruction on the Shuttle life support and communication systems in both nominal and off-nominal situations, payload activities, Earth observation and photographic activities. On one mission (Krikalev on STS-60), cosmonaut duties included the use of the Shuttle Remote Manipulator

System (RMS), and on another flight (Titov on STS-86), the cosmonaut was trained in American EVA operations. All cosmonauts conducted egress and emergency egress operations to ensure personal and crew safety under all conditions. This included trips to KSC, Florida, for on-site training. Cosmonauts also received training on the operation and scope of crew equipment, hardware and facilities (meal preparation, stowage, exercise equipment, waste management, etc.) and trained as crew members of T 38 aircraft for flights to and from JSC and KSC as required, with a member of the NASA astronaut corps piloting the aircraft.

Where cosmonauts were only serving as passengers to and from Mir (Dezhurov, Strekalov, Onufriyenko, Usachev, Budarin and A. Solovyov), their training was reduced and mainly focused on keeping the cosmonauts safe. This included training in the life and crew support systems.

For Mir crews that only visited the Shuttle while in orbit during docked phases, training focused on a general familiarity with the Shuttle crew compartments, life support systems and transfer operations between the two vehicles. This generally averaged out at about 36 hours for each Mir resident cosmonaut crew member. Some of the payload training for cosmonauts also occurred during joint sessions in the United States.

The first Russian Mission Specialists

In the articles of agreement between NASA and the Russian Space Agency, dated 5 October 1992, Article I (Description of Cooperation)[4] stated that 'An experienced cosmonaut will fly on the Space Shuttle on the STS-60 mission, which is currently scheduled for November 1993. The cosmonaut will become an integral member of the crew and will be trained as a Mission Specialist on Shuttle systems, flight operations, and manifested payload procedures, following the existing Shuttle practices. In addition, the RSA will nominate two cosmonauts for approval by NASA as 'candidates' for STS-60. One would be designated prime, the other as back-up. Both would receive NASA Mission Specialist astronaut training until the time STS-60 begins dedicated mission training.' Once the crew had been assigned, the back-up cosmonaut would complete as much training as was practical. Their arrival at NASA JSC Houston was scheduled for October 1992. Article IV (Selection of Candidates), stated that the selection of suitable flight candidates would be based on mutual agreements, but that all candidates would be flight experienced. The cosmonauts were also expected to have 'a sufficient knowledge of English in both verbal and written forms.' Under Article V (Training), the cosmonauts would be based at NASA JSC, assigned to the Astronaut Office (Code CB) in the Flight Crew Operations Directorate. At the start of training, each candidate would be required to enter into a Standard of Conduct Agreement with NASA, which included safety and security matters, prohibition to use the position for private gains, recognising the authority of the Mission Commander and limiting the use of information gained or used from the training or mission.[5]

Table 12 Cosmonaut Shuttle-Mir Training – Systems, procedures and hardware (hours)

Cosmonaut Name	Int. Ascent	Int. Entry	Int. Orbit	Orbit Ops	Ascent Entry	Orbit Systems	Crew Systems	EVA	PDRS	Payloads	Rndz/ Prox ops	Spacelab	Spacehab	Total
Krikalev*	1	15	75	53	63	9	70	24	151	70	70	0	16	617
Titov V.**	17	30	162	117	178	10	103	137	75	28	34	0	46	937
Kondakova	1	7	50	60	21	8	70	13	0	6	22	0	27	285
Sharipov	1	7	50	0	4	0	50	0	0	2	0	0	3	117
Ryumin	0	7	40	36	8	8	74	0	0	0	15	0	12	200
Dezhurov	0	0	7	0	4	8	12	0	0	0	0	1	0	32
Strekalov	0	0	7	0	4	0	12	0	0	0	0	2	0	25
Onufriyenko	0	0	7	9	4	0	50	0	0	2	0	2	0	74
Usachev	0	0	0	9	0	0	30	0	0	2	0	2	0	43
Budarin	0	0	7	9	8	0	61	0	0	2	0	2	0	89
Solovyov, A.	0	0	12	9	4	0	49	0	0	2	0	2	0	78
TOTALS	20	66	417	302	298	43	581	174	226	114	141	11	104	2497

* The table reflects only formal training for each cosmonaut. The number of hours spent with workbooks during initial preparation training while still in Russia is not recorded

** Vladimir Titov completed two training cycles for STS-63 and STS-86. He was also back-up to Krikalev on STS-60, and vice versa on STS-63. It was not clear in the report if the training times included back-up training.

Sergey Krikalev practices operating the Shuttle RMS during a training exercise at the Shuttle mock-up and integration laboratory, Bldg 9 at JSC September 1993. (Courtesy NASA)

First cosmonauts selected for NASA training

'Of course I would like to fly on the Space Shuttle. It's a chance to fly on new technology,' stated veteran Mir cosmonaut Sergey Krikalev (through an interpreter), during a visit to the NASA Kennedy Space Center in Florida on 18 August 1992. Krikalev, accompanied by his Soyuz TM 12 commander, Anatoliy Artsebarskiy, was in the United States to attend a space conference in Washington DC and visited KSC for a goodwill tour of facilities and to meet with dignitaries.[6] A few weeks later, on 6 October 1992, his wish came closer when he was named with Colonel Vladimir Titov as the two Russian cosmonauts who would train as Shuttle Mission Specialists, with one flying on STS-60. Both were experienced cosmonauts, with Titov having logged over 367 days in space, including a single 365-day residence on Mir. Krikalev has also lived on Mir and his two missions logged in excess of 463 days. They would arrive at JSC in late October.[7] On 28 October, the NASA crew for STS-60 was named as Charles Bolden (Commander), Kenneth Reightler (Pilot), and Franklin Chang-Diaz, Jan Davis and Ron Sega as Mission Specialists. The cosmonauts spoke to the American press (through an interpreter) for the first time on 10 November. Krikalev stated, 'We have already started the training and learning the materials necessary for the accomplishment of the mission.' Titov added, 'Hopefully, the experience of cooperation we gain here will lay the foundation for the future work between our cosmonauts and scientists and will be followed by other endeavours.'

Their families would not arrive until 21 November and both cosmonauts recognised that their wives and children would experience some difficulties in adjusting to life in Houston.[8]

Initial Mission Specialist training flow for cosmonauts

In the first Russian MS training flow (for STS-60 only), NASA planned 574 total hours for each cosmonaut. This included:

Table 13: Russian MS Training Flow (STS-60)

Course	*Hours*
Ascent flight operations	16
Orbit flight operations	10
De-orbit and entry flight operations	09
Payload deployment and retrieval systems (RMS)	51
Orbiter systems	84
Rendezvous and proximity operation	25
Data Processing systems/software	16
Crew systems	65
Flight support	14
Informal training	186
Total	476

In addition, there were 98 hours planned for training as a crew member on the T 38 jet. Several adjustments had to be made to the training programme to take into account the shortage of time (twelve months from announcement to flight), to include a full Ascan MS training programme (twelve months), and for specific mission training time (9–12 months). The basic Ascan training programme was abandoned, largely because the cosmonauts were already experienced and qualified. There was also an adjustment to accommodate the language barrier and because very little of the repetitive training was needed.

Work books were used only as reference, with one-hour briefings given before each lesson, for which the cosmonauts had an interpreter. The lessons were conducted separately for each cosmonaut, with no interpreter. By early January 1993, the basic training was on schedule, but at a point where the Payload Deployment and Retrieval System (PDRS) and Rendezvous (RNDZ) training flows were starting, which were the most difficult parts of the remaining course work to master. These would also provide the best demonstration of which cosmonaut was beginning to master the English language most effectively. In the first six weeks of 1993, there was little time for the Russians to participate in crew activities other than completing basic MS training. The NASA training team was very sensitive about overloading the cosmonauts, with time included for study and reflection, so it was important that gaps in the training course remained free. As the basic course wound down, the decision about which cosmonaut would fly and which would be back-up

Sergey Krikalev prepares to operate an M-113 armoured personnel carrier as part of emergency egress training for STS-88 at KSC in Florida, November 1998. Fellow STS-88 astronauts Bob Cabana and Fred Sturckow are at rear

had to be made, but discussions were already being held about the feasibility of flying the second cosmonaut as well, enthusiastically supported by the training team.[9]

Basic training lasted from late October 1992 to early February 1993, and mission training from then until launch in February 1994 (the mission was delayed from November 1993 as a result of launch delays and priorities from other 1993 Shuttle missions). In December 1993, the STS-60 training manager and instructors responsible for training the prime and back-up cosmonauts for STS-60 compiled a set of memos, listing lessons learned from training the cosmonauts for the mission. These had important implications for the forthcoming flights of other Russian cosmonauts on the Shuttle, under Shuttle-Mir or the future ISS programme.[10] The decision was made early in the planning of cosmonaut assignments to a Shuttle mission that they would fly as a Mission Specialist 'with substantial onboard duties', as opposed to Payload Specialist with very few. The training department and astronaut office stated that, preferably, the cosmonauts would be fluent in English and should begin their training in the United States several months prior to normal crew assignment (3,000-hour level, or about nine months), which for STS-60 was February 1993 for a November 1993 launch. However, it appears that no one at JSC had any input into who the Russians selected for the mission and found out 'at the last minute' who they were. The cosmonauts themselves had very little time to prepare for the assignment before being rushed over to Houston in November 1992, with only three months to complete the equivalent of the pilot pool flow.

Originally, it was planned to use training manuals in written English, but it soon became evident that the lack of technical English ability and the time limitations meant that a series of introductory briefings with translators was required to prepare the cosmonauts for the Single System Trainers (SST). An already modified training flow had to be modified again in real-time to accommodate the steep learning curve for both the cosmonauts and the instructors. It was already recognised that this would be a major challenge and as a result, the SMS team and team leaders were hand picked. That same team would provide as much of the cosmonauts' non-SMS training as possible.

STS-60 Training feedback notes

In the feedback from the STS-60 training team, the following points were observed:

T. Capps, STS-60 Training Manager

Notes dated 15 November 1993 identified a variety of pointers for the future training of cosmonauts in the United States:

Personal Provision of furnished apartments only, with all bills paid. No houses to be used, as bill paying and associated decisions proved too complicated. Provision of an info package, in simple terms, for living and working in Texas and working at JSC, for such everyday needs as driving, parking, logistics, etc.

Lesson materials Use workbooks as reference only material; clearly mark 'Reference Only' on the books and turn as many of the workbooks into practical briefings as possible. Use visuals in all classes and provide a dictionary of applicable NASA terms and acronyms arranged in areas of concentration or systems. Provide audio-visuals for teaching English and personal computer interfaces directly to TsPK in Russia.

Teaching skills Separate the cosmonauts into different classes and individual lessons and reinforce the purpose and application of materials to motivate the cosmonauts. Instructors must inform the cosmonauts where any information does not appear to be learned and why this appears to be the case. Provide formal status to the cosmonauts to reveal their progress. (It was clear the cosmonauts were goal driven and needed to be informed of their level to reach a given goal). Designate back-ups to prime instructors to limit the number of interfaces with training staff. Limit confusion, and screen and limit the number of different presentations on the same subject. Tailor requirements to the individual and provide flexibility in training flows to accommodate each individual and their personal skills and shortcomings. Make decisions and stick with them and do not present too many options to the cosmonauts.

Crew assignment It became clear that crew assignments and flight positions were identified early in the training flow and changed only if the performance of the cosmonauts warranted such a change. Focus on crew assignments and concentrate their curriculum on that position. Teach only the subject matter that is required.

General remarks Channel all PAO activities through the training manager and/or the Commander. Lobby NASA Headquarters to get future cosmonauts to JSC two months earlier in the training cycle than the STS-60 selection, in order to concentrate on English language, 'NASA-ology', culture and family logistics. Build time into the schedule for the gymnasium, limit T 38 training and accept these limitations. Simplify the logistics and decision making and try to provide and tell at the earliest point. Formularise team member's sharing of experiences and approaches that work.

Bill West, the SMS Team 6 Training Instructor
Notes dated 4 November 1993 stated that training for both cosmonauts began in November 1992 and continued for Krikalev as he prepared for STS-60. West trained Titov up to his inclusion in the crew for STS-63. The initial training consisted of a series of briefings that provided a basic overview of the Shuttle flight control, propulsion and guidance systems. This was followed by SST and Shuttle Mission Simulator (SMS) familiarisation sessions prior to specific STS-60 mission training. West summarised the lessons learned from training Krikalev and Titov and put forward recommendations for further training of cosmonauts. He explained that, prior to initial briefings, he obtained a copy of the ASTP Russian-to-English and English-to-Russian glossaries and a general Russian-to-English and English-to-Russian dictionary and used these in briefings and SST classes. When pointing out unfamiliar words in English, written right next to it in the dictionaries were the

STS-60 back-up crewmember cosmonaut Vladimir Titov practices his skills in a small life raft during bailout training at the JSC WET-F, Bldg 29, during February 1993. (Courtesy NASA)

Russian equivalents, and when the cosmonauts were unsure of an English word, they pointed to the Russian listing with an English definition beside it. It was very basic, but it seemed to work well. When presenting briefings, West used simple diagrams of Shuttle systems, but labelled them in Russian using words copied from the technical glossaries. Wherever possible, he compared US systems to Russian ones, such as orbiter entry guidance vs. Soyuz entry guidance, or Soyuz propulsion systems vs. orbiter propulsion systems, and presented all numbers in both imperial and metric units. West found that the briefings took much longer than scheduled, with one briefing taking four hours instead of the normal two. Models and diagrams helped significantly in getting the point over to the cosmonauts.

From his experiences on console training the STS-60 cosmonauts, West suggested that if a decision was made to train future cosmonauts as 'full-up Mission Specialists', then more than a year should be allocated to training them. The information load thrown at them in the early weeks after arriving at JSC required several refresher sessions later on. West suggested, 'It may be desirable to select several cosmonauts with an Ascan class and have them start (Shuttle) training from the beginning.' There also needed to be more time allocated to classes and briefings and the class size should be small. Titov, for example, was found more likely to try and speak English in a small group than a large one. West also stated, 'I feel that it is a waste of time and money to send (training) personnel off to learn to speak Russian, unless they are going to Russia.' He felt it was much easier to have cosmonauts learn English than for an instructor to learn Russian. This raised the question of which language would be priority on ISS and implied that learning the language as soon as possible would be an advantage to future cosmonaut selections for American crews.

Barbara Severance, STS-60 Ops/Nav Instructor

In undated notes, she explained that the training was a compressed Ascan flow, and remarked, 'It should be noted that no one really knew what to expect and the instructors had little or no time to prepare for this task.' The trainers were tasked with training Krikalev and Titov to a level where they could make CRT (Cathode Ray Tube) screen interface inputs, could read, understand and implement the Flight Data File, and could be conversant with orbiter systems and workings to be able to contribute to flight activities, as well as their assigned tasks on orbit, such as RMS operator. Again, initial meetings via an interpreter were at times challenging, with old training schedules discarded and rewritten from scratch. Severance observed, 'There were quite a few times when the student and interpreter would have a lengthy conversation in Russian and eventually present a question to me. I never knew if the interpreter was attempting to answer the question on his own, or if the student was trying to explain enough technical information so that the interpreter could then more accurately ask me a question.'

Though their command of English improved, Severance wrote that, at first, Krikalev was easier to teach because his command of English was better than Titov's, whose initial skill was extremely limited. At the start of training, they started lessons with Krikalev acting as interpreter for Titov. Krikalev also demonstrated the most hands-on approach to simulator time, whereas Titov wanted to observe at first and thus was much slower. This did change later, as the absence of an interpreter forced him to learn English much faster. Severance also observed that briefings needed to be streamlined to the essentials and recognised that the extensive space flight experience and cosmonaut training of both men allowed them to demonstrate 'their extensive capabilities and intelligence once the language barrier was overcome.' This resulted in the cosmonauts offering great insight into the training protocol and coming up with questions that tasked the trainers. Severance observed that speaking slowly, clearly, and without too many acronyms or slang words helped the cosmonauts understand the sentence. Presenting 'the big picture' also generated more detailed, and sometimes difficult, questions from the cosmonauts, but she noted that the trainers should be prepared to encourage such questions. She also found that reading the question to the interpreter helped with the translation and watching facial expressions closely revealed more than words could in several cases. She suggested developing a standard training flow for future Mission Specialists from foreign countries, standardising lessons and developing handouts for instructors as a reference to provide them with training methods and suggestions. She also suggested that the US/Russian technical library should be expanded.

Dave Pitre, Comm Instructor

His undated notes observed that the information that both cosmonauts could speak some English did prove a little optimistic, although 'to their credit, it did not take them long to improve (their) skills. But it was not without pain.' He suggested that the ASTP technical directories were very useful and that a similar Shuttle Russian-English dictionary should be developed, perhaps on a computer disc which would be more accessible and portable. During single system training, it was found that free-form instruction was better than following the rigid training plan, and overall patience was important. In the SMS, with the student removed from the direct view of the instructors, the Russians were apprehensive at first and needed some encouragement to participate, but identifying responsibilities, and training and encouraging them, helped improve their confidence and abilities.

Lisa Reed, SMS Team 6 Systems Instructor

In her (undated) notes, she suggested that, in addition to the language barrier, the compressed training time added to the pressure for the cosmonauts to succeed. 'There was so much more I would have liked to teach Titov and Krikalev but I just did not have time. Keeping the pace was hard on the instructors; I can only imagine what it must be like for Titov and Krikalev. I commend their efforts.' In general, she stated that, given the time and resources they had, the training team did an excellent job of training, but with a little more advanced notice, much more could have been done. 'If we commit to training Russian cosmonauts or any other foreign crew, I highly recommend that (NASA) management look (at) the lessons learned and the recommendations given by the instructors.' Reinforcing the opinion of other instructors, Reed added that no information was passed to the SMS team on who would be trained and that they had to source a PAO press release to find out who was scheduled and what space flight experience they had. They were not informed exactly how to train the cosmonauts and the suggestion was made that 'if the foreign students were introduced to our training processes before they ever walked through the door, this might help them understand more of the what, how and why we train as we do.'

Reed also expressed frustration in the way the basic Shuttle training system was compressed into two months. The training team were asked to waive prerequisites and trim back their requirements as much as possible, but still provide a detailed overview for someone flying as a fully-fledged Mission Specialist. She explained that, taking the simulator element of a nominal NASA Ascan training as an example, it could take classes of eight hours, on consecutive days, spread over several weeks or months to develop understanding, even with students that could speak English. With the Russians, these classes were compressed into one working day![11] It became obvious to the team that to schedule basic training early would be more beneficial and productive to all concerned. The Ascan training programme was 'modified to fit' by trimming sections or adding abbreviated lessons, but to then try to use the formal workbooks with NASA jargon and abbreviations complicated the delivery process. The suggestion was to create a training flow for future international students that is suited to their particular instructional needs. What did work was assigning one SMS training team to work with the Russians, using familiar faces over time to bring out the confidence and elements of teamwork and team sprit from both sides. Reed observed that, for her team, interpreters or translators were cumbersome and time consuming, and it would be far better to teach English to the students first, and make translation reference documents more available for current systems. Class time frequently overran, but this should also be expected, and care should be taken to select instructors with proven communication skills and patience, especially in dealing with new foreign students.

In summary, STS-60 Flight Section Head, Dennis Beckman, praised both his team and the cosmonauts for adapting a difficult situation to achieve excellent results, and cited the skills and abilities of both cosmonauts in persevering to overcome their training hurdles. 'They are all to be commended for a difficult job well done.'

Sergey Krikalev, the first Russian Shuttlenaut (STS-60)

Krikalev was assigned to the prime crew of STS-60 to begin mission training with effect from 1 February 1993, with Titov as back-up also being prepared for his own flight aboard STS-63 (for which Krikalev would be back-up). The official announcement was made on 2 April 1994.[12] STS-60 was flown between 3 and 11 February 1994, carrying the retrievable Wake Shield Facility (a free-flyer disc designed to generate new semiconductor films for advanced electronics, although technical problems prevented its deployment on this mission) and the second flight of the commercially developed research facility, Spacehab. The mission was the 60th of the series and utilised Discovery (OV-103), on its 18th mission. Krikalev flew as MS4 and became the first Russian cosmonaut not to leave Earth for space from the Baykonur cosmodrome, and the first to launch, orbit and return solely on a foreign spacecraft and land outside Soviet or Russian territory. He logged just over 8 days and 7 hours in space and, ironically, also became the first cosmonaut who had received some limited Soviet Buran shuttle training to fly on a 'Shuttle mission' – even if it was an American one!

Krikalev flew into space on the middeck (Seat 6) and swapped to Seat 3 on the flight deck for entry. According to the STS-60 crew task list, Krikalev was assigned a variety of responsibilities on the crew, reflecting the scope and depth of his training for the flight in such a short time. His orbiter tasks included responsibilities for photo and TV issues (back-up), in-flight maintenance (primary), and Earth observations (back-up). He was also assigned back-up responsibilities for both the Remote Manipulator System and for Spacehab systems. In payloads, he was back-up for the Wake Shield Facility, and in the Getaway Special (GAS) experiments, and held primary responsibilities for the Capillary Pumped Loop Experiment and GAS Bridge Assembly. Krikalev was also primarily responsible for the SAREX-II Shuttle amateur radio experiment and Aurora Photography Experiments (APE-B). Spacehab-2 consisted of twelve investigations, of which Krikalev was primarily responsible for four (Space Acceleration Measurement System (SAMS); Three-Dimensional Zero-gravity Accelerometer (3-DMA); Bio serve Pilot Lab (BPL); and Immune Response studies (IMMUNE-01)) and back-up on one (Astroculture Experiment ASC-3). For the joint US-Russian medical investigations (Detailed Supplementary Objectives, or DSO), Krikalev was a primary on DSO 204 (visual observations) and back-up on DSO 200 (radiological) and DSO 201 (sensory).[13]

Titov flies on STS-63

It was agreed that Titov would fly on STS-63, a Mir rendezvous mission as a dress rehearsal of future Shuttle docking missions. Originally planned for June 1994, the mission was delayed until February 1995 because of the inability to fill the Spacehab module with experiments. The crew was named on 8 September 1993[14], and consisted of James Wetherbee (Commander), Eileen Collins (the first female pilot assigned to a Shuttle crew), Janice Voss (Payload Commander/MS1 identified on 3 August 1993), Mike Foale (MS2), Bernard Harris (MS3) and Titov as MS4. The delays to the mission probably helped Titov improve his Shuttle skills and become more familiar with his assigned responsibilities and tasks. Known as the 'Near-Mir' mission, STS-63 was flown between 3 and 11 February 1995, the 67th flight of the series and the 20th for Discovery. As well as carrying the Spacehab-3 payload, the crew deployed the Orbital Debris Radar Calibration System-II (ODERACS–II), deployed and retrieved the SPARTAN 204 free flyer, and closed to within ten metres of the Mir space station on 6 February. Foale and Harris also completed a four-hour EVA on 9 February. For the ascent, Titov rode on Seat 6 on the middeck, swapping to flight deck Seat 3 for re-entry.

Titov's orbiter tasks included a primary role on in-flight maintenance (IFM) and Earth observations and a back-up role for communications and instrumentation, photo and TV, and the RMS. His payload responsibilities included back-up duties for the Spacehab-3 module, and back-up for the secondary payloads IMAX Cargo Bay Camera (ICBC), and the Solid Surface Combustion Experiment (SSCE), located on the middeck. Titov held primary responsibility for four Spacehab-3 experiments and back-up responsibilities for a further seven. His primaries were Astroculture-IV, Charlotte, Chromex-06 and Commercial Generic Bio-processing Apparatus (CGBA-6), and his back-up roles were for Commercial Protein Crystal Growth – Vapour Diffusion Apparatus (CPCR-VDA), CREAM-06, ECLIPSE-Hab 3, IMMUNE-02, Protein Crystal Growth – STES-03, 3-DMA and Window Experiment-01 (WIND-EX-01). In addition, he held responsibilities for 11 DSOs: 200B Radiobiology Effects; 201B Sensory-Motor; 204 Visual Observations; 484B Circadian Shift; 486 Physical Exam; 487 Immune Assess; 491 Microbial Transfer; 377 Shuttle-Mir VHF, responsible for communication with the resident crew (EO-17, Viktorenko, Kondakova, Polyakov); 901 Documentary TV; 902 Documentary Movies; and 903 Documentary Still Photography.[15]

Cosmonauts on the Shuttle-Mir dockings

In the three years between June 1995 and June 1998, there were nine Shuttle-Mir docking missions.[16] Of these, five carried Russian cosmonauts as members of the crew and their training for these missions is reflected in Table 14. By capitalising on the lessons learned from training Krikalev and Titov, a pattern was developed to assist in preparing cosmonauts for the role of MS on these visiting missions to the space complex. This also provided an invaluable database of training experience for direct application to ISS Shuttle missions that might include members of the cosmonaut team.

STS-71 On 3 June 1994, NASA named the crew of STS-71 (Atlantis – OV-104), the first Shuttle docking mission.[17] Previously assigned to the flight for ascent only were Anatoliy Solovyov and Nikolay Budarin (the Mir 19 resident crew), who would remain aboard Mir when Atlantis departed and returned the Mir 18 crew of Vladimir Dezhurov, Gennadiy Strekalov and American astronaut Norman Thagard. Since they would only occupy the Shuttle for ascent or descent, no crew responsibilities were assigned to the cosmonauts for this mission. For launch, Solovyov was designated RC-1 (Russian Cosmonaut) and occupied Seat 6 on the middeck, while Budarin was designated RC-2 and occupied Seat 7 on the middeck (cosmonauts Onufriyenko and Usachev also trained as back-ups for these positions). For the descent, Dezhurov served as RC-1 in Seat 7, and Strekalov as RC-2 in Seat 6 (Thagard served as MS 4 and occupied Seat 8). The returning Mir 18 crew rode in specially designed Recumbent Seating Systems, to help them endure re-entry and landing loads following their extended duration mission (a method used later by returning American Mir resident crew members and ISS crews on the Shuttle). This was only the second time an eight-person crew had ridden on the Shuttle (five returning STS-71 crew and three returning Mir 18 crew); the first was STS 61-A (Spacelab D1) in 1985. STS-71 flew between 27 June and 7 July 1995.

STS-84 Veteran Mir cosmonaut Yelena Kondakova was named as MS4 for the sixth Shuttle docking crew on 22 August 1996,[18] six weeks after the NASA crew had been identified and presumably after she had completed basic Shuttle training. A report issued by the BBC on 8 November 1996 suggested that the only criteria for her

Table 14 Russian cosmonauts as crewmembers on US Space Shuttle missions 1994–2002

Cosmonaut	*Mission*	*Position*	*Orbiter*	*Launch*	*Duration*	*Landing*	*EVA*	*Notes*
Shuttle Mir Programme 1994–1998								
Krikalev	STS-60	MS-4	Discovery	1994 Feb 03	08:07:09:22	1994 Feb 11	–	1st cosmonaut on Shuttle; Spacehab 2
Titov V.	STS-63	MS-4	Discovery	1995 Feb 03	08:06:28:15	1995 Feb 11	–	Near-Mir rendezvous mission; Spacehab 3
Solovyov A.	STS-71	MS-4	Atlantis	1995 Jun 27	approx 2 days	(Soyuz TM21)	–	Up only – transferred as Mir 19 crew
Budarin	STS-71	MS-5	Atlantis	1995 Jun 27	approx 2 days	(Soyuz TM21)	–	Up only – transferred as Mir 19 crew
Dezhurov	STS-71	MS-4	Atlantis	(Soyuz TM21)	approx 3 days	1995 Jul 07	–	Down only – former Mir 18 crew
Strekalov	STS-71	MS-4	Atlantis	(Soyuz TM21)	approx 3 days	1995 Jul 07	–	Down only – former Mir 18 crew
Kondakova	STS-84	MS-4	Atlantis	1997 May 15	09:05:19:56	1997 May 24	–	Shuttle Mir (SM) docking mission 6
Titov V. (2nd)	STS-86	MS-1	Atlantis	1997 Sep 25	10:19:20:50	1997 Oct 06	1 (05:01)	SM docking mission 7; 1st Russian Shuttle EVA (EV2)
Kadenyuk (Ukraine)	STS-87	PS-1	Columbia	1997 Nov 19	15:16:34:04	1997 Dec 05	–	Science mission; Kadenyuk flew as Ukrainian not Russian
Sharipov	STS-89	MS-4	Endeavour	1998 Jan 22	08:19:46:54	1998 Jan 31	–	Shuttle Mir docking mission 8
Ryumin	STS-91	MS-4	Discovery	1998 Jun 02	09:19:53:54	1998 Jun 12	–	Shuttle Mir docking mission 9 (last of series)
International Space Station Programme 1998–2002								
Krikalev (2nd)	STS-88	MS-4	Endeavour	1998 Dec 04	11:19:17:57	1998 Dec 15	–	ISS-2A; first Shuttle assembly mission
Tokarev	STS-96	MS-5	Discovery	1999 May 27	09:19:13:57	1999 Jun 06	–	ISS-2A.1; second ISS assembly mission
Usachev	STS-101	MS-5	Atlantis	2000 May 19	09:21:10:10	2000 May 29	–	ISS-2A.2a; third ISS assembly mission
Malenchenko	STS-106	MS-4	Atlantis	2000 Sep 08	11:19:12:15	2000 Sep 20	–	ISS 2A.2b; fourth ISS assembly mission

Cosmonaut	*Mission*	*Position*	*Orbiter*	*Launch*	*Duration*	*Landing*	*EVA*	*Notes*
Morukov	STS-106	MS-5	Atlantis	2000 Sep 08	11:19:12:15	2000 Sep 20	–	ISS 2A.2b; fourth ISS assembly mission
Usachev (2nd)	STS-102 up	ISS-2	Discovery	2001 Mar 08	approx 4 (2+2) days	(STS-105)	–	ISS-2 resident crew member returned on STS-105
Gidzenko	STS-102 down	ISS-1	Discovery	(Soyuz TM 31)	approx 2 days	2000 Mar 21	–	ISS-1 resident crew return only
Krikalev (3rd)	STS-102 down	ISS-1	Discovery	(Soyuz TM 31)	approx 2 days	2000 Mar 21	–	ISS-1 resident crew return only
Lonchakov	STS-100	MS-5	Endeavour	2001 Apr 19	11:21:31:14	2001 May 01	–	ISS- 6A; 9th ISS assembly mission
Dezhurov (2nd)	STS-105 up	ISS-3	Discovery	2001 Aug 10	approx 4 days (2+2)	(STS-108)	–	ISS-3 resident crew member returned on STS-108
Tyurin	STS-105 up	ISS-3	Discovery	2001 Aug 10	approx 4 days (2+2)	(STS-108)	–	ISS-3 resident crew member returned on STS-108
Onufriyenko	STS-108 up	ISS-4	Endeavour	2001 Dec 05	approx 4 days (2+2)	(STS-111)	–	ISS-4 resident crew member returned on STS-111
Korzun	STS-111 up	ISS-5	Endeavour	2002 Jun 05	approx 7 days (2+5)	(STS-113)	–	ISS-5 resident crewmember returned on STS-113
Treshchev	STS-111 up	ISS-5	Endeavour	2002 Jun 05	approx 7 days (2+5)	(STS-113)	–	ISS-5 resident crewmember returned on STS-113
Yurchikhin	STS-112	MS-4	Atlantis	2002 Oct 07	13:18:48:38	2002 Oct 18	–	ISS-UF-2; 15th ISS assembly mission
Budarin	STS-113 up	ISS-5	Endeavour	2002 Nov 23	approx 2 days	(Soyuz TMA 1)	–	ISS-5 crew returned on Soyuz TMA 1

The STS-91 crew are briefed prior to entering the Systems Integration Facility at JSC. Standing at rear wearing the training version of the Shuttle launch and entry ('pumpkin') suit is veteran cosmonaut Valeriy Ryumin. (Courtesy NASA)

selection was her previous experience in orbit aboard Mir. Kondakova was the wife of former cosmonaut, and leading Energiya official, Valeriy Ryumin, and according to one source, she was invited to join the STS-86 crew by Charles Precourt after he had met her during a trip to Russia. The obvious political advantage of this was quickly seized upon by NASA and it was officially approved, even though this meant one of the NASA astronauts would lose their seat on this mission. This echoed comments from Irina Pronina, who had stated that no female cosmonaut ever received a mission assignment without influence![19]

For both launch and landing, Kondakova remained in Seat 6 on the middeck. Her crew responsibilities included opening and closing the payload bay doors and radiators inside the payload bay doors, as well as photo and TV documentation during the mission, contingency EVA crew member IV2, radio communications with the EO-23 Mir crew (Tsibliyev and Lazutkin) during rendezvous and docking of the Shuttle, supporting in-flight maintenance, the post-orbit insertion operations, and de-orbit preparations. She also supported the set up of Spacehab shortly after orbital insertion, and the operation of refrigeration and freezers in the module, and was responsible for the deactivation and stowage of equipment in Spacehab in preparation for the return to Earth. On Spacehab, she participated in the operation of thirteen Bio rack experiments and the CREAM risk mitigation experiments on the middeck of the orbiter, as well as two DSOs. During the transfer operations at Mir, she supported the transfer of Russian and powered hardware between the spacecraft, assisted Mike Foale in joint US-Russian science experiments during the docked phase, and was crew representative for Russian language issues.[20] STS-84 was flown between 15 and 24 May 1997.

STS-86 Vladimir Titov remained at JSC after his flight on STS-63, while Krikalev (supporting Shuttle-Mir activities as a consultant) returned to TsPK to begin training as a member of the first ISS crew, to which he was named on 30 January 1996.[21] In April 1996 Titov supported the flight of Kondakova on STS-84 and was identified as a crew member for STS-86 in November 1996. On 6 December, he was named MS1 for STS-86, the seventh docking mission, with Jim Wetherbee (Commander), Mike Bloomfield (Pilot), Scott Parazynski (MS2) and French astronaut Jean-Loup Chrétien, a veteran of both Salyut 7 and Mir, as MS3. Titov and Wetherbee had flown together on STS-63 and Wetherbee invited him to fly on STS-86, apparently supported by the Americans, even though the Russians did not want to assign a cosmonaut to this flight. Persuasion from the Americans allowed Titov to fly a second Shuttle mission (capitalising on his previous training) and to perform an EVA.[19] His assignments reflected his growing capability during MS training.

Titov sat on the flight deck (Seat 3) for ascent and on the middeck (Seat 5) for entry. As MS1, he had more ascent responsibilities than during his first mission on the Shuttle, supporting the 'orbiter crew' during phases of the ascent to orbit. He was also responsible for the transfer of logistics across to Mir, activation and deactivation of Spacehab science and systems, including in-flight maintenance on the module, and Earth observation. Titov was the crew representative for Russian language issues and was responsible for the communications and instrumentation, as well as prime on opening the payload bay doors and supporting closing of the doors towards the end of the mission. He ensured that the External Tank (ET) connection doors were closed at separation from the fuel tank and photographed the separated tank as the orbiter moved into orbit. His other orbiter responsibilities included cable configurations, and supporting IFM and Waste Collection System (WCS) issues. He also worked on post-insertion activities and was prime for de-orbit preparations. During Mir operations, he handled communication with the EO-24 crew (A. Solovyov and Vinogradov) and operated the hand-held laser for range finding during docking. On the science objectives, Titov was assigned to support four Detailed Text Objectives (DTO), one DSO, three Risk Mitigation Experiments (RME) and one GAS Canister payload. His major contribution to the mission was the 5-hour 1-minute EVA on 1 October 1997 with Parazynski, wearing US EMU suits to deploy and retrieve equipment and experiments on the US Docking Module. The STS-86 mission was flown between 26 September and 6 October 1997.[22]

Valeriy Ryumin is seated in the Crew Compartment Trainer (CCT-2) in the Systems Integration Facility at JSC. At rear is astronaut Franklin Chang-Diaz. The simulation of launch and landing operations was completed during April 1998. (Courtesy NASA)

STS-89 The NASA astronauts named to STS-89 (the eighth docking mission) on 4 March 1997 were Terrence Wilcutt (Commander), Joe Edwards (Pilot) James Reilly (MS1) Michael Anderson (MS2) and Bonnie Dunbar (MS3). [23] On 15 October, rookie cosmonaut Salizhan Sharipov was added to the crew as MS4. Three candidates (Sharipov, Morukov and Kotov) were identified a couple of months previously.[24] Apparently, NASA had now dropped the veteran cosmonaut rule for Shuttle assignments, but with only two docking missions to Mir remaining on the schedule, the Americans wanted a native Russian speaker on the last two missions, both as an aide to the transfer operations and as a safety element in case of further difficulties with the ageing station. It also would help to ease the transition from Mir over to ISS. Sharipov became the only rookie cosmonaut to fly on a Shuttle during the Shuttle-Mir programme. He was not in line for any crew assignment, had recently completed routine water egress and survival training in the Black Sea and was about to go on a family vacation during the summer of 1997 when Yuri Glazkov, director of cosmonaut training at TsPK, informed him that he was going to Houston for six month's Shuttle training and his holiday had been cancelled. The happy but surprised cosmonaut would be entering space probably three to five years before he expected to. He left for JSC in August 1997 (his wife would join him in December) and apparently found the English language lessons the hardest hurdle to overcome (in addition to Houston's heat and humidity), but he made excellent progress despite the short training period and again, NASA allowed for the cosmonaut's previous training at TsPK in lieu of full Ascan training.[25]

Sharipov flew the ascent on Seat 6 and descent on Seat 7 on the middeck. He had minimal assignments on the crew, mainly configuring the middeck after post-orbit insertion and for de-orbit preparations. He was primarily responsible for the ergometer and assisted with photo documentation and TV activities throughout the mission. During Mir transfer operations, he assisted in the transfer of water supplies, the IELK, cold stowage, gyrodome and batteries across to the station, supported a few middeck science payloads, and participated in Earth observations and the photo survey of Mir. STS-89 was flown between 22 and 31 January 1998. NASA archives reveal that Sharipov had very little input into crew tasks – probably because of his late assignment to the crew (although he still logged 140 hours crew training), his difficulty in mastering English, or politics.[26]

STS-91 The final Shuttle docking mission was flown between 2 and 12 June 1998. The NASA crew assigned on 23 October 1997 was Charles Precourt (Commander), Dominic Gorie (Pilot), Franklin Chang-Diaz (MS1), Wendy Lawrence (MS2), and Janet Kavandi (MS4). When NASA offered a seat on the final Shuttle docking mission with Mir to the Russians, Valeriy Ryumin nominated himself (with the support of Energiya). On 5 September 1997, he began NASA MS training, having been identified to the crew in August 1997. This would be a challenge, because he could not speak English, was seriously overweight (a potential safety risk for emergency situations, although by January 1998 he had lost 25 kg in weight) and had been out of space flight training for eighteen years. On the other hand, he had helped design the Salyut and Mir stations, had been a cosmonaut since 1973 and had been involved with Energiya (formerly OKB-1) since 1966. A three-time veteran cosmonaut, he had also accumulated over a year aboard the Salyut 6 space station and, in logged time, was more experienced than all his American crewmates put together. His primary role on the flight would be to personally examine the condition and status of Mir after twelve years in space and to apply these findings to future space stations. He was also to supply information to support the decommissioning of Mir as soon as possible to make way for ISS, but after NASA took him to the station, he reported that it was in fine condition and proclaimed it should continue to be occupied (which it was for another three years), the exact opposite of what NASA wanted to hear![19]

Vladimir Titov is seated on the middeck of Discovery on Pad 39B at the Cape during January 1995. He was participating the Terminal Countdown Demonstration Test (TCDT) dress rehearsal for the launch the following month

Ryumin's crew assignments on the mission included supporting opening and closing of the payload bay doors, IFM, photo and TV documentation objectives, post-insertion and de-orbit preparations, and radio communications with Mir resident crew EO-25 (Musabayev and Budarin). He also supported the activation, operation and deactivation of the Spacehab payload and experiments, the transfer of ambient and Russian hardware across to the station, the retrieval of Mir water samples, and Earth observation objectives, and was support (IV-2) for contingency EVA operations. After this mission, the way was clear to begin ISS flight operations, which began five months later with the launch of the first element from Russia, the Zarya module.[27]

AMERICANS AT TSPK

Following the visits of American astronauts to TsPK during 1973–1975 as part of the ASTP training preparations, there was a gap of nearly twenty years before the next American astronauts undertook training sessions at the Cosmonaut Training Centre. This time however, they did not receive simple familiarisation training on Russian space hardware, but a full training programme to qualify them for resident missions on space station Mir under the joint Shuttle-Mir programme. As with the Russians training in the United States, the Americans encountered language and social hurdles to overcome, and periods of adjustment, isolation, and at times, frustration, during long missions onboard the station between 1995 and 1998. Though each mission was a challenge in different ways, the experience was vital for preparing Americans to work on international crews on ISS. The only experience the Americans had in space station operations were the three Skylab missions of 1973/4, lasting 28, 59 and 84 days.

COSMONAUT TRAINING FOR ASTRONAUTS

All eleven NASA astronauts who were selected for long duration flights to Mir were trained as Flight Engineer-2 cosmonauts in two stages. Phase 1 included training as part of a group of astronauts, with Phase 2 as part of a specific crew.[3] The eleven astronauts were Norman Thagard, Bonnie Dunbar, Shannon Lucid, John Blaha, Jerry Linenger, Michael Foale, James Voss, David Wolf, Andrew Thomas, Wendy Lawrence and Scott Parazynski (who also completed some training for a long duration mission but his training hours were not recorded in the report). They pioneered long-term American international space flight training and also moved international cosmonaut training into a new era at TsPK.

Astronaut Mir assignments

The first astronauts assigned to Mir training were Norman Thagard (prime) and Bonnie Dunbar (back-up), who were officially announced on 3 February 1994. They had been preparing for departure since the end of 1993, taking language studies at the US Army's Defence Language Institute (DLI) in Monterey, California. They departed for Star City later in February 1994 and, after completing the Phase 1 training programme, they joined the cosmonauts assigned to Mir 18 and 19 for crew training.[28]

Mir 18	*Crew*	*Cdr*	*FE-1*	*NASA*
	Prime	Dezhurov	Strekalov	Thagard
	BUp	Solovyov A.	Budarin	Dunbar
Mir 19	*Crew*	*Cdr*	*FE1*	*FE2*
	Prime	Solovyov A.	Budarin	–
	BUp	Onufriyenko	Usachev	–

The first NASA astronaut launched (14 March 1995) on a mission outside of the United States (Soyuz TM 21), and the only NASA Mir astronaut to be launched from Baykonur, was Thagard, who flew a 115-day mission as part of the Mir 18 crew. Dunbar flew on the STS-71 mission in July 1995, the first Shuttle docking mission that returned Thagard and his colleagues and delivered the Mir 19 cosmonauts (the only Mir crew to be launched to the station by the Shuttle). In early plans, Dunbar could have remained on Mir replacing Thagard, but this option was not pursued.[29]

The second Shuttle-Mir docking mission (STS-74 in November 1995) did not include any American resident astronauts, but from STS-76 in March 1996 until the ninth and final Shuttle docking mission (STS-91 June 1998), all NASA Mir residents flew to and from the station aboard an American Shuttle. The NASA-2 training group was named as Shannon Lucid and John Blaha (the only 'pilot' astronaut selected for Mir residency crew training) on 3 November 1994 and by 30 March 1995, the group was expanded to prepare for a further three-mission plan announced on 14 December 1994.[30]

Mission	*Prime*	*Back-up*	*Launch*	*Land*	*Duration*
1	Thagard	Dunbar	(TM 21)	STS-71	3 months
2	Lucid	Blaha	STS-76	STS-79	5 months
3	Linenger	Parazynski	STS-79	STS-81	4 months
4	Blaha	Lawrence	STS-81	STS-84	6 months
5	Parazynski	Lawrence?	STS-84	STS-86	4 months

The training plan projected Russian language studies at DLI, then a two-phase cosmonaut training programme at Star City, with Shuttle training (mainly ascent and decent) at JSC and KSC in America. The back-up position to Mission 5 was a dead-end role and was most likely to have been filled by Lawrence in order to cut training costs. On 14 October 1995, NASA announced that Parazynski would discontinue Mir training following concerns over his ability to fit safely in the Soyuz decent vehicle due to his height. A preliminary evaluation had cleared him for training, but further evaluation into height increases caused by zero-g (2 cm average), deceleration loads, sitting height issues, and the safety margins against injury meant that the module could not accommodate Parazynski comfortably or safely in a Sokol suit. According to preliminary training reports, the astronaut had scored highly in performance and had earned the respect of the Russian training staff. A second blow befell the group on 24 October 1995 when, just prior to her departure for Russia, NASA announced that Lawrence would discontinue Mir residency crew training while NASA discussed the problem that she did not meet the minimum height requirements for Soyuz training with the Russians.

The system that was originally developed and approved by both the Russians and Americans stipulated that the flight of an astronaut on a residency crew would be preceded by back-up training, which ensured a longer joint training programme. However, the decision not to fly Dunbar (who probably would have been backed up by Lucid), the removal of Parazynski, and subsequent alterations in the training cycle and dates of arrival at Star City, prevented the full back-up system from being

NASA astronauts Thagard (l) and Dunbar inside the Soyuz TM simulator at TsPK, September 1994. Both are wearing Sokol pressure garments. (Courtesy NASA)

implemented. As a result, NASA reassigned the training group to accommodate the departure of Parazynski and Lawrence, with Mike Foale and Jim Voss brought in to replace them.

Mission	*Prime*	*Back-up*	*Launch*	*Land*	*Duration*
2	Lucid	Blaha	STS-76	STS-79	5 months
3	Blaha	Linenger	STS-79	STS-81	4 months
4	Linenger	Foale	STS-81	STS-84	6 months
5	Foale	Voss	STS-84	STS-86	4 months

On 30 January 1996, NASA announced an expansion of the Shuttle-Mir programme, with two further Shuttle dockings and the decision to select and train two further astronauts for residency missions.

The second astronaut training group was paired with their Russian colleagues in February 1996 for mission training. The Russian crew had been in training since February 1994:

Mir 21	*Crew*	*Cdr*	*FE-1*	*NASA*
	Prime	Onufriyenko	Usachev	Lucid
	BUp	Manakov	Vinogradov	Blaha

Lucid was to work with the Mir (EO)21 crew until replaced by Blaha, returning home on STS-79. Blaha would work with the EO-21 crew for a short time until replaced by the EO-22 crew (Manakov and Vinogradov), with whom he had trained.

However, technical problems delayed STS-79 resulting in a change of resident crew before Blaha arrived. In addition, a slight heart irregularity had grounded Manakov, so the EO-22 crew was replaced by their back-ups, Kaleri and Vinogradov. Blaha had to endure a long mission on Mir with cosmonauts he hardly knew and had not trained with. Lucid flew a 118-day mission, while Blaha flew a 'difficult' 128-day mission.

On 16 August 1996, NASA re-instated Wendy Lawrence to Mir training for the new sixth mission. She would be backed-up by Dave Wolf, who would fly the seventh and final long duration mission to the station. Andy Thomas was assigned as the final back-up but, as with Voss, without the prospect of a flight to the Mir station, with operations handing over to ISS from the end of 1998.

Mission	*Prime*	*Back-up*	*Launch*	*Land*	*Duration*
3	Blaha	Linenger	STS-79	STS-81	4 months
4	Linenger	Foale	STS-81	STS-84	6 months
5	Foale	Voss JS	STS-84	STS-86	4 months
6	Lawrence	Wolf	STS-86	STS-89	4 months
7	Wolf	Thomas	STS-89	STS-91	4 months

Crews for the next missions were paired on 1 July 1995 (Mir 22) and 21 February 1996 (Mir 23):

Mir 22	*Crew*	*Cdr*	*FE-1*	*NASA*
	Prime	Korzun	Kaleri	(Blaha) Linenger
	BUp	None	None	Foale
Mir 23	*Crew*	*Cdr*	*FE-1*	*NASA*
	Prime	Tsibliyev	Lazutkin	Foale
	BUp	Musabayev	Budarin	Voss JS

Linenger flew a 132-day mission, including the first American EVA from a Russian space station. During the mission, a fire broke out on the station, causing some concern and requiring the crews to put into practice the fire emergency drills they had trained on. Then, following a collision on Mir later in 1997, during which Foale's emergency training was also called upon (including an unplanned EVA), astronaut participation on the station was reviewed and a joint decision was made that all NASA astronauts should be trained and qualified in operating the Orlan EVA suit. Lawrence did not fit in this suit and was not EVA trained at NASA, and was again removed from the long duration residency group. However, she still flew on STS-86 to assist in the change over of NASA 5/6 (taking advantage of her training and familiarisation with both the Shuttle crew and Mir systems and experiments). As a result, on 10 October 1997, Thomas was named as the final astronaut to fly a long duration mission to Mir, while Jim Voss was reassigned to fill the dead-end back-up assignment, utilising his previous Mir training rather than assigning a new astronaut to the group. Voss was already in preliminary training for an early assignment to ISS.

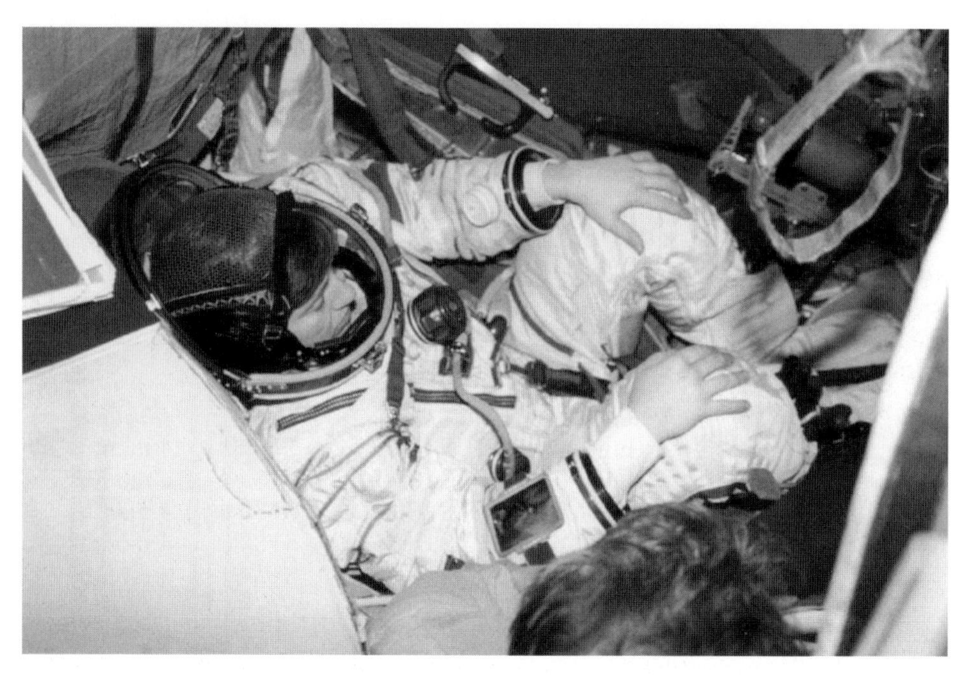

Astronaut Wendy Lawrence wearing a Sokol pressure garment inside the Soyuz TM simulator at TsPK. This September 1996 training session was part of her preparations for a subsequently cancelled long duration flight on Mir. (Courtesy NASA)

Mission	Prime	Back-up	Launch	Land	Duration
6	Wolf	Thomas	STS-86	STS-89	4 months
7	Thomas	Voss JS	STS-89	STS-91	4 months

Russian crews for the final missions with Americans on Mir were paired up in October 1996 (Mir 24) and March 1997 (Mir 25). The lack of integrated crew training for these missions was reflected in the comments of the astronauts and cosmonauts assigned (see page 289).

Mir 24	Crew	Cdr	FE-1	NASA
	Prime	Solovyov A.	Vinogradov	(Wolf/Thomas)
	BUp	Padalka	Avdeyev	(Thomas/Voss)

Mir 25	*Crew*	*Cdr*	*FE-1*	*NASA*
	Prime	Musabayev	Budarin	(Thomas)
	BUp	Zalyotin	Kaleri	(Voss)

Wolf flew a 127-day mission that included the final American EVA from Mir, while Thomas completed a 140-day mission to close out the programme.

NASA ASTRONAUT MIR TRAINING

The Americans' personal experiences in Russia and their activities during these flights have been recorded in several sources (see bibliography), but details of the training for these missions is not so widely reported. Therefore, a review of NASA astronaut Mir resident training follows.

Training as part of a group – phase 1

The first stage of training for these astronauts at TsPK included technical training for the Soyuz TM transport vehicle; practical classes and training sessions in the Soyuz TM simulators and part-task trainers; technical training for the Mir orbital complex; practical classes and training sessions on the station and module simulators; a programme of medical/biological training, including sessions in flying laboratories to simulate weightlessness; medical examinations and physical training programmes; survival and wilderness training under 'severe conditions'; independent training; and continued studies of the Russian language.

The scope of this training allowed each astronaut to acquire fundamental knowledge of the design, layout and operation of onboard systems of the Mir complex, develop skills in control and servicing of onboard systems, gain confidence in using and understanding the concepts, terms, and abbreviations of the Russian space programme (which allowed them to use a series of Mir flight data files and other flight documentation, all in Russian), and continue to expand their individual Russian language skills. Upon achieving these targets, each astronaut would then be qualified for inclusion into a specific crew assigned to the Mir residency crew training flow. NASA astronaut training for Mir is detailed in Table 15.

Training as part of a crew – phase 2

Once assigned to a crew, the astronauts then began the second phase of their cosmonaut training syllabus. This included technical training for the Soyuz TM spacecraft; practical classes and training sessions using the Soyuz TM simulator and system mock-ups; technical training; practical classes and training sessions on the Mir station and its modules using the simulator; continued medical and biological training; training for the programme of American and joint scientific research under the NASA-Mir agreement; training for the EVA programme (where appropriate); pre-flight training as a member of a Mir resident crew; and independent and Russian language training. This training allowed the Americans to develop the necessary skills to function as FE-2 during a mission. These included crew safety, including contingency situations using the Soyuz TM spacecraft for emergency return to Earth (even though launch to and recovery from Mir was assigned to a NASA Shuttle mission); supporting the nominal operation of onboard systems and equipment of Mir and its modules; operating various workstations; establishing regular communications and exchange of information with the NASA consultant group at NASA mission control in Houston; performing research and experiments; and using onboard facilities to perform personal hygiene, physical exercise and household procedures to ensure the health and hygiene of the astronaut and fellow crew

Table 15a Scope and dates of Mir training for NASA astronauts

NASA Mission	*Prime* Astronaut	*Backup* Astronaut	*Phase 1 (Group)* Start/Complete	*Phase 2 (Crew)* Start /Complete	*Total Hrs in Group*	*Total Hrs in a Crew*	*Total Hrs as backup*	*Total Training hrs*	*Launch date*	*Launch Mission*
NASA 1	Thagard	Dunbar	1994 Mar 01–1994 Oct 07	1994 Oct 10–1995 Feb 21	883	845	–	1728	1995 Mar 16	Soyuz TM 20
NASA 2	Lucid	Blaha	1995 Jan 03–1995 Jun 24	1995 Jun 26–1996 Feb 26	795	1127	–	1922	1996 Mar 24	STS-76
NASA 3	Blaha	Linenger	1996 Feb 23–1996 Jul 01	1995 May 29–1996 Jul 19	795	503	959	2257	1997 Jan 15	STS-79
NASA 4	Linenger	Foale	1996 Sep 23–1996 Dec 06	1995 Nov 29–1996 Dec 20	765	605	1054	2424	1997 Jan 15	STS-81
NASA 5	Foale	Voss J.S.	1997 Jan 13–1997 Apr 09	1996 Apr 03–1997 Apr 30	899	408	840	2147	1997 May 17	STS-84
NASA 6	Wolf	Lawrence	1996 Sep 02–1997 Aug 27	1996 Sep 02–1997 Aug 12	1081	614	–	1695	1997 Sep 30	STS-86
NASA 7	Thomas A.	Voss J.S.	1997 Jan 16–1997 Dec 05	1997 Sep 08–1997 Dec 05	982	553	–	1535	1998 Jan 12	STS-89

Table 15b NASA astronaut Mir training hours as part of a Group

NOTE: Hours for both prime and back-up together as 'a group'

NASA Mission	*Prime Astronaut*	*Backup Astronaut*	*Soyuz TM training*		*Mir training*		*Medical/ biological*	*EVA training*	*Independent training*	*Russian language*	*Total hours*
			Technical	*Simulators*	*Technical*	*Simulators*					
NASA 1	Thagard	Dunbar	134	173	120	50	170	–	86	150	883
NASA 2	Lucid	Blaha	20	50	114	60	122	–	161	268	795
NASA 3	Blaha	Linenger	20	50	114	60	122	–	161	268	795
NASA 4	Linenger	Foale	26	21	114	34	132	–	152	286	765
NASA 5	Foale	Voss J.S.	50	23	108	40	156	93	154	275	899
NASA 6	Wolf	Lawrence	77	91	54	22	153	–	172	349	918
NASA 7	Thomas A.	Voss J.S.	49	165	60	13	180	32	147	336	982

Table 15c NASA astronaut Mir training hours as part of a crew

NOTE: Prime/backup hours given where known

NASA Mission	*Prime/ Backup*	*Soyuz TM training*		*Mir training*		*Medical/ biological*	*EVA training*	*Science training*	*Pre-flight Training*	*Independent Training*	*Russian language*	*Total hours*
		Technical	*Simulators*	*Technical*	*Simulators*							
NASA 1	Thagard/Dunbar	35	90	128	68	94	4	311	80	11	24	845
NASA 2	Lucid/Blaha	80/–	130/–	141/–	142/–	180/–	–	266/–	24/–	76/–	88/–	1127/–
NASA 3	Blaha/Linenger	6/79	29/172	16/139	81/141	100/147	–	239/209	–	–	32/72	503/959
NASA 4	Linenger/Foale	13/49	20/206	26/84	81/97	60/153	46/75	303/230	–	–	56/160	605/1054
NASA 5	Foale/Voss J.S.	14/18	22/50	22/102	46/78	62/110	4/57	142/339	–	41/48	55/38	408/840
NASA 6	Wolf/Lawrence	10/–	82/-	126/–	100/–	71/–	96/–	121/–	–	–	8/–	614/–
NASA 7	Thomas A./ Voss J.S.	18	58	78	77	144	64	104	–	6	4	553

NOTE: Foale received 137 hours additional group science training; Thomas received 93 hours additional group science training; /– = no training figures available.

members. After integrated training sessions, it was determined whether the training was sufficient for the astronaut to be cleared for space flight and a science programme on the Mir space station.

Mir systems and Soyuz TM technical training

The objective of technical training for NASA astronauts was to develop the knowledge and skills to enable them to operate onboard systems for both the Soyuz TM transport vehicle and Mir space station, so that they could complete a training session in simulators within the limits of their duties as Flight Engineer-2. This training focused on systems that had an impact on crew safety, such as life support, thermal mode control, and the motion control system.

Mir Technical training included:

Life Support System Training on the control and servicing of the LSS complex of Mir.

Thermal mode control systems Development of practical skills for vital operations in servicing and maintaining the system, as well as monitoring and control.

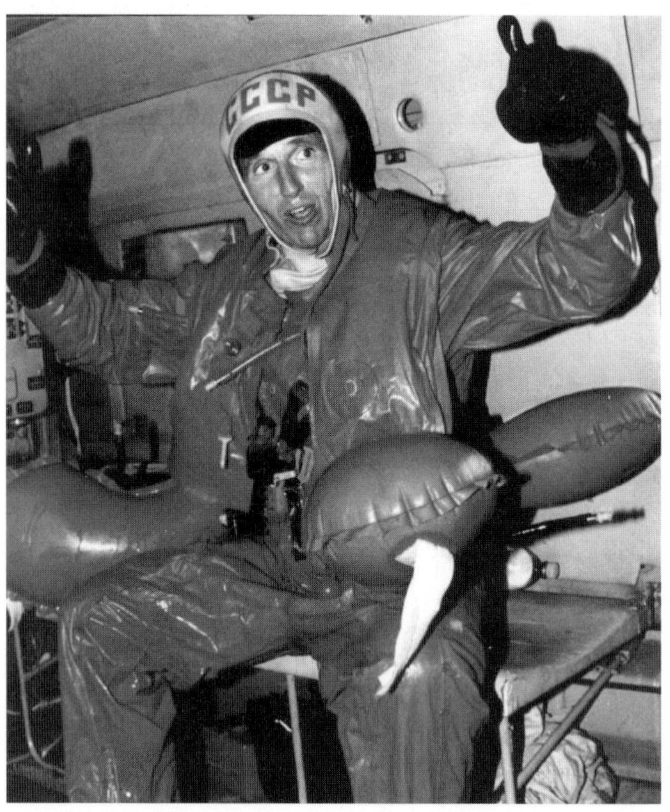

Astronaut Jerry Linenger participates in a simulated helicopter rescue during water survival training in May 1996, as part of his Mir resident crew training programme at TsPK. (Courtesy NASA)

Motion control systems Theoretical training for off-nominal situation recognition, including a series of practical courses held at RKK Energiya's control and test station for servicing and repair, replacement of units, and exchange of electrical cables.

In addition to these sub-systems, theoretical and practical courses were completed for other onboard systems.

Soyuz TM technical training took into account the astronaut's function as a cosmonaut researcher during operations for an ahead-of-schedule or emergency descent from orbit. Each astronaut received a general overview of the Soyuz TM and its onboard systems, plans for the decent from orbit, practical skills for self-help using the spacecraft's life support system, conducting radio communications with mission control in Moscow, evacuating the spacecraft after landing (splashdown), and survival training.

Training in the Soyuz TM Integrated simulator

Only Norman Thagard would reach orbit aboard a Soyuz TM transport spacecraft and its R-7 launch vehicle, so his training included sessions as part of the Mir 18 crew (and with Dunbar and the Mir 19 crew) for the launch and docking phase. Subsequently, all NASA astronauts resident on Mir were transported to and from the station aboard the Shuttle, so they only completed Soyuz training for the descent from Earth orbit (including emergency situations).

This training required each astronaut to become familiar with the design layout and onboard systems of Soyuz TM and execute emergency evaluation from Mir as part of a resident crew, following correct procedures in the event of fire, depressurisation, or exceeding specific flight rules. They had to be familiar with specific emergency flight data files, and perform several practical tasks including hatch opening/closing, operating personal protection gear (Sokol, Forel suit, etc.), operating specific valves and input commands on the right control panel of Soyuz TM, and familiarising themselves with Soyuz sub-systems such as radio communication systems, the water supply system, the waste water system, and Soyuz survival equipment.

The training included practical courses on the Soyuz TM Integrated Simulator, including training as part of a crew for integrated control of the transport vehicle during descent from orbit, which included off-nominal situations. The astronauts also studied relevant flight data file sections, flight programmes and profiles, including decent vehicle ballistics.

The competence of each astronaut was then verified by a board during a test training session on the TM integrated simulator, focusing on the descent flight programme and vehicle ballistics as part of a 'typical' training programme (nominal and off-nominal).

Table 16 NASA Astronaut Soyuz TM Simulator Training

Name of exercise	*Exercises*	*Hours*
Practical exercise training	3	6
Practical exercises in TM integrated simulator	3	12
Part crew training sessions for integrated control of Soyuz TM during descent from orbit	5	10
Crew training sessions for integrated control of Soyuz TM during descent from orbit	5	20
Class study of FDF, flight programme, and ballistics	10	20
TOTALS	26 exercises	68 hours

Astronaut training for Mir Orbital Complex simulator and system mock-ups

Flight Engineer-2 duties in the operation of onboard systems of Mir and in the execution of the science programme required all NASA astronauts assigned to resident crews to be trained on station stimulators and mock-ups, using the documentation and methods used to train Russian Mir resident crews for almost a decade.

Initially, the classes were for individual practical instruction, without the participation of the full crew, where the astronauts used the Mir simulators to

Astronaut Mike Foale participates in a fit check inside a Soyuz TM simulator at TsPK during training for a residency mission to Mir. (Courtesy NASA)

develop their skills in operating onboard systems. This helped to evaluate the individual astronauts prior to inclusion into a resident crew. The astronaut training programme for Mir crew assignment focused on the ability to 'accomplish the entire mission onboard the station and to take action in emergency and off-nominal operations.' Under guidance from an instructor, the crew would work as a team, practicing basic elements of the flight programme, including operating several onboard systems and science hardware simultaneously. This included still camera and video filming inside the simulator and radio and TV communications with a simulated mission control centre.

The next phase of training focused on organising work aboard the station. Here, the crews encountered problem situations associated with rescheduling tasks and refresher training with onboard systems, science hardware, and certain flight modes or operations (Soyuz redocking, EVA operations, preparing for and conducting Progress cargo transport vehicle remote operator docking, etc.). This phase was under the integrated control of Mir onboard systems and science hardware.

The process of Mir crew training represented a completely different philosophy of space mission training for the NASA astronauts. Since the mid-1970s, US astronaut training had focused on the Space Shuttle and Spacelab science module missions of up to 17 days, where a packed flight plan meant intense orbital operations. With assignment to Mir missions, they not only had to cope with the cultural shock of working in Russia and mastering the language, they also had to come to terms with a training regime that was the complete antithesis of Shuttle training. Now, the focus was on basic system familiarisation in a general form, because the flight plans would frequently be amended during the course of the mission.

This approach required a work pattern that encompassed total knowledge of the skills necessary to complete the mission programme, as well as finding optional solutions to both planning and organising work on the station. There was also a significant amount of crew safety training, in particular the emergency evaluation of the station following depressurisation and fire (which Linenger and Foale both had to call upon during their residencies in 1997). In addition to practical classes and training sessions in the simulators, there were classroom sessions on the Flight Data File (FDF), various flight simulations using these FDFs, technical classes updating the changing technical status of the station, studies of the function of the MCC in Moscow, and classes in the specific mission programme. As with the Soyuz training, astronaut residents were examined by a board of specialists during a session in the Mir Integrated Simulator (DON-17KS), where the crew completed a 'standard day flight programme' and tests of their mission programme. Graduating from these tests enabled the astronaut to be considered for flight assignment to the appropriate Mir resident crew.

Table 17a Practical Classes and Classes on FDF, Mir technical studies, MCC-Moscow and Mission Programme

Class	*Code*	*Class topic*	*hours*	*location*	*notes*
1	PZ-1	Development of practical skills in operating onboard consoles	2	DON-17KS	Conducted with crew
2	PZ-2	Development of practical skills in operating onboard systems	2	DON-17KS	Conducted with crew
3	PP-1	Technical status of Mir orbital systems and science hardware	2	Classroom MCC-M	
4	PP-2	Flight Data Files (FDF)	2	Classroom	Conducted with crew in preparation for session
5	PP-3	Analysis of Mir mission progress	2	Classroom TsPK	
6	PP-4	Mir Shuttle joint operations	2	Classroom TsPK	Jointly with STS crew
TOTAL		7 Classes	14 hours		

Table 17b Integrated Training Sessions

Class	*Code*	*Class topic*	*hours*	*location*	*notes*
1	Tr-1	PDS operation, experiments	6 (2 + 4)	DON-17KS	only PDS operation
2	Tr-2	PDS operation, experiments	6 (2 + 4)	DON-17KS	only PDS operation
3	Tr-3	PDS operation, experiments	6 (2 + 4)	DON-17KS	
4	Tp-4	PDS operation, experiments, fire	6 (2 + 4)	DON-17KS & TDK-7ST	as part of Mir crew
5	Tp-5	SP-ZO depressurisation	6 (2 + 4)	DON-17KS	
6	Tp-6	SP-ZO depressurisation	6 (2 + 4)	ZU-734	as part of Mir crew
7	TPS	standard flight days	10 (2 + 8)	DON-17KS	as part of Mir crew
8	ZKT	standard flight days	10 (2 + 8)	DON-17KS	as part of Mir crew
TOTAL:		8 classes	52 hours		

Astronaut training for the Mir Science programme

The preparation and training of NASA astronauts was important for successful completion of the scientific investigation programme. The preliminary American training plans were based upon information supplied on each experiment and were refined by the Russians to ensure that, while the objectives of the American experiments were met, they also stayed within Mir safety and operational guidelines. The successful implementation of the science programme also depended upon the organisation of crew system training, the availability and accuracy of science hardware training models, the time lines of the flight data file, and the development of training systems, as well as the experience and proficiency of the assigned crews and Principle Investigators (PIs), hardware construction contractors, crew trainers, flight controllers and support teams.

The organisation of science training was developed by both sides. This began with a three-week training session at NASA JSC that involved basic training in each experiment assigned to a NASA long duration mission, and familiarisation with each

Mike Foale prepares to enter a training version of the Orlan M EVA suit prior to a training session in the WET-F facility at JSC in Houston. (Courtesy NASA)

item of science hardware. The next phase of science training was accomplished by TsPK instructors at the Cosmonaut Training Centre, with the support and participation of associated organisations where appropriate. Six months prior to the planned launch, a second three-week session was held at NASA JSC, which included practical training and meetings with PIs and experiment suppliers. The final training stage was back at TsPK, using current sets of FDFs.

Documentation on experiment methods and crew training hardware was delivered to TsPK and RKK-E, together with supporting dimensional installation drawings, electrical diagrams, developmental requirements and technical descriptions, as well as FDFs and training method documentation. Most important was the active participation of the astronauts (and cosmonauts) in the assigned science investigations and experiments. By not restricting training to the work books and acquiring fundamental knowledge about the research objective, a broader understanding of the science hardware, its design and function was gained, allowing for fuller understanding of its operation once on orbit. This also made crew participation much more interesting.

Mike Foale, wearing a specially weighted Orlan M training suit, is lowered into the WET-F at JSC for an evaluation of the Russian EVA pressure garment. (Courtesy NASA)

A number of crew tasks and functions were defined during the training planning, including crew tasks in preparatory operations (such as circuitry assembly) and the execution of experiments and investigations in accordance with onboard instructions and procedures; recording experiment results using onboard recording systems and associated hardware; operation, maintenance and repair of the science hardware as required; and the storage and delivery of ground materials with the results of all experiments and investigations.

Experts from TsPK participated in blending the science programme with the development of the experimental procedures and updating the FDFs for flight from those documents used during crew training. Science training for practical and theoretical sessions at TsPK integrated the Mir simulators and mock-ups, specialised science hardware stands (duplicating operator workstations), and science hardware training models. Crews from both the US (resident astronauts, their back-ups and STS assigned crews) and Russia (Mir resident prime and back-up crews), and American and Russian instructors, participated in training sessions. Experiment suppliers, curators of the hardware and librarians involved with the FDFs also initially participated, and graded training sessions determined the crew's readiness to perform the scientific investigations.

A quantity of scientific training hardware was transferred from NASA to TsPK to facilitate the training there. Experts at TsPK developed and utilised simulator models for science experiments, simulators for crew work stations, and specialised databases, and incorporated a number of modern technologies into the training programme.

To improve training facilities, laboratories at TsPK were adopted or developed for all science fields in the experiment programme. These included:

- Training in technical experiments (k.106–3; k.107–3)
- Cosmonaut training in biotechnical and biological experiments (k.313.KMU)
- American hardware installed in k.225–2 (astrophysical and technical experiment cosmonaut training laboratory) and k.208–2 (cosmonaut training laboratory for geophysical experiments)

In addition, experiment procedures and onboard instructions had to be developed. A crew training programme was also evolved to conduct medical investigations and in the first stage of training, the cosmonauts and astronauts were instructed on how to draw human blood samples. The first familiarisation sessions were held at NASA JSC and continued at TsPK and the training materials included video, which detailed the requirements of the World Health Organisation for medical personnel regarding compliance with safety procedures when working with biological material. TsPK instructors were faced with a dilemma at this stage of training, given the lack of volunteers on which cosmonauts and astronauts could practice. In most cases, training was accomplished using the TsPK physician instructor and NASA flight surgeon as guinea pigs.

Astronaut comments on training

NASA training protocol gave operational responsibility to the crew, with hands-on control of their spacecraft and almost constant communications with ground control for immediate advice and assistance. In contrast, Russian spacecraft are generally automated, with only short communication passes of about ten minutes per ninety-minute orbit. As expeditions on Russian stations could last many months, they needed to be able to respond to both sudden and evolving contingency situations. NASA training relies on detailed simulations ('sims') of varying situations and astronauts are not given formal written examinations at the end of their training programme to qualify for flight. The Russians train on the use of equipment in simulators, but do not simulate many situations or activities, with cosmonauts spending a lot more time in classroom activities and the oral examinations at the end than the American astronauts were used to.[31]

In his oral history transcript, John Blaha recalled that courses in Russia during training were not actually on the language but on a specific element of space flight hardware, systems or operations (communications, flight control, etc.) 'in the Russian language and Russian cultural context.' He noted the difference between America, with the use of simulators, and Russia, with the use of chalk and chalk boards. In a class of only one or two students, a Russian instructor teaches a specific system in Russian using the chalk board and the student takes notes and asks questions. After the course, a team administrator sets an oral examination. This administrator is an expert or member of the design team for that item of hardware or system. The student is then graded and either passes or fails. Blaha noted that although the systems were very different, the results were the same – cosmonauts fly well on their space station; astronauts fly well on the Shuttle. The hurdle to overcome therefore, was to fly well on each other's spacecraft, which was one of the objectives of Shuttle-Mir.

Shannon Lucid noted that colleagues assumed she would go over to Russia and train with the crew she would fly in space with, but this was not the case, at least not immediately. With Blaha (her back-up, who would follow her on Mir), she sat in classrooms all day and that was all. 'We didn't interface with anybody else. Only towards the end did we do just a very few sims with the Russians, but it was very minimal. There wasn't the training with a crew like you would think.'

Dave Wolf recalled how similar Russian training was to American training in many cases, to achieve a certain type of proficiency. After mastering the language, he did not even notice he was speaking and thinking in Russian. The training was long and intense and the survival training north of the Arctic Circle was extremely serious, with temperatures more than forty degrees below zero outside for several days. This led to excellent camaraderie, learning how to respond under stress as well as live off the land. To Wolf, this training raised important side issues, such as how to get through the next hour and then the one after that during a particularly difficult task. But all of this was transferable to longer missions on ISS.

Andy Thomas recalled that training in Russia was 'a fascinating experience.' It was hard work and 'a big undertaking to do all that in a year to prepare for the flight on (Mir).'

LEARNING NEW LESSONS FROM SHUTTLE-MIR

Feedback from the training for the first Shuttle-Mir missions in Russia suggested that it would be more effective if fundamental knowledge of Russian theoretical training could be increased and the level of space flight experience and quality of NASA astronaut training was taken into greater account. It was clear that more intense study of the Russian language and its technical application was needed before technical training on Russian systems and hardware. The Phase 1 Joint Programme Report cited that, 'An optimum combination of theoretical knowledge and the independent work of NASA astronauts should be provided during the initial stages of training when the level of Russian language study is not high enough.' A limit of four hours was suggested for these theoretical classes and that for the rest of that day, the astronauts should complete 'independent work', consultation, or physical training. During the early stages of training, there should also be source material available in both Russian and English.

A unique aspect of the Shuttle-Mir missions was the shift rotations required onboard Mir to utilise the Shuttle for crew delivery and pick up, rather than the Soyuz TM. The rotation of Russian resident crews differed from that of the NASA astronauts, which meant that it was not always possible for the NASA astronaut to train as part of the crew they would be working with on orbit. It was the opinion of both the cosmonauts and astronauts that, during the phased training for long duration missions, greater attention needed to be paid to the psychological compatibly of crew members. To achieve this, a longer training period was suggested for each resident crew with which an astronaut would be an integral team member, and it was suggested that joint survival training sessions under extreme conditions could contribute to this. This was raised because most of the survival training for the Americans on Soyuz TM was for emergency descent conditions. However, several Americans could not train with both the crews they would work with in orbit, as their training times were different, which may have led to an infringement of crew safety due to the lack of familiarity with each specific Russian crew's methods.

The American system of assigning a back-up crew member was officially phased out early in the Shuttle programme (1982) when a pool of trained astronauts was available to replace any injured or ill crew member should the need arise. The flight profile of the Shuttle allowed astronauts to easily fit into new crews shortly before commitment to the final mission training phase, because of the short nature of the missions (up to ten days on average), the distinct roles of each crew member, their cross-training and the 'series of flights' nature of the missions (satellite deployment, EVA, RMS work, Spacelab missions, retrieval and repair etc.). For Mir, the Americans adopted the Russian system of always selecting prime and back-up crews from a training group to allow extended training for those assigned to fly long duration missions. Thus a back-up on one flight would normally be assigned as prime on a later flight. The longer training programme allowed for familiarisation with Russian crews and flight regimes, but was prone to disruption, as with Parazynski and Lawrence.

As the replacement of Russian cosmonauts on Mir did not coincide with the replacement of the astronauts, the joint training sessions also did not usually coincide. In one case (John Blaha), a series of flight delays and crew changes through medical disqualifications meant that he spent his entire mission in space with cosmonauts he had not trained with. The Phase 1 Report noted that, 'the result was that in some flights, the crew commander, without knowing the actual proficiency level of the astronaut, did not always trust the astronaut to perform individual FE-2 operations, even when the latter was adequately trained to do so.' Looking ahead, the report also stated, 'Joint training of all members of a specific ISS crew should be conducted as frequently as possible.' Adopting this policy would raise the effectiveness and productivity of the crew on the station, and the language problems between crew members and between flight crews and ground controllers would be significantly reduced, as would the need for interpreters in the training cycle.

In the case of science training, much of the science hardware was not forthcoming and in many cases, rather than having adequate simulators installed in the Spektr and Priroda simulator models at TsPK, the crews only had face panels, or even photos of the science hardware. It would have been much more beneficial for the science programme to be worked out earlier and information on the equipment needed for crew training confirmed with TsPK. The fact that only three to four weeks worth of science hardware training was completed in the US during the training programme and that training models were not available at TsPK both disrupted the continuity of training and deleted science classes during integrated training sessions on the Mir simulator. This not only limited proficiency training by the Americans, but also made the flight experiments unfamiliar to the Russian crews once in orbit, which made it difficult to work together on the research programme. This prompted the comment that future (ISS) science training must be planned in formal training sessions, with direct interaction between flight crews, science experts, flight controllers and trainers. Where possible, the time between final training sessions on the science hardware and its implementation in space should be reduced. On Shuttle-Mir, this time lag sometimes reached six months. It would also have been useful to update science programme training for both cosmonauts and astronauts with the results from earlier missions and thus amend the hardware for training and flight equipment accordingly. The Flight Data Files and science briefing material would also have to be updated as required. For a number of experiments, there was no Russian cosmonaut participation in the operation or collection of data, and this meant that no cosmonaut training was planned, despite the fact that the cosmonauts had to participate in almost all the experiments and science hardware repair tasks once in orbit.

In general, the Shuttle-Mir programme afforded the opportunity for both the Russian Space Agency and NASA to work with one another and with the system and specifics of training cosmonauts for space flights in Russia and the USA. This promoted further mutual cooperation and improvements to both countries' methods of space flight training, flight planning and implementation, as well as selecting and training crews for ISS assembly and operation.

INTERNATIONAL SPACE STATION TRAINING

The dream of creating a large space station in Earth orbit had featured in the long-term plans of both the Soviet and American manned space programmes for decades, and in November 1998 after years of development and discussion, the first element of ISS (the Zarya Functional Cargo Block) was launched into space by Proton rocket. Over the next two years, the station was expanded to include the Zvezda Service Module and the American Unity Node, with the facility for docking a Shuttle and Progress unmanned re-supply craft, as well as the assembly of the first of the large solar array trusses designed to power the station. By October 2000, the station could

Cosmonauts Sergey Zalyotin and Nadezhda Kuzhelnaya during a break between training sessions for the Soyuz TM 33 mission, for which they served as back-up crew

support a resident crew without needing a Shuttle docked to it, and with the delivery of the US Destiny Laboratory in February 2001, the first expanded science programme could begin. During 2001 and 2002, the station expanded further, with more solar array trusses, the Quest and Pirs airlocks, the robotic arm systems, and a host of internal equipment, experiments and supplies. This expansion was to continue during 2003 to the point where the station could support large scale science activities by an international crew, but was halted when Shuttle Columbia was lost at the start of February. While NASA worked towards returning the Shuttle to flight after the second fatal accident in its 22-year history, the Russians continued to support ISS operations with Progress re-supply flights and Soyuz ferry missions, maintaining a minimal human presence until the resumption of Shuttle flights. With the Shuttle planned to retire by around 2010 (and, possibly, the venerable Soyuz), the future of ISS over the next decade or so is to support long duration manned space flight programmes targeted for 2020–40, supported by new vehicles. This expanded programme should include the creation of a semi-permanent/permanent human presence on the Moon and the first manned flights to Mars. While the training flow for Shuttle will be decreased over the next five years, that of the ISS has continued to gain pace, and a significant percentage of ISS resident crew training, as well as for Soyuz ferry training, is focused at TsPK.

ISS training group created

To train for a long duration mission in space can take up to several years so, despite the fact that the first element of ISS was yet to launch, the first resident crew was named on 30 January 1996.[32] Originally named as Russian cosmonauts A. Solovyov (Soyuz Commander), Sergey Krikalev, (Flight Engineer) and American NASA astronaut Bill Shepherd (ISS commander), Solovyov stood down when it was clear that Shepherd would be commander once the crew docked to the station, not him. The Russians' experience was vast compared to the American. Solovyov had logged over 651 days in space from five Mir flights, as well as sixteen EVAs (82 hours). Krikalev had logged 484 days in space, including seven EVAs totalling over 36 hours and two Shuttle flights. Shepherd's eighteen days on Shuttle missions, with no EVAs, paled in comparison. Solovyov's pride took over and he stood down in October 1996, to be replaced by Yuri Gidzenko, who had 179 days (and one 2.5-hour EVA)

ISS-1 Commander Astronaut Bill Shepherd exits the forward hatch of a Soyuz DM training mock-up during winter/wilderness survival training near TsPK. He is assisted by ISS-1 FE Sergey Krikalev. (Courtesy NASA)

aboard Mir. Though the experiences of both the Russians and Americans would be of benefit, training for ISS involved working in both countries in greater depth than for Shuttle-Mir. When the crew was launched in October 2000, they had been in training for almost five years.

On 17 November 1997, a year before the first element was launched, the crews for ISS-2, 3, and 4 were identified, with crews 3 and 4 backing up the flights of 1 and 2. The Russians were drawn from a pool of cosmonauts who had been identified for ISS training in 1996.[33] With the training programme designed to be conducted at both TsPK and JSC, a support crew team was added to help with non-training issues for each resident crew and plans were being developed, as the station grew, to include training in Canada (for robotic arm operations), Europe (Columbus module) and Japan (Kibo Research Facility). The International Space Station was also creating an international space training programme, and a challenge for even the most veteran Russian cosmonauts.[34]

ISS Offices in Houston and Moscow

In order to support the expanded and prolonged ISS operations, NASA created an office at TsPK and the RSA created a similar office at JSC Houston. These offices support the training and oversee the operational (and personal) requirements of their astronauts/cosmonauts in the respective country. From 1998, NASA also appointed a senior manager (Director of Human Space Flight – Russia) based in Moscow to support NASA operations there.

The ISS-2 crew participate in winter/wilderness survival training in March 1998 near TsPK. (from left) Susan Helms, Yuri Usachev, Jim Voss. (Courtesy NASA)

Table 18 Astronaut and Cosmonaut Service Manager Roles (ISS) 1998–2005

Nasa Director of Operations Russia (ISS) TSPK 1998–2005

From	*To*	*Astronaut*
1998 Aug	1999 Apr	Terence Wilcutt
1999 Apr	1999 Aug	Joe Edwards
1999 Aug	2000 Jun	Donald Thomas
2000 Jun	2000 Dec	Scott Kelly
2000 Dec	2001 Jun	William McArthur
2001 Jun	2003 Jan	Chris Hadfield
2003 Jan	2004 Jan	Kenneth Cockrell
2004 Jan	2005 Jan	Kevin Ford
2005 Jan	date	Douglas Wheelock

RSA Director of Operations – Houston

From	*To*	*Cosmonaut*
2000 Jun	2000 Nov	Yuri Lonchakov
2000 Nov	2001 Jun	Konstantin Valkov
2001 Jun	2001 Dec	Roman Romanenko
2001 Dec	2002 Nov	Maksim Surayev
2002 Nov	2003 Aug	Aleksandr Skvortsov
2003 Aug	2004 Apr	Dimitriy Kondratyev
2004 Apr	2004	Sergey Volkov
2004	date	Konstantin Valkov

Director of Human Spaceflight – Russia 1998–2005

From	*To*	*Astronaut*
1998 Jan	2001 Aug	Mike Baker
2001 Aug	2002 Dec	Robert Cabana
2002 Dec	2005	James Newman

Table 19 ISS Resident Crew Assignments 2000–2005

NOTE: Dates are from NASA News releases but they are selected and often train for some months (or years) before the official announcement

Exp	*Named (NASA)*	*Pos.*	*Prime*	*Back-up*	*Support TsPK*
One	1996.31.01	ISS Cdr	Shepherd	Bowersox	Melvin
		Soyuz Cdr	Gidzenko*	Dezhurov	
		Flt Eng	Krikalev	Tyurin	

* Gidzenko replaced A. Solovyov from Oct 1996; Bowersox also completed support roles on this mission

Two	1997.17.11	ISS/Soyuz Cdr	Usachev	Onufriyenko	Fuglesang (ESA)
		FE-1	J.S. Voss	Walz	Robertson (deceased)
		FE-2	Helms	Bursch	
Three	1997.17.11	ISS Cdr	Culbertson**	Whitson	Nespoli (ESA)
		Soyuz Cdr	Dezhurov	Korzun	Creamer
		FE-1	Tyurin	Treshchev	

** Culbertson replaced previously named Bowersox

Four	1997.17.11	ISS/Soyuz Cdr	Onufriyenko	Malenchenko	C. Anderson
		FE-1	Walz	S. Robinson	Wheelock
		FE-2	Bursch	Fincke	
Five	2001.23.03	ISS/Soyuz Cdr	Korzun	Kaleri	Wilmore
		FE-1	Treshchev	Kondratyev	Caldwell
		NASA Sc Officer	Whitson	S. Kelly	
Six	2001.23.03	ISS Cdr	Bowersox	Noriega	Chamitoff
		NASA Sc Officer	D. Thomas***	Pettit***	Nyberg
		FE	Budarin	Sharipov	

*** Thomas was replaced by Pettit and Pettit by Fincke

Also on 2001.23.03, two further three-person crews were named: (Expedition 7) Malenchenko, Lu and Kaleri and (Expedition 8) Foale, Tokarev, McArthur. The loss of Columbia on 1 February 2003 during the ISS-6 residency changed future crew assignments to a two-person core crew from ISS-7

Seven	2003.31.03	ISS/Soyuz Cdr	Malenchenko	Krikalev	Woodward
		NASA Sc. Officer	Lu	S. Volkov	Boe
Eight	2003.25.07	ISS Cdr/Sc Officer	Foale	McArthur	Patrick
		Soyuz Cdr/FE	Kaleri	Tokarev	

On 2003.21.11, the original Expedition Nine crew was named as Tokarev and McArthur, but this was changed early in 2004

Nine	2004.06.02	ISS/Soyuz Cdr	Padalka	Sharipov	Virts
		Sc. Officer/FE	Fincke	Chiao	
Ten	2004.06.02	ISS Cdr/Sc. Officer	Chiao	McArthur	Drew
		Soyuz Cdr/FE	Sharipov	Tokarev	Stott
Eleven	2004.23.11	ISS/Soyuz Cdr	Krikalev	Tyurin	R. Romanenko
		Sc Officer/FE	Phillips	Tani	S. Volkov
					Coleman
					Feustal

Probable crewing 2005–2006 (to be confirmed)

Twelve		ISS Cdr/Sc. Officer	McArthur	J. Williams	Lucid
		Soyuz CDR/FE	Tokarev	Lazutkin	Kopra

Planned return to three person resident crews (dependent upon Shuttle Return-to-Flight)

Thirteen	ISS/Soyuz Cdr	Vinogradov	Yurchikhin
	Sci. Officer	Tani	Grunsfeld
	FE-2	Kondratyev	Kotov
Fourteen	ISS Cdr/Sc. Officer	J. Williams	Lopez-Alegria
	Soyuz Cdr/FE	Lazutkin	Treshchev
	FE-2	Anderson	Reisman

ISS general training flow

A general ISS crew training flow was created in 1998 as a foundation for the development of resident crew training as part of the ISS programme. The training was based on the experiences of the Salyut and Mir programmes and was amended to incorporate the new hardware, systems and foreign elements as required. This training flow features three main training blocks:

Basic This element is partner-specific, where each partner retains responsibility for training its own candidates in generic space flight and science knowledge, flying, scuba diving, survival training, and necessary language training (the ISS common language is English – resulting in English lessons taught to Russian cosmonauts, although most foreign resident crew members also learn Russian) to reach the minimum requirement for basic training set by the international partners. NASA Ascan training can take one to two years and Russian basic cosmonaut training can also take about two years to complete.

In Orbiter Processing Facility 1 at KSC, members of the STS-88 crew examine equipment they will use on their upcoming flight as part of the Crew Equipment Interface Test held in October 1998. (from left) Jerry Ross, Sergey Krikalev and Jim Newman. (Courtesy NASA)

STS-88 MS Sergey Krikalev (l) and Jim Newman sit inside Endeavour's middeck during the TCDT, which included mission familiarisation activities, emergency egress training and a simulated launch countdown. (Courtesy NASA)

Advanced This provides generic ISS knowledge and skills which are not mission related, but which are required for understanding and mastering onboard systems and operations, irrespective of which mission is flown. These include nominal and malfunction system training, emergencies, crew systems and facilities. This programme takes about one to two years to complete and includes a number of support roles prior to being assigned.

Increment Specific A one-and-a-half-year training programme designed to prepare an assigned crew for a specific mission to ISS. This includes all the skills and knowledge required to complete the flight objectives, science programmes, and EVAs, and to refresh malfunction, systems and survival training begun under the advanced phase. Here, training facilities at TsPK can be used (for the Russian segment elements) in addition to training facilities around the world. The added challenge is the increased time away from home and the amount of travel across the time zones that is now required as part of long duration cosmonaut training for international ISS mission. According to a recent interview with Dutch ESA astronaut, André Kuipers[35], there is currently talk of forming a non-Russian, non-American group of astronauts (Europe, Canada and Japan) to train for long duration missions. This would give them a chance to do as much training as possible, as soon as possible, so that they can return home more frequently once specific mission training began. Apparently, the Americans had expressed dismay about the fact that training in Russia meant they weren't home half the time, to which the Europeans reacted by saying that they were in Russia half the time and Houston for the rest. In other words, they weren't home at all!

ISS Advanced training at TsPK began in 1997/1998 (original contract effective from 1 August 1994) with about 800 contracted hours per year for a series of Soyuz lectures and practical training, ISS Russian segment lectures and practical training, a physical training programme, EVA lectures and Neutral Buoyancy Simulator (NBS) training, and ongoing Russian language training.

In detail, this 'Advanced' training for each assigned crew included:

ISS Russian Segment lectures	36 hours
ISS Russian Segment practical training	14 hours
EVA training	28 hours
EVA lectures	16 hours
Scuba practical training	8 hours
Orlan training	4 hours
Russian/English language training	120 hours
Soyuz lectures	125 hours
Soyuz training	33 hours
Total approximately	356 hours

ISS Expedition training – an insight

The most practical application of training time and funding is to assign a crew member in a support role, progressing to a back-up and then a prime role, and recycling this person to a subsequent crew further down the manifest. This system helps to progress training smoothly and efficiently. American back-up and support crew use was discontinued from 1982 for most Shuttle missions, but was reinstated for NASA Mir and ISS crew assignments due to the complexity and length of training. The Russians have always adopted a primary / back-up / 2nd back-up training group.

Crew Support Role If a crew member is assigned to an American crew support role, it only involves a couple of short visits (up to two weeks) to TsPK for more orientation classes, with no formal training sessions. The role mainly focuses on 'issues' that occur during launch preparations.

ISS Training Flow Americans assigned to the ISS training flow usually complete four or five one-month visits to TsPK each year, concluding with assignment to a formal back-up crew (after about eighteen months) and leading, after a further year or so, to a flight crew assignment and a resident crew trip to ISS. A detailed insight

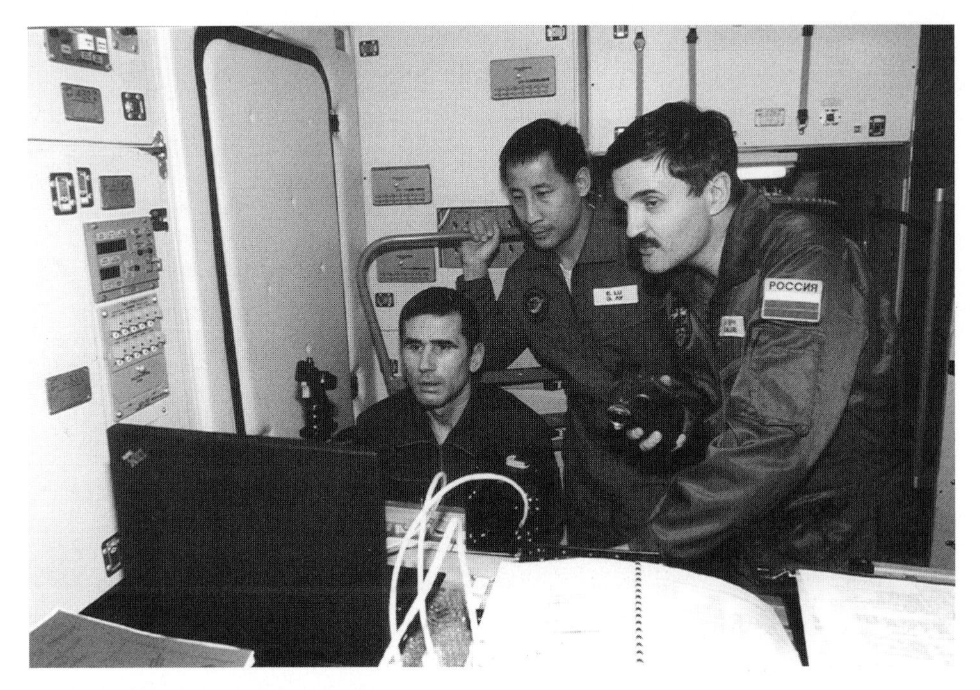

Yuri Malenchenko, Ed Lu and Aleksandr Kaleri, the original ISS-7 crew, participate in crew training in the Zvezda simulator at TsPK

into NASA astronaut assignments at TsPK in this role has been provided by Clayton Anderson, in a series of postings on the NASA Human Spaceflight web pages.[36]

Anderson detailed the initial academic classes as daily sessions divided into four periods (1 hour 50 minutes each) between 9 am and 6 pm, with a lunch break from 12.50–2 pm. These are mostly theoretical, but there is some practical training in the ISS modules (SM and FGB) where some of the end of training exams are held, with the candidate showing the examiners what they know (and equally, what they do not know). The weekly programme also includes two four-hour Russian language training sessions.

Survival training is completed over two days, after two days of preparation, as part of a three-person Soyuz crew (the Soyuz 'commander' is normally a veteran cosmonaut to pass on their experience to the rookies). Day one of preparation at TsPK is an eight-hour session about Soyuz emergency kit contents, how to construct shelters, survival clothing and either winter or summer survival techniques. The next day is spent donning and doffing Sokol pressure garments (also at TsPK), climbing into a previously flown capsule now used for training, and getting changed into survival gear to practice exits in a training building, as they would 'in the field'. The three-day field exercise begins at lunch time on Day 1 and is completed by lunch time on Day 3. It is conducted in a heavily wooded area eight miles from TsPK, during which the 'crew' constructs a lean-to from local trees (which also supply the firewood) and parachute cloth. A signal fire is prepared, but not lit until a rescue

The original STS-101 crew are joined by their training team near the Service module/ Functional Cargo Block mock-up at TsPK. (rear from left) trainer, Jeff Williams, Ed Lu, Jim Halsell, Boris Morukov and three trainers. (front from left) trainer Mary Ellen Weber, Scott Horowitz, Yuri Malenchenko and trainer. (Courtesy NASA)

team is clearly sighted, and a shelter fire is used for warmth while eating from the survival rations. A day's rations consists of freeze dried yoghurt, freeze dried chocolate, a fig bar, cookies and two tea bags with sugar and lemon. At night, two crew members sleep and the third takes watch to keep the fire going and watch out for rescue teams. After two days of this, the crew is 'rescued', sometimes with one of them feigning an injury or broken limb for their team mates to construct a stretcher and drag their injured colleague to the rescue chopper in a nearby clearing.

The crew receives training on the Russian systems of ISS (Zarya, Zvezda, Progress, Soyuz, Pirs, Orlan and Sokol suits), then the Americans go home after four weeks to receive training on US systems (Unity, Destiny, Quest, power systems, EMU systems). Some sessions include training on three or four different systems at the same time. The Russian system continues to be theory in a classroom, followed by practical sessions on mock-ups of actual hardware, followed by study of text books and documentation provided to support the training. The day before the exams, a review session is held, during which the student can ask any question he or she likes in order to clarify certain issues or queries. The exams, which are oral or practical in nature, usually last one hour. Examiners ask various questions and the student has to demonstrate their knowledge or experience on a set system or function, conducted in a simulator, or on the station via laptops. At the end of the session, the examiners grade the student, with an unofficial scale of 1 for a 'fail' up to 5 for 'excellent pass'. According to Anderson, grade 5 is a 'pretty normal occurrence' for the astronauts, having already gone through the TsPK training.

STS-106 MS Ed Lu and Yuri Malenchenko conduct an EVA simulation on a training mock-up of the SM in the Hydrolaboratory at TsPK in August 1999. (Courtesy NASA)

Physical training is a must, and the Americans use a personal gym set up in the American Houses basement. This includes free weights, weight machines, and typical aerobic equipment of various types, including 'astronaut specials' such as a cycle that you pedal with your hands. They also run a well-beaten track around Star City and in nearby woods, which is about a 6 km round trip. The runners are often accompanied by a former TsPK trainer, who carries an old fishing pole to fend off any stray dogs in the woods!

Upon completion of the basic and advanced training, assignment to a back-up crew is a step closer to assignment to a specific flight crew although, as with any crew assignment, nothing is certain until the vehicle leaves the pad.

TAXIS AND TOURISTS

Roughly every six months the Soyuz spacecraft at ISS needs to be exchanged for a fresher vehicle. If a resident crew rotation is not scheduled, a visiting Soyuz crew can be launched to take the next Soyuz to the Station and bring home the older craft. With limited numbers of experienced cosmonauts available and the requirement for a Soyuz commander to be flight experienced, the Russians have trained European astronauts as Soyuz Return Commanders, qualified in undocking, separation, de-orbit and landing activities. This has amended the traditional Russian cosmonaut roles and responsibilities and reflects the international flavour of the current

The ISS-6 crew (Ken Bowersox, Don Pettit and Nikolay Budarin) receive congratulations from Maj.-General Pyotr Klimuk on completion of their training and passing their exams

programme (and the need to obtain funds for the national programme by bartering seats). One of the first ESA astronauts to complete this type of training was Thomas Reiter.

The opportunity also arose to 'market' the third seat on Soyuz to those able and willing to pay for it (currently US$20 million), continuing the plan that originated towards the end of the Mir programme. The general rule in the Soviet / Russian programme since the 1977 Soyuz 25-Salyut 6 docking failure was to have at least one member of the crew flight experienced. The flight profile of a taxi mission to and from the ISS allowed for the flight of two Russian rookies for flight experience, or one rookie and an ESA or Canadian astronaut if no 'fare paying' passengers could be found. Of course, selling seats to ESA or Canada was an option when Expedition crews were being rotated as part of a Shuttle assembly mission. Offering seats to 'millionaire' individuals or organisations caused resentment in many unflown Russian cosmonauts, who lost their only chance of flying in space, just as they did during the Interkosmos missions in the late 1970s and early 1980s.

When the inadequately prepared, but Russian trained Dennis Tito flew on a Soyuz visiting mission to ISS, he was prevented from accessing the American elements of the station because NASA felt he was not sufficiently skilled, especially in NASA safety procedures. As a result, the ISS partners developed a series of requirements for any 'guest' cosmonaut, who now have to pass safety and familiarisation training at NASA JSC before they can fly in space.[37] The Russians

The ISS-4 crew lay flowers at the Korolyov statue at RKK Energiya during a post-flight ceremony. (from left) Dan Bursch, Yuri Onufriyenko, and Carl Walz

demanded a minimum six-month training programme for any tourist candidate, but paying several million dollars for a short training programme and a flight as a passenger on a Russian mission with nothing to do during the flight was fine, as long as it did not involve the American elements or astronauts. Of course, on ISS this was impossible, so rules had to be agreed, and for the only other millionaire passenger to date (South African Mark Shuttleworth), the flight had to include a programme of experiments or investigations that made the investment worthwhile. For European astronauts (ESA or a national space agency), qualification as Flight Engineer and Soyuz Return Commander was combined with an extensive science programme to be carried out during a few days of exchange onboard the station, which would not interrupt the flow of resident operations.

Table 20 ISS Soyuz 'Taxi flights' and Visiting Missions (2000–2005)

ISS Soyuz	*Soyuz mission up/down*	*Science Mission*	*Commander Back-up*	*Flight Engineer Back-up*	*3rd Seat Back-up*
S1	TM 31/(STS-102)	ISS-1	Gidzenko	Krikalev	Shepherd (NASA FE-2)
			Dezhurov	Tyurin	Bowersox (NASA)
S2	TM 32/TM 31	–	Musabayev	Baturin	Tito (US SFP)
			Afanasyev	Kozeyev	–
S3	TM 33/TM 32	Andromeda	Afanasyev	C. Haigneré (CNES)	Kozeyev
			Zalyotin	Kuzhelnaya	–
S4	TM 34/TM 33		Gidzenko	Vittori (ESA)	Shuttleworth (S.A SFP)
			Padalka	Kononenko	–
S5	TMA 1/TM 34	Odissea	Zalyotin	De Winne (ESA)	Lonchakov
			Lonchakov	Lazutkin	–

The loss of Columbia in February 2003 changed future crew assignments to a two-person core crew and delivery to and from the station via Soyuz TMA craft, restricting the seats available for ESA or Space Flight Participants. TMA 1 brought back the ISS-6 resident crew of Bowersox, D. Thomas and Budarin.

ISS Soyuz	*Soyuz mission up/down*	*Science Mission*	*Commander Back-up*	*Flight Engineer Back-up*	*3rd Seat Back-up*
S6	TMA 2/TMA 2	ISS-7	Malenchenko	Lu (NASA)	–
			Kaleri	Foale (NASA)	–
S7	TMA 3/TMA 3	ISS-8/ Cervantes	Kaleri	Foale (NASA)	Duque (ESA)
			Tokarev	McArthur (NASA)	Kuipers (ESA)
			Duque returned with the ISS-7 crew aboard TMA 2		
S8	TMA 4/TMA 4	ISS-9/ DELTA	Padalka	Fincke (NASA)	Kuipers (ESA)
			Sharipov	Chiao (NASA)	Thiele (ESA)
			Kuipers returned with the ISS-8 crew aboard TMA 3		
S9	TMA 5/TMA 5	ISS-10	Sharipov	Chiao (NASA)	Shargin
			Tokarev	McArthur(NASA)	–
			Shargin returned with the ISS-9 crew aboard TMA 4		
S10	TMA 6/TMA 6	ISS-11/ Eneide	Krikalev	Phillip (NASA)	Vittori (ESA)
			Tyurin	Tani (NASA)	Thirsk (CSA)
			Vittori returned with the ISS-10 crew aboard TMA 5		

Paying for the pleasure – Space Flight Tourist Cosmonauts

The chance to buy a seat on a Soyuz has been a contentious issue and has resulted in only two flights to date, although many have been rumoured, planned and attempted until the finances fail to materialise. In recent years, there has been the opportunity to sample elements of 'cosmonaut training' via such organisations as Space Adventures where, for a fee, tourists can sample a flight on a MiG jet to the edge of space, weightlessness in parabolic flight, suiting up in an Orlan EVA suit, a ride on the centrifuge or tours of the space facilities. This is about as close to space as most people can get, but for a few million dollars more the ultimate trip is possible.

Roberto Vittori exiting the Soyuz TMA simulator. He is an Italian ESA astronaut who has flown two missions to ISS

The second entrepreneur to experience a Soyuz flight to the ISS was South African Mark Shuttleworth in April 2002, the year after Dennis Tito. Upon hearing that an opportunity to fly in space was open to him, Shuttleworth's first hurdle was the mountain of bureaucracy he had to go through to convince the Russian space authorities that he was genuine and committed in his application to fly. Shuttleworth reasoned that volunteering for a medical experiment programme and conducting an extensive scientific research programme offered a different reason for his flight into space, other than just being able to afford it. Working with medical specialists early in training also helped him prepare for the rigours of space flight and get the most from a short, abbreviated training programme. Negotiations were part of the 'training' in order to work with the Russians, and it was a challenge for him to get as much 'training' out of them as possible. Shuttleworth has said that a third of his training at TsPK was on Soyuz systems, a third on ISS systems and a third on survival training.

During water survival training on the Black Sea, he was told that recovery took around thirty minutes and was asked how long he could hold his breath – just in case. Shuttleworth was assured that, even after a space flight, the Soyuz would not sink and that a water landing was unlikely. He was also told that the training model he was to practice in also 'probably would not sink.' For one test, it was planned to induce violent motion sickness in the crew inside the cramped, hot Soyuz, in a strong swell, but that day the water was calm and a team of divers had to sway the spacecraft in the water to achieve the desired effect. Shuttleworth commented that the best way to prepare for a space flight was by daily conditioning on the vestibular chair, not the most popular cosmonaut training device, but which apparently worked very well. He would certainly use this method again if he took a second space flight. He noted that the training for a Soyuz flight was excellent and well thought out practice for preparing crews for flights into space, but it was still nothing like achieving the real thing. As cosmonauts have often stated, the best training for space flight is in space itself.[38]

There have been two or three others who have undergone 'tourist' training, but did not progress to a flight:

Mark Shuttleworth in the centrifuge at IMBP as part of his medical examination prior to his selection as a tourist cosmonaut. (Courtesy Mark Shuttleworth)

- Lance Bass – an American pop singer, who completed some initial sessions but failed to generate the required funding to pay for the mission.
- Lori Garver – a former NASA manager who completed a lot of PR work but no training.
- Greg Olsen – the American industrialist, who failed a medical test but did complete some ISS training. However, in July 2005, he resumed training, for a flight in October 2005.

The US$ 20 million fee includes half a million for the training at TsPK. The biggest cost is for the launch, which goes to Energiya.

Future cosmonaut training for ISS

With the gradual winding down of the Shuttle programme, the opportunity for flights to ISS will focus on Soyuz and its planned replacement, currently known as Klipper (although this programme remains funded only by Energiya, not the Russian government, and it is unclear whether there has been any cosmonaut involvement as yet).

NASA has said that its work on ISS will be completed by 2016, and it is possible that operational activities on board ISS will be handled by Russia and European nations (as well as Japan), while American astronauts focus on certifying the new Crew Exploration Vehicle (CEV) that will lead to a return to the Moon and provide baseline data for possible flights to Mars. Presumably, the opportunity for crew seats to ISS will be expanded for cosmonauts to draw upon the decades of experience on Salyut and Mir space stations. Whether the facilities at TsPK will be expanded to incorporate training equipment and simulators that will allow cosmonauts to fly on the CEV is at this time uncertain. What is clear is that the training of Russian cosmonauts for ISS is already encompassing training facilities around the world.

An indication of future operations was revealed in reports of continuing difficulties in reaching agreements over future crewing of ISS. On 3 May 2005, Aleksey Krasnov, head of the RSA, indicated that the Russians would launch only Russian cosmonauts to the ISS on Soyuz, and once the Shuttle resumes flight, the American resident crew members would be launched on that and not Soyuz (though Soyuz training for emergency landing would still have to be completed). From 2006, crewing will focus on Russian cosmonauts working on expanding the Russian segment. However, the Americans desire to launch one American on *every* Soyuz, and will agree to deliver a Russian resident crew member on the Shuttle. Comments from the Europeans, Japanese and Canadians were not reported at the time.[39]

REFERENCES

1 D.J. Shayler: The proposed USSR Salyut and US Shuttle docking mission c.1981, *Journal of the British Interplanetary Society*, **44**, no. 11, 1991

2 US-Russian Cooperation in Space, Office of Technical Assessment, Congress of the United States, OTA-ISS-618, April 1995

3 Phase 1 Programme Joint Report, George C. Nield and Pavel Mikailovich Vorobiev Editors, NASA SP-1999–6108 (in English), January 1999, Section 7 – Crew Training
4 'Implementing the Agreement between the National Aeronautics and Space Administration of the United States of America and the Russian Space Agency of the Russian Federation on Human Spaceflight Cooperation', (undated) copy in the STS-60 Mission Archive File, Charles F. Bolden Collection, NASA JSC History Collection, University of Clear Lake, Houston, Texas; photocopy in AIS Archives
5 Russian Space Agency (RSA) sponsored Mission Specialist Standards of Conduct Agreement, (dated 11 October 1992); original copy in the STS-60 Mission Archive File, Charles F. Bolden Collection, NASA JSC History Collection, University of Clear Lake, Houston, Texas; photocopy in AIS Archives
6 Cosmonauts tour KSC and dream of flying on the Space Shuttle, *Spaceport News*, NASA KSC Space Centre, 28 August 1992, p 4
7 Cosmonauts selected to fly on a Space Shuttle, *NASA News* 92–166, 6 October 1992
8 Cosmonauts begin training at JSC, Kelly Humphries, JSC Space News Round Up, **31**, No 44, p 113, November 1992
9 STS-60 Russian MS Basic Training Status, 4 January 1993, DG6/CAPPS, NASA JSC, presentation by Dennis Beckman, DT48/STS-60 Flight Section Head; original copy in the STS-60 Mission Archive File, Charles F. Bolden Collection, NASA JSC History Collection, University of Clear Lake, Houston, Texas; photocopy in AIS Archives
10 NASA memo DT48–93–121, dated 3 December 1993, Subject: Lessons Learnt from Training the Russians for STS-60, from Dennis D. Beckman, DT48/STS-60 Flight Section Head; original copy in the STS-60 Mission Archive File, Charles F. Bolden Collection, NASA JSC History Collection, University of Clear Lake, Houston, Texas; photocopy in AIS Archives
11 For a comparison of a typical Shuttle crew's training hours *after* completing Ascan training, see Women in Space: Following Valentina, by David J. Shayler and Ian Moule, p 210, Springer-Praxis 2005
12 NASA News, 93–061, 3 August 1993
13 STS-60 Press Kit, February 1994, p 15–16; also, Crew Task Assignments, STS-60 Query Book, PAO Archives, NASA JSC
14 NASA News 93–070, 8 September 1993
15 STS-63 Press Kit, February 1995; also, Crew Task Assignments, STS-63 Query Book, PAO Archives, NASA JSC
16 American Flights to Mir (Space Shuttle), David J. Shayler, in The History of Mir 1986–2000, pp 71–85, Editor Rex Hall, British Interplanetary Society, 2000
17 NASA News 94–039, 3 June 1994
18 NASA News 96–171, 22 August 1996
19 Star-Crossed Orbits, p 160, James Oberg, McGraw Hill 2002
20 STS-84 Crew Assignments, Query Book, PAO archives, NASA JSC; Women In Space: Following Valentina, already cited, pp 278–279
21 NASA News 96–18, 30 January 1996
22 STS-86 Crew Assignments (dated 11 August 1997), Query Book, PAO Archives, NASA JSC
23 NASA News 97–33, 4 March 1997
24 *Novosti Kosmonavtiki*, 28 July-10 August 1997
25 Ref 19, p 162
26 STS-89 Crew Assignments, Query book, PAO Archives, NASA JSC
27 STS-91 Crew Assignments, Query book, PAO Archives, NASA JSC

28 Mission Data for All Manned Missions, pp 33–37, in History of Mir 1986–2000, already cited
29 Women in Space: Following Valentina, already cited, p 320–322
30 Phase One Overview, Tom Holloway, Programme Director presentation handout, 14 December 1994
31 Shuttle-Mir history website, http://space flight.nasa.gov/history/shuttle-mir/history NASA Human Spaceflight website with related links
32 NASA News (JSC) 96–18, 30 January 1996
33 NASA News (JSC) 97–269, 17 November 1997
34 The International Space Station, From Imagination to Reality, Editor Rex Hall, British Interplanetary Society, 2002
35 Telephone interview with André Kuipers by Bert Vis, May 2005
36 ISS Expedition Journal – Training, by Astronaut Clayton Anderson, http://space flight.nasa.gov/station/crew/andersonjournals/index.html
37 Principles Regarding Processes and Criteria for Selection, Assignment, Training and Certification of ISS (Expedition and visiting) Crew members, ISS Multi-lateral Crew Operational Panel, Revision A, November 2001
38 Shuttleworth's Space Experience, by David J. Shayler, an account of the L.J. Carter Memorial Lecture 2004, BIS/RAeS London, 24 March 2004, Spaceflight **46**, No 6, June 2004, pp 261–263, British Interplanetary Society; interview with Mark Shuttleworth by D. Shayler, London, 28 April 2004
39 RIA News Agency, Moscow 3 May 2005

Zvyozdnyy Gorodok – the town

The Central Committee of the Communist Party agreed the establishment of a cosmonaut training centre on 11 January 1960. This was soon followed by a decision by the Head of the Air Force that the first commander of the training centre would be Colonel Yevgeniy Karpov and that he would have a staff of 250 specialists. The training centre would be a military unit, with the responsibility to house the staff and their families. To that end, the town of Zvyozdnyy Gorodok was constructed.

The monument on the main road to Star City. This is about a kilometre from the main entrance

An Aerial view of Star City circa 1970. On the left are Dom 2 and 4 and the smaller block 5. At the end of the main pedestrian road is the House of Cosmonauts. Also visible are blocks 10 and 11, in the foreground right. The old Orbita hotel is the small block at right angles to these two blocks

STAR CITY IS BORN

Zvyozdnyy Gorodok (or Star City) is a military (or closed) town, which means it is completely surrounded by a perimeter fence with guarded gates for cars and pedestrians. It has no civilian mayor, and it is the responsibility of the commander of the training centre to look after the town and all its inhabitants. For many years, it was a secret location that was not shown on any maps, nor signposted on the road system. The soldiers, many of whom are conscripts, guard the gates, clear up the snow, and do general maintenance work around the town. There is an armed officer and guards operating at night and the police also have a station in the town. A pass is required to enter the town, either on foot or by vehicle, and this can only be obtained from a current resident or on an official visit. Security is tight and has increased in recent years.

The decision about where the training centre should be located was taken at the highest level within the Soviet political system. A number of factors were taken into account, according to officials who were around in 1960. It had to be close to Moscow, given that all the major establishments involved in the development of the human space effort were based there and the initial training of the 1960 selection was

An aerial view circa 1970. In the foreground are Dom 2, 4 and the smaller block 5. In the bottom right of the photo, the site of the Gagarin memorial is being cleared but the statue has not yet been constructed. The photo also shows the school, the shop complex and the House of Cosmonauts

taking place in Moscow at various medical and military establishments. It also had to be close to an airfield for access to Baykonur and other parts of the Soviet Union, and near a railway link with access to central Moscow. But it had to be off the beaten track and not in the public gaze, on a green field site that had the capacity for development. The decision on the site was made in the summer of 1960, when the 'Green Village' site was identified. It met many of the criteria laid down by the Air Force, particularly that it was very remote and anonymous.

The site chosen was close to the military air base at Chaklovskiy and was already in use by the military, although it only had a few buildings located on it. The site was close to the single railway line from Moscow to the town of Monino, which was the site of a leading Air Force academy and another military air base. It was a healthy site, with lakes and a marsh, and with plenty of opportunity to develop as demand directed.

Responsibility for construction of the town and the centre was given to one of Karpov's deputies, E. Cherkasov, a deputy director for administration. The construction of Zvyozdnyy Gorodok was undertaken by engineers and construction troops and the cosmonauts and senior managers came to live in nearby Chaklovskiy

The school, which was constructed in 1967, is named after Vladimir Komarov

in late 1961 while they waited for their homes and the centre to be constructed. They commuted to Moscow for most of their training sessions.

The first major tower block, Dom 2 (Russian for house), was ready for occupation by cosmonauts of the 1960, 1962 and 1963 selections on 6 March 1966. It also became the home for some senior figures in the centre's administration. The first housing blocks were completed in 1964 (Dom 10) and 1965 (Dom 11) and the cosmonauts moved in immediately. Plans to build three more similar blocks were dropped, as it was felt this would affect the country feel of the town.

The school opened on 3 September 1965, the beginning of the Soviet academic year, with a very small number of pupils. The teachers, some of whom were the wives of cosmonauts, were all living in Star City. On 15 November, the railway station which would serve Zvyozdnyy Gorodok was opened by cosmonaut Pavel Popovich with a big celebration. The station is called 'Tsiolkovskiy', named after the father of Soviet cosmonautics.

The pendant block, Dom 4, was occupied by cosmonauts in 1967. By Soviet standards, the apartment blocks were very large and luxurious, with their own concierge system. Linking the two buildings was a meeting hall and reception centre, which is no longer used for this type of activity. The blocks today are in need of some repair, but are still largely occupied by retired cosmonauts and officials, although a number of the flats have been leased to NASA and ESA for their astronauts and officials. The blocks are twelve storeys high and neither has a Flat 13, reflecting a Soviet superstition. The decision to build tower blocks might have been influenced by a statement from a book on the life of Gagarin: 'When Zvyozdnyy was being built, a discussion arose as to whether to build separate cottages for the cosmonauts or a big block of flats in which they would all live. Gagarin argued in no uncertain

The Gagarin memorial was constructed in 1971. The picture shows blocks Dom 2 and 4 where the cosmonauts live. The trees in the background are young, but had matured by the time the image on page 322 was taken

terms for the block of flats: 'Why cottages? Let's all live together. It will be all the merrier'.' It is fair to say that many of the decisions made about the centre and the town have been attributed to Gagarin. It is likely that he did have an influence, but is not clear exactly how much so.

On 26 July 1967, the school was renamed after cosmonaut Vladimir Komarov, who had been killed a few months earlier during the mission of Soyuz 1.

The House of Cosmonauts and the Museum

Construction of a cultural centre started on 6 October 1966 and on 14 October 1967, the House of Culture (called Dom Kosmonavtov, or House of Cosmonauts) was opened. The first major event was held on 6 November, when they marked the 50th Anniversary of the October Revolution. This centre was and is a major building in the cultural life of the town. It has a theatre which doubles as a cinema and can seat about 600 people. All major events are held here, including the welcome back of crews, events to mark Cosmonautics Day, and major mission anniversaries. It also has a function room, called the 'Hall of Mirrors', which has a number of different roles. This is where couples are married, where officers who died are laid out for the regiment and friends to pay their respects, it acts as a dance hall for classes and discos, and it is also the venue for banquets after major events in the theatre. Another room, on the second floor, is used as an aerobic studio and the building also

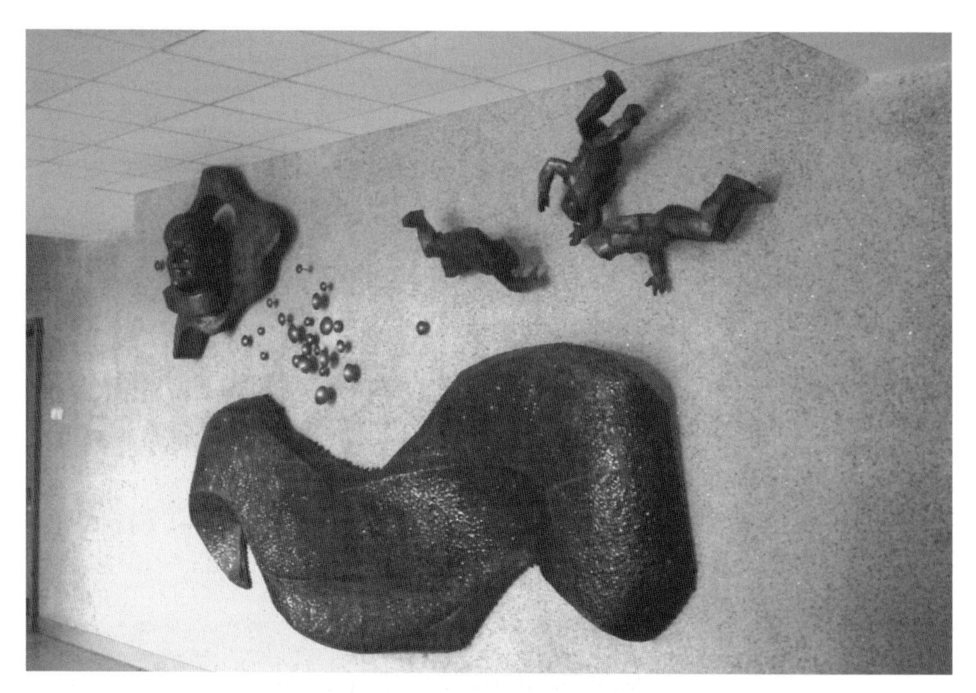

The Memorial to Flight in the lobby of the House of Cosmonauts

The House of Cosmonauts which houses the museum, theatre and reception hall. The memorial to Leonov's flight is in the centre of the photo. This building is where all major celebrations occur

houses a library, where pupils do homework (there is a very large collection of space books for loan), a cafeteria which is open to the public but also holds wedding parties and funeral events, a souvenir shop (which sells china, glass and small souvenirs of the training centre such as badges, posters, clocks and calendars), a small number of shops selling items such as films and drinks, and an administration office for the building, whose occupant is appointed by the commander of the training centre.

The museum of the cosmonaut centre is located in the House of Cosmonauts. It formally opened on 12 April 1968 and has expanded over the years, so that it now occupies four rooms on two floors in the building. The collection was started in 1961 by staff in the political department of the centre. According to the training centre website, the idea for a museum came on 24 June 1961 from Gagarin, who said that an idea had crossed his mind to create a museum in Star City. 'Let's create our own museum. It will be a kind of report on our work and actually, we need a museum for ourselves, for history for our followers.' He presented the first exhibit, which he had obtained while he was visiting Prague soon after his mission. The museum occupied its initial space in the House of Cosmonauts on 6 November 1967 and the first room is now dedicated to the origins of Soviet cosmonautics. The second room is a memorial room to Yuri Gagarin and contains his uniform and medals, which were donated by his wife. The third room has exhibits on international space missions and long duration orbital stations, describing the flights under the Interkosmos which opened on 17 September 1969. This is Gagarin's original office, laid out as it

Gagarin's memorial office, which is part of the Museum in the House of Cosmonauts. It was moved to this location in 1969 from the training centre.

This is a view of the interior of the Museum. It occupies a number of rooms on the ground and first floor of the House of Cosmonauts. The uniforms of Gagarin and Belyayev are in the glass cases on the right

programme. The fourth room is the Gagarin memorial working study, or cabinet, was on the day of his death on 27 March 1968. It has his desk, letters, phones, cupboard and coat hanging on a hook and the wall clock is stopped at the time of his death. This office is where all crews and their back-ups go to sign the memorial book before going to Baykonur for their own launch.

The museum has an extensive collection of personal mementos of the Air Force cosmonaut team and a number of unique artefacts, including uniforms, documents and flight suits donated by the families of cosmonauts. It also has a flown Soyuz capsule, space suits and other hardware and sends out exhibitions to other cities, both in Russia and abroad. The museum has its own director and staff, with the director also appointed by the commander of the training centre. There were plans to build a large new museum close to the training centre in the mid-1970s, and a site was identified and steel purchased, but the money ran out and the site lies derelict. The current museum needs more space, but lacks the resources for such expansion. All important visitors to the training centre are brought to see the history of human space flight and the museum has won some awards for its work.

Close to the House of Cosmonauts, there is a small number of shops, a bank and a post office which opened in 1968. They are currently being refurbished, with a new private enterprise café and ice cream parlour being new features.

The first US astronaut who visited the centre and Zvyozdnyy Gorodok was Frank Borman, on 5 July 1969. The following year, Neil Armstrong toured the centre and held a reception at the House of Cosmonauts on 1 July 1970. Dave Scott was the first astronaut to sit in a simulator and be shown how it worked, by Leonov, when he visited the centre in 1973.

The 'Orbita' Hotel opened in 1969 and new housing blocks were started, located on either side of the hotel. The hotel was built to house new workers at the centre who did not have allocated accommodation and the 1970 cosmonaut selections lived there for some months while their own permanent housing was under construction. The Orbita is not a hotel in the Western sense of the word, but more like a long stay hostel for incoming workers. Around the time of its construction, a number of key decisions were taken about the role and future of the training centre, and major expansion plans for the town were put in place.

A monument to Gagarin

On 18 August 1971, the statue of Gagarin was unveiled, located close to Dom 2 and 4 where Gagarin and his fellow cosmonauts lived. It is positioned so that he looks at the training centre and the site is well tended. All crews, before and after missions, lay flowers at the statue, as do all visiting dignitaries. The image on page 322 shows how the trees have matured since the statue was placed on the site. The statue was sculpted by B. Dyuzhev and its architect was A. Zavarzin. One nice story was that the walkways in the town were more or less made by the inhabitants during the winter. Aerial photographs had been taken from a helicopter which showed the paths that had been formed by people walking through the snow and when the snow disappeared, these paths were made permanent.

A group of cosmonauts lay flowers at the Gagarin memorial. Note absence of the trees which now surround the memorial. The picture was taken in the winter of 1971

The Gagarin memorial as it is today, surrounded by mature trees

A town is created

In 1973, Zvyozdnyy Gorodok had a population of around 3,000. A new commercial centre, including shops and a cafeteria, was planned and constructed, and was opened in the centre of the town in 1978. Its inception came at a time of major construction within the town and the training centre and in 1974, the Profilactorium was opened, as part of the preparations for the upcoming joint Soviet-American ASTP programme. This was located next to an artificial lake, which was especially constructed to retain the open air feel to the town. The American crews visited the centre in November of the same year and were impressed by the range of facilities available to support the mission.

In 1975, the kindergarten opened. The need for a secondary school meant that a new extension was built on the old school, opening in 1979 and reflecting the growth of families in the town. A new kindergarten was started, but never completed, and its foundations are located close to Dom 47.

The next major expansion was in the late 1970s when, due to the demand for housing, it was decided to build ten blocks in a new part of the centre's grounds. These ambitious plans were toned down, and three blocks (numbers 43, 44 and 45) were built and occupied in the period 1979 to 1980. A further block (number 46) was

Housing blocks Numbers 47, 48 and 49. These blocks are where many of the specialists who work at the training centre live

This is the new tennis hall that was constructed in the year 2003 following the old one being burnt down

built in 1983 and additional blocks (numbers 47, 48 and 49) were built between 1987 and 1989. All these were tower blocks and reflected the responsibility of the military unit to house current and retired officers and their families. Further blocks were built outside Zvyozdnyy in 1993.

There was a lot of pressure from residents to increase the recreational facilities in the town, but in the 1980s, there was no money for such facilities and a bureaucratic system running the country. The main task had been to build living quarters for the people, but when the centre asked for funding for a building which was officially described as 'a Facility for Stationary and Kinetic Investigation', the money came without a problem. The facility was constructed under the project title of the 'Chamber of Extreme Humidity' and opened in 1983 as a sports complex. Next to it was an indoor tennis court, and tennis is a very popular sport in the town. In 2001, this burned down, but it was replaced by a new structure in 2003. Plans were also drawn up for a football field, but this was never built. In addition to these facilities, the school has a small swimming pool.

In 1984, the authorities built a shop for the elite of Star City. It was reserved for Hero Cosmonauts, senior managers and others and was well stocked in terms of food and other goods. This was a fairly normal practice in the Soviet Union at this point, but the shop has since closed down and the building has been occupied by the Militsiya (Police) station, a trade union office, and the Cosmonauts' Post Office. This office was established on 21 April 1965 to deal with mail addressed to the cosmonauts that was coming in from all over the Soviet Union and from the rest of the world. The post office moved from the administration centre to the House of Cosmonauts, before moving again to its current office in the early 1990s. It is part of the information group of the training centre and holds archives of all the post sent to cosmonauts from 1961. These are kept in books and in files, and would certainly be a valuable source of research information. There is also a bar and pool room on the ground floor of the building, which opens late at night.

A new Orbita hotel was finished in 1990 and the old one became a block of flats. The new hotel was used, like the old one, as temporary housing for military officers, contractors and cosmonauts who were posted here from other parts of Russia. Flats for families were on the top floors and in the mid-1990s, these were rented out to foreigners visiting the centre. In 2002, an extension of the hotel was completed, whose rooms are available for guests to rent. They are quite large and modern. The hotel itself has a small café and hairdressing salon.

In the mid-1990s, when NASA astronauts started training for long duration missions, they occupied flats in Dom 2 and 4. But they found life very lonely and took a long time to adjust. As it became clear that more NASA personnel and astronauts would be staying in Star City, three houses were built, which in fact were six separate dwellings, close to the Profilatorium and the lake. They were prefabricated Canadian condominiums, capable of surviving Russian winters. The Russians call these 'American cottages' and there is a bar in the basement of one, set up by astronaut Bill Shepherd, which is called 'Shep's bar' in his honour. NASA and ESA have offices in the Profilactorium and a number of foreign astronauts have lived in this building. It is also used as the quarantine area for crews before and after

The new 'Orbita' hotel, which is where the authors have stayed, usually hosts visiting specialists

These are the houses which were built for US astronauts, who were spending long periods training for missions on Mir and ISS at the training centre

returning from a mission. They stay here under medical supervision for up to three weeks.

Since the break up of the Soviet Union, the character of who lives at Zvyozdnyy Gorodok has changed. Many officers were stood down due to cuts in the staff, but continued (as was their right) to live in the town. Many of the veteran cosmonauts also retired and were also due to be offered housing. This meant that there was serious pressure on the housing stock and many of the new Air Force officers and their families had nowhere to live, which meant living in the Orbita or on the base. There was also no space in Zvyozdnyy within the perimeter of the town for major construction projects, but the authorities had to build and began doing so outside the perimeter fence. They are currently building four blocks between Zvyozdnyy and the adjacent village of Leonikha. These are all nine storeys tall and have been built close to the back security gate and to a block built in the mid-1990s. The new buildings are Dom 61, 62 and 63, with Dom 64 still under construction and likely to be occupied in 2005. A number of the younger cosmonauts live in Dom 61, 62 and 63.

There is now a minibus service which leaves from the main gate and takes passengers into Moscow. The terminal is a Metro station and the service was started in the mid-1990s. There is no hospital in Star City, so people use a military hospital in Chkalovskiy, where there is also a big market, cafés and a wide variety of shops. Many residents of Zvyozdnyy Gorodok have dachas, or summer cottages, some of which are in the grounds of the town while others are within walking distance. Here, they spend evenings and weekends and grow vegetables and fruit. Zvyozdnyy Gorodok is located in the middle of a large forest and there is a lot of walking around the town. The lakes, both the natural and the artificial ones, are clean, and

A view over the lake of the Profilactorium on the left and the swimming pool and public gym on the right. The lake is stocked with fish and is used for swimming in the summer

people bathe in them during warm weather. It is fair to say, however, that there is little to do socially in the town and many young people hang around at night or go into Moscow. Drinking is also an issue, but activities held in the House of Cosmonauts, such as dancing and aerobics, are very popular.

The cemetery for Zvyozdnyy Gorodok is in Leonikha and is also used by the residents of Chkalovskiy, so many of the graves are those of military officers. A number of cosmonauts and their families are buried in the cemetery, which is about a twenty-minute walk from Zvyozdnyy Gorodok, located in a wood made up of silver birch.

Zvyozdnyy Gorodok is a town that was created to help send people into space. It has achieved its goal.

In 2000, when two of the authors attended the celebrations for the 40th Anniversary of the formation of the Cosmonaut Training Centre named for Yuri Gagarin, we realised the ever-changing road it had taken. When it was formed, its existence and even its location was one of the most guarded secrets within the Soviet Union. Since those days, it has created new machines and methodologies to enable humans to work and live in space. As we edit this book, Sergey Krikalev will pass the cumulative total of two years in space during his sixth flight. The Training Centre has now become an international centre for astronauts and cosmonauts to gain the skills and insights needed to work in the hostile environment of space. It has helped humans to learn the physical, medical, psychological and scientific knowledge which is now required to fly ever-longer and more demanding missions, but the Centre faces new challenges in a market-led Russia. Space flight is seen by more and more as a luxury, but every country needs its pride and the flights of cosmonauts still do inspire many. All this has led to the transformation of the Centre from one focused solely on Soviet space missions to one where the international crewing of missions is the norm.

Under the leadership of Generals Tsibliyev and Korzun, it will meet these challenges and will lead the way to longer and longer missions as humans stretch out to the Moon and Mars. We wish them and their dedicated staff continued success.

REFERENCES

Much of the information gathered for this chapter was collected during the numerous visits to Star City made by the authors. Specific additional material was obtained from:

Starry published 1982 in Russian
Moscow Rabochin Chronicle Star City p 198–206, by Nikolay Kopylov
Interview with Colonel Gennadiy Ivanovich Sokolov (who was in charge of construction at Star City between 1969 and 1989) by Bert Vis and Rex Hall, Star City, August 2004
Kamanin Diaries
'Soviet Space Centre being expanded', *Aviation Week*, 25 June 1973, p 18–20
Cosmonaut Training Centre named for Yuri Gagarin 2002, second edition, produced by the cosmonaut training centre
The official website of the training centre; http://www.gctc.ru
Our Gagarin, published by Publisher Progress

Appendix 1 – Biographies of Key Personnel

In this appendix are the short biographies of key figures involved in the command structure of the Cosmonaut Training Centre named for Yuri Gagarin and the training of the cosmonauts. Also included are short biographies of the commanders of the various cosmonaut teams selected by RKK Energiya, The Institute of Medical Biological Problems, NPO Mashinostroyeniya, LII and the Buran Air Force contingent.

Commanders of the Cosmonaut Training Centre named for Yuri Gagarin.

Karpov, Yevgeniy Anatoliyevich (Major-General) was born on 19 February 1921 in Kiev. He attended the Kirov Military Medical Academy, graduating in 1942, and was later appointed to an air division which was flying bombers on long range missions to Germany. After the war, he was appointed to the Institute of Aviation Medicine, where he developed a range of tests and knowledge which would be used to screen and test the first group of cosmonauts. Karpov was appointed as the first commander of the Cosmonaut Training Centre on 24 February 1960 and in the press, was given the title of Chief Doctor (being a Colonel in the Medical Services). He served for three years as Director and was replaced because the centre had acquired a special significance, so the leadership felt that it should be headed by an Air Force general with a reputation and fame comparable to the cosmonauts. He returned to the Institute and was promoted to Major-General in 1966. In 1973, he was appointed as a supervisor in the Ministry of Civil Aviation State Scientific Research Institute and in 1978 he joined the Federation of Cosmonautics as one of its permanent staff. He retired and went to live in Kiev in the Ukraine, where he died on 25 May 1990.

Odintsov, Mikhail Pyotrovich (Colonel-General) was born on 18 November 1921 in the Perm Region. He was one of the most outstanding fighter pilots of the Great Patriotic War (WWII) and was twice awarded the title of Hero of the Soviet Union, on 2 April 1944 and 27 June 1945. After the war, he attended the Lenin Military-Political Academy and the General Staff Academy and also held a number of senior positions in the Air Force. Odintsov was appointed commander of the Cosmonaut Training Centre for only a few months in 1963. He was probably a Major-General at the time. The reason for his departure has never been disclosed, although it is known that he did not get along with the cosmonauts and that people like Kamanin were not very enthusiastic about him. He continued to serve in the Air Force until his retirement.

Kuznetsov, Nikolay Fedorovich (Major-General) was born on 26 December 1916 in Pyotrograd (Leningrad). He trained at an Air Force school named for Kalinin, served as a fighter pilot in World War II and was made a Hero of the Soviet Union on 1 May 1943, having shot down thirty-seven German aircraft during the conflict. He held a number of command positions, including a period in Korea during the early 1950s, and attended the General Staff Academy in 1956. Kuznetsov was appointed to command the training centre on 2 November 1963, with Yuri Gagarin as his deputy. He remained in command until May 1972, when he went to work at NPO Energiya until he retired in 1987. He continued to live at Star City until his death on 5 March 2000, and is buried in a cemetery on the outskirts of Moscow.

Beregovoy, Georgiy Timofeyevich (Lieutenant-General) was born on 15 April 1921 in the Poltava Raion of the Ukraine. He went to work in a steel plant before going to the Lugansk Air Force School, from which he graduated in 1941. He flew 185 combat missions against the Luftwaffe and was shot down three times. In April 1944, Beregovoy became a Hero of the Soviet Union and after the war, he became a test pilot, eventually flying sixty-three different types of aircraft and logging over 2,500 flying hours. He also undertook a correspondence course at the Red Banner Air Force Academy. In 1963, the heads of the manned space flight programmes decided that the team needed more experienced candidates, and Beregovoy joined the team on 17 January 1964. He underwent some initial training and in April 1965, was selected as the back-up commander of Voskhod 3. He joined the Soyuz training group and was assigned to fly the first Soyuz mission after the Soyuz 1 crash that killed Komarov in April 1967. He flew Soyuz 3 in October 1968, but was unsuccessful in his attempt to dock with the unmanned Soyuz 2. He retired from the team in April 1969, having been promoted to Major-General after his mission, and in June 1972, he was appointed Director of the training centre, remaining in charge until January 1987. In 1977 he was promoted to the rank of Lieutenant-General and after his retirement, he worked for the Academy of Sciences. Beregovoy died on 30 June 1995. He was a twice Hero of the Soviet Union and Pilot Cosmonaut of the USSR. He is buried at the Novodeviche cemetery in Moscow.

Shatalov, Vladimir Aleksandrovich (Lieutenant-General) was born on 8 December 1927 in Pyotropavlovsk, Kazakhstan. He attended the Kacha HAFP School, graduating in 1949, and stayed at the school as an instructor before attending the Red Banner Academy, from which he graduated in 1956. He had a number of assignments, rising to senior inspector pilot. He joined the cosmonaut team on 10 January 1963 and was a Capcom on the Voskhod missions, as well as one of the back-ups for Voskhod 3. He became a member of the original Soyuz training group and was a Capcom on Soyuz 1, then backed up Beregovoy on Soyuz 3. He commanded three early Soyuz flights in 1969 and 1971 including being the command cosmonaut on the 1969 group flight (Soyuz 6, 7 and 8). He left the cosmonaut team on 26 June 1971, was promoted to the rank of Major-General, and replaced Kamanin as Director General of Cosmonaut Training in the Soviet Air Force High Command. He was promoted to the rank of Lieutenant-General in 1977 and became a member of the Supreme Soviet. In January 1987, he succeeded Beregovoy as

Commander of TsPK (and his former post was abolished), holding the post until his removal in September 1991. He is now retired, but still lives at Star City. Shatalov is a twice Hero of the Soviet Union and a Pilot Cosmonaut of the USSR.

Klimuk, Pyotr Ilyich (Colonel-General) was born in the Brest Raion of Byelorussia on 10 July 1942. He attended a pilot school and after graduating, flew MiG 15s. He became a cosmonaut on 28 October 1965 and was immediately assigned to the lunar training group. He also worked on the Kontakt Soyuz missions in 1970 and commanded the Soyuz 13 Orion mission in 1973. He flew a second mission in 1975 to the Salyut 4 space station, and a third on an Interkosmos mission in 1978 with a Polish cosmonaut. He retired in 1978 and was promoted to the rank of Major-General. He then became the TsPK political chief and graduated from the Lenin Military-Political Academy in 1983. In September 1991, he succeeded Shatalov as Director of the training centre, and was promoted to the rank of Lieutenant-General in 1992 and Colonel-General in 1998. He retired as Director on 25 September 2003 and subsequently retired from the Air Force. Klimuk is a twice Hero of the Soviet Union and a Pilot Cosmonaut of the USSR.

Tsibliyev, Vasiliy Vasiliyevich (Lieutenant-General) was born in the Kirov district of Russia on 20 February 1954. He graduated from the Kharkov HAFP School in 1975 and served as pilot in the Air Force, flying MiG 21s, before graduating from the Gagarin Air Force Academy in 1987. He joined the cosmonaut team on 26 March 1987 and qualified as a Soyuz TM commander. He served on a number of back-up crews before commanding two missions to the Mir station, Soyuz TM 17 and TM 25. He logged over 381 days in space. Tsibliyev left the cosmonaut team in 1997, and in 1998, he was appointed to work at the training centre. In 2000, he was made deputy director in charge of cosmonaut training and was promoted to the rank of Major-General. On 25 September 2003, he succeeded Klimuk as TsPK commander. He is a Hero of the Russian Federation and a Pilot Cosmonaut of the Russian Federation. In late 2004, he was made a Lieutenant-General in charge of Cosmonaut Training and Selection, based at the Ministry of Defence.

Directors of Cosmonaut Training

Kamanin, Nikolay Pyotrovich (Colonel-General) was born, according to official records, on 18 October 1908 (his actual birthday was on 18 October 1909). He was one of the first Heroes of the Soviet Union, which was awarded in 1934. In 1958, he was appointed to be the Deputy Chief of the Soviet Air Force responsible for manned space flight. The post was redesignated Director of Cosmonaut Training in 1960 and then renamed Aide for Space Matters of the Air Force C in C. He remained in this post until July 1971, when he was replaced by cosmonaut Shatalov. Kamanin died on 11 March 1982 and is buried at the Novodeviche Cemetery in Moscow.

Goreglyad, Leonid Ivanovich (Major-General) was born on 13 April 1916 and served in the Air Force during the Second World War. He was made a Hero of the Soviet Union on 23 February 1948 and attended the General Staff Academy in 1950. He was deputy to Kamanin for over ten years, with the official title of 'Inspector

General for Spaceflight'. He was present at the landing site when the Soyuz 11 crew returned to Earth. Goreglyad retired in 1976 and died on 17 July 1986.

Biographies of selected Deputy Directors of Star City

Gagarin, Yuri Alekseyevich (Colonel) was born on 9 March 1934 in Klushino, in the Smolensk region. He became a factory worker before joining the Air Force in 1955, where he served with the Northern fleet. Gagarin was selected as a cosmonaut in 1960 and flew the first manned space mission on Vostok in April 1961. He was then grounded on the order of the Soviet leadership due to his status. In 1963 he was made deputy commander of the Cosmonaut Training Centre and later persuaded the commanders to reinstate him to flight status. Gagarin became a member of the original Soyuz training group and in 1966, was selected as back-up to Komarov on Soyuz 1. He was involved in the investigation of Komarov's accident, but was himself killed in an air crash on 27 March 1968. He is buried in the Kremlin Wall.

Nikolayev, Andriyan Grigoryevich (Major-General) was born on 5 September 1929 in the Chuvash Autonomous Republic. He initially became a lumberjack before becoming an Air Force pilot in 1951, where he flew MiG 15s. Nikolayev was selected as a cosmonaut in 1960, made his first flight in 1962 on Vostok 3 and was a member of the original Soyuz training group selected in 1966. He backed up a number of early Soyuz missions before commanding the Soyuz 9 long duration mission on which he set a world duration record of eighteen days. He retired in 1970 and worked at TsPK in a number of senior training and command positions. He retired from the Air Force in 1992 and lived at Star City near Moscow until he died from a heart attack on 3 July 2004. He was a twice Hero of the Soviet Union and a Pilot Cosmonaut of the USSR.

Popovich, Pavel Romanovich (Major-General) was born on 5 October 1930 in the Kiev region of the Ukraine. He served in the Soviet Air Force as a pilot of MiG 17 aircraft. He was the first cosmonaut selected in 1960 and acted as Quartermaster to the first group. He flew his first mission on Vostok 4 in 1962 and was then assigned to the military Soyuz VI program before joining the lunar training group. Both programs were cancelled. In 1972, he was given command of the first Soyuz ferry mission (Soyuz 14) to fly to a military space station, Salyut 3 – Almaz. He then retired and served in a number of senior positions within the command structure of the training centre before retiring to live in Moscow in 1989. He is a twice Hero of the Soviet Union and a Pilot Cosmonaut of the USSR.

Leonov, Aleksey Arkhipovich (Major-General) was born in the Kemerovskoye Region on 30 May 1934. He joined the Soviet Air Force and was selected to join the cosmonaut team in 1960. He made his first space mission on Voskhod 2, making the world's first space walk in March 1965, and was a member of the original Soyuz group selected in 1966. He was also the head of the lunar training group and would have been the first Soviet citizen on the Moon if the programme had not been cancelled in 1970. He was assigned to command a Salyut mission in 1971 (but was stood down when one of his colleagues failed a medical) and subsequently trained for a number of missions which, for a variety of reasons, were all cancelled. However, he commanded the ASTP Soyuz 19 mission in 1975 before becoming a senior commander at the Cosmonaut Training Centre. He retired in 1992, but still lives close to Star City, and is currently a president of a Russian bank. He is a twice Hero of the Soviet Union and a Pilot Cosmonaut of the USSR.

Commander of the air regiment attached to Star City

Seryogin, Vladimir Sergeyevich (Colonel-Engineer) was born on 7 July 1922 in Moscow. He was a graduate of the Tambov school for military pilots and was sent to the Front during WWII. He was awarded the title of the Hero of Soviet Union on 26 June 1945, and after the war, he went to the Zhukovskiy AF Engineering Academy as a test pilot. Seryogin was commander of the cosmonaut flight training squadron based at Chkalovskiy Air Base and one of his roles was overseeing the planes that cosmonauts used to keep their flying skills up to par, as well as the weightless training aircraft. He was overseeing such a flight with Gagarin when they were killed when their MiG 15 crashed on 27 March 1968. Seryogin died instantly and was buried in the Kremlin Wall. The air regiment which is still based at Chkalovskiy is named after him.

Commander of the Vostok training group

Gallay, Mark Lazarevich was born on 16 April 1914 in Saint Petersburg. He was a famous aviator and test pilot who served with distinction in WWII. He was a test pilot of the USSR working at LII, and was awarded a Hero's Star on 1 May 1957. In 1960, he was assigned to support the training of the Gagarin selection and worked with the 'group of immediate preparedness'. He then went on to work at Moscow Aviation Institute with Korolyov. He died on 14 July 1998 and is buried in a cemetery in Moscow.

The organisation and command structure of the cosmonaut team of RKK Energiya

Heads of the Flight Test Department of the design bureau of Energiya.

Anokhin, Sergey Nikolayevich was born on 19 March 1910 in Moscow. He initially became a railway worker before attending a Higher Air Force School in the 1930s. In the 1940s, he set a number of world records as a glider pilot and during the war, he commanded a regiment in the Air Force on the Belorussian front and in support of partisans. In 1943, he became the leading test pilot for the Soviet jet plane and was one of the leading test pilots in the country. On 3 February 1953, he was awarded the title Hero of the Soviet Union and in 1959, he became the first Merited Test Pilot of the Soviet Union. In 1964, Korolyov asked him to head the Flight Methods section of the design bureau responsible for the selection and training of civilian cosmonauts, which was named Department 90. Anokhin was an early cosmonaut himself, but in May 1966, the State Commission formed Department 731, which

Anokhin headed until 1978. He died on 15 April 1984 and is buried at Novodeviche Cemetery in Moscow.

Makarov, Oleg Grigoryevich was born in the Tver Region of the USSR on 6 January 1933. He graduated from the Bauman Higher Technical School in 1957 and joined Korolyov's design bureau, where he was involved in the development of Vostok and helped design Soyuz. He joined the cosmonaut team in May 1966 and was assigned to the prime crew of the first lunar mission with Leonov. He worked on this project until 1970 when it was cancelled. He flew in 1973 as the Flight Engineer on Soyuz 12, the first flight of the ferry craft, and was then involved in the launch abort of 1975, but went on to fly on Soyuz 27 in 1978 and Soyuz T 3 in 1980. He retired from the team in 1987 and died from an heart attack on 28 May 2003. At the time of his death, he held a senior position in NPO Energiya. Makarov lived in Moscow and is buried close to the Ostankino TV tower. He was a twice Hero of the Soviet Union and a Pilot Cosmonaut of the USSR.

Sevastyanov, Vitaliy Ivanovich was born on 8 July 1935 near the city of Sverdlovsk. He graduated from the Moscow Aviation Institute in 1959 before joining the Korolyov design bureau where he was part of the team that created Vostok. He lectured to the first groups of cosmonauts, and joined the team himself in January 1967, where he was immediately assigned to the lunar program working on a Zond lunar mission. When it was cancelled in late 1968, Sevastyanov was paired with Nikolayev and trained for a number of missions before flying the Soyuz 9 long duration mission in 1970. He went on to work on Salyut stations and flew his second mission, Soyuz 18, in 1975. He then headed his own department in Energiya before beginning training for a TM Mir mission in the late 1980s. Sevastyanov retired in 1993 and is now a member of the Duma (the Russian parliament) for the Communist Party. He lives in Moscow and is a twice Hero of the Soviet Union and a Pilot Cosmonaut of the USSR.

Kubasov, Valeriy Nikolayevich was born in Moscow on 7 January 1935. He graduated from the Moscow Aviation Institute in 1958 and went to work in the Korolyov design bureau, where he worked on spacecraft trajectories. He was a candidate to join the cosmonaut team in 1964, but formally joined in May 1966. Kubasov was a member of the original Soyuz training group and served as a back-up to a number of early Soyuz flights, before being the Flight Engineer on Soyuz 6 in 1969. He was involved in a number of other prime crews whose missions were cancelled, before completing two more Soyuz flights in 1975 and 1978. On his third mission, he acted as the commander. In 1987, he took up a senior engineer post at Energiya and was then made a deputy director of a branch of Energiya concerned with life support systems. He lives in Moscow and is a twice Hero of the Soviet Union and a Pilot Cosmonaut of the USSR.

Aleksandrov, Aleksandr Pavlovich was born in Moscow on 20 February 1943. He initially enrolled in the Rocket Forces, but left and joined Energiya in 1964. He tried to join the cosmonaut team but failed a medical test in 1967. In 1969, he graduated from the Bauman Higher Technical School and finally joined the team in December

1978. He flew two Soyuz missions, in 1983 and 1987, involving Soyuz T and TM craft. In 1993, he retired and became Head of Department 291 in Energiya, which is the manned space flight directorate with responsibility for all cosmonauts and training in the bureau. Aleksandrov lives in Moscow and is a twice Hero of the Soviet Union and a Pilot Cosmonaut of the USSR.

Strekalov, Gennadiy Mikhailovich was born in Moscow on 28 October 1940. He worked as a coppersmith in OKB-1 on the construction of Sputnik 1. In 1965, he graduated from the Bauman Higher Technical School and joined the Korolyov design bureau, where he was involved in the design and development of Soyuz. He joined the cosmonaut team 1973 and went to work as a controller at TsUP (mission control) prior to his first assignment. He served on a number of back-up crews before his first mission, on Soyuz T 3 in 1980. He was a flight engineer on four more Soyuz flights, including the launch abort in 1983, and served as the Head of NPO Energiya's cosmonaut team. Strekalov retired as a cosmonaut in 1995 and as head of the team in 2003. He died from cancer on 25 December 2004 and is buried in central Moscow. He was a twice Hero of the Soviet Union and a Pilot Cosmonaut of the USSR.

Head of the Chelomey OKB-52 Mashinostroyeniya cosmonaut team

Makrushin, Valeriy Grigoryevich was born on 14 January 1940. In 1963, he graduated from the Leningrad Institute of Aviation Instrumentation and joined the Chelomey design bureau at Reutov, near Moscow. He joined the cosmonaut team on 22 March 1972, one of the first cosmonaut group selected from this design bureau, and became its Head until the group was disbanded on 8 April 1987. He worked on the Almaz military program, but was so angry at the group disbanding that he left the design bureau and became a trolley bus driver in Moscow, where he still lives.

Head of the cosmonaut team of the Institute of Medical and Biological Problems

Polyakov, Valeriy Vladimirovich was born on 27 April 1942 in Tula. He graduated from the Sechenov First Moscow Medical School in 1965, devoted himself to space medicine and applied to join the cosmonaut team. He was enrolled in the team on 22 March 1972 and became the IMBP team head until he retired in 1996. He served on a number of crews as crew physician and then as a back-up cosmonaut, and flew two long duration missions to the Mir Space Station; on Soyuz TM 6 where he stayed for 240 days, and on Soyuz TM 18 in 1994 when he stayed for 14 months, establishing the world record for an individual mission. He retired on 1 June 1996 and is currently deputy director of IMBP in charge of manned flight. He lives in Moscow and is both a Hero of the USSR and a Hero of the Russian Federation. He is also a Pilot Cosmonaut of the USSR.

Head of the group of cosmonauts of LII

Volk, Igor Pyotrovich was born in the Ukraine on 12 April 1937. He became a bomber pilot in the Soviet Air Force and in 1962, he joined the Moscow Aviation

Institute, graduating in 1969. Also in 1962, he became a test pilot for the Mikoyan Aircraft design bureau, flying the Soviet equivalent of the X-20 (50–50). He has logged over 7,000 hours and has flown over 80 different types of aircraft. In December 1978, he was enrolled in a group of pilots to test and fly the Soviet Buran shuttle. He became the group leader on 10 August 1981 and they became known as the 'wolf pack', because 'Volk' means 'wolf' in Russian. He flew on Soyuz T 12 in 1984 and also commanded the test program of the Buran analogue (BTS-02) between 1985 and 1988. Volk is a Merited Test Pilot of the Soviet Union, a Hero of the USSR and a Pilot Cosmonaut of the USSR, one of the top pilots of the Soviet Union. He stood down when the Buran program was finally cancelled in 2000.

The group of cosmonauts of GKNII VVS named for Chkalov

Bachurin, Ivan Ivanovich (Colonel) was born on 23 January 1942 in the Ukraine. He joined the Air Force in 1959 and became a test pilot in 1967. He has flown a large number of aircraft, including fighters and bombers. On 7 August 1987, he was enrolled in an Air Force group to test Buran and commanded a group of Air Force cosmonauts. He flew the Buran analogue BTS-02 in 1988 and also trained at Star City, where he trained to fly a Soyuz craft. He retired in December 1992 and is a Merited Test Pilot of the Soviet Union.

There were other selections, but with the exception of the Academy of Science group, they were single representatives of design bureaus or were attached to existing groups for training. One example of this was the journalist selection of 1991.

Single cosmonauts were selected from the design teams or organisations of Paton Institute, the Khrunichev Rocket Design Bureau, Zvezda, and the Russian Rocket Forces. None of them were deemed to constitute a full group, but they were included in a general classification of cosmonauts.

Appendix 2 – Current Deployment of the Russian Cosmonaut Team *(updated to July 2005)*

This is the current deployment of the Russian team, as at July 2005.

ISS Expedition Crews

MKS 11 – In space
Sergey Krikalev (RKK Energiya) and John Phillips (NASA)
They were due to be joined by Reiter (ESA-Germany), launching on STS-121 and returning on TMA 8, but as the Shuttle will be further delayed, he is likely to join expedition 12.

MKS 12 – due October 2005
Prime: William McArthur (NASA) and Valeriy Tokarev (Air Force), with Greg Olsen (Tourist)
Back-up: Jeffrey Williams (NASA) and Aleksandr Lazutkin (RKK Energiya), with Sergey Kostenko (Space Adventures Russian Tourist)
It was also planned for Sunita Williams (NASA) to join this crew on a Shuttle flight. Her back-up is Clayton Anderson (NASA)

MKS 13 – due April 2006
Prime: Pavel Vinogradov (RKK Energiya), Dmitriy Kondratyev (Air Force) and Daniel Tani (NASA)
Back-up: Fyodor Yurchikhin (RKK Energiya), Oleg Kotov (Air Force) and John Grunsfeld (NASA)

MKS 14 – due October 2006
Prime: Jeffrey Williams (NASA) and Aleksandr Lazutkin (RKK Energiya)
Back-up: Michael Lopez-Alegria (NASA) and Mikhail Tyurin (RKK Energiya)
Garrett Reisman (NASA) is in training as an increment crew member.

MKS 15 – due April 2007
Prime: Peggy Whitson (NASA), Fyodor Yurchikhin (RKK Energiya) and Oleg Kotov (Air Force)
No back-up crew confirmed

ISS Pool assigned for training purposes only

RKK Energiya:	Aleksandr Kaleri, Oleg Kononenko and Mikhail Korniyenko
Russian Air Force:	Yuri Lonchakov, Roman Romanenko, Sergey Volkov, and Maksim Surayev
NASA:	Sandra Magnus, Greg Chamitoff, Michael Barrett, Tim Kopra, Bob Behnken and Nicole Stott
ESA:	Leopold Eyharts, Frank De Winne and André Kuipers
CSA:	Bob Thirsk
JAXA:	Koichi Wakata

Training Groups

'MKS Group 1'	Yuri Lonchakov, Viktor Afanasyev, Yuri Baturin, Yuri Malenchenko, Sergey Treshchev, Gennadiy Padalka and Konstantin Kozeyev
'MKS Group 2'	Aleksandr Skvortsov, Maksim Surayev, Roman Romanenko and Sergey Volkov
'MKS Group 3'	Sergey Revin, Sergey Moshchenko, Mikhail Korniyenko, Oleg Kononenko, Oleg Skripochka and Yuri Shargin

The candidate group passed their examinations in late June 2005 and became available for training assignment in July 2005

Candidate Group

Russian Air Force:	Aleksandr Samokutyayev, Anton Shklaperov, Anatoliy Ivanishin and Yevgeniy Tarelkin
RKK Energiya:	Mark Serov, Andrey Borisenko and Oleg Artemyev
IMBP:	Sergey Ryazanskiy
RosKosmos:	Sergey Zhukov
Kazakhstan:	Aydyn Aimbetov and Mukhtar Aymakhanov

Cosmonauts not currently occupied with mission preparation

Salizhan Sharipov is currently undergoing a course of rehabilitation after prolonged space flight
Konstantin Valkov has been the representative of RGNII TsPK at NASA's Johnson Space Center in Houston, Texas, since September 2004
Aleksandr Kaleri and Mikhail Tyurin are working in the division of cosmonauts at RKK Energiya
Boris Morukov is working at IMBP

Thus, as at 31 July 2005, there are 38 cosmonauts in training in Russia. ESA, CSA and JAXA are preparing a series of three-month flights to ISS to support their modules. They assume that all the flights will be on Soyuz

Group E (Europe)

E1	Leopold Eyharts (prime) and Frank De Winne (back-up). Thomas Reiter's mission does not support any European development and is not included in the sequence.
E2	De Winne (prime). The back-up role is not yet announced but is likely to be Christer Fuglesang in recognition of his bad luck with Space Shuttle missions. André Kuipers is currently filling this role as well as backing-up Bob Thirsk
E3	Fuglesang or Kuipers (prime)

Group C (Canada)

C1	Bob Thirsk (prime) Kuipers (back-up) in return for Thisk backing-up Roberto Vittori. Canada is supported by ESA astronauts

Group J (Japan)

J1	Koichi Wakata (prime) with no back-up currently named

Appendix 3 – The Volga Stratospheric Balloon programme

During the 1930s, the Soviet Union pursued an active programme of research by balloon to support both military (strategic) and civilian (scientific) studies of the stratosphere. This research also had practical applications for the creation of pressure garments, life support systems, and the launch, tracking and recovery systems that would help to pioneer manned space flight three decades later in the Vostok and Voskhod programme. These studies were suspended in the 1940s as the Great Patriotic War (World War II) engulfed the Eastern Front and plunged the Soviets into a war with Nazi Germany. Following the war significant information, material, hardware and personnel was acquired by the Soviet authorities, which helped support their own development of ballistic missiles and, ultimately, the quest for space. The development of the Cold War between the US and USSR gave rise to the strategic build up of military might and the scientific and technical expertise to apply this powerful resource to the exploration of space. One programme that seems to have been linked to early human space exploration research and support was the development of a new manned stratospheric balloon programme 'Object SS' which became known as Volga in the West. It could also have been connected directly to the training (or at least in the support of training) of Soviet cosmonauts, or for evaluating techniques for future space flights that were then incorporated into the cosmonaut training programme.

In response to high-altitude US reconnaissance balloon flights over Soviet territory, which were flying too high to be shot down, design bureau OKB-424 was created on 24 December 1956, essentially from the 13th Laboratory of the Central Aerohydrodynamics Institute (TsAGI), and with Mikhail Gudkov as its first General Designer. Assigned to the development of Soviet high-altitude balloons for both military and scientific studies, the bureau was spread over three locations until, in 1959, all its facilities were centralised in Dolgoprudnyi, Moscow. On 30 April 1966, it was renamed DKBA (Soviet acronym for Dolgoprudnyi Design Bureau of Automatics) and continued to work on atmospheric balloon research and development, including the 'Saturn' balloon for high-altitude studies during the 1960s.

It appears that the Object SS programme was created to support development of high-altitude aircraft and the first manned Soviet spacecraft – Vostok. Work began on development of the gondola in 1958 and unmanned test flights commenced in 1959. The chief designer was Gennadiy F. Chekalin. There seems to have been a series (at least 5–6, maybe as many as 10) gondolas constructed for ground tests, unmanned flights and manned flights. The flights were dispatched from a launch site near the Volga River, and hence the 'public' name 'Volga' for the one known flight which tragically ended in the death of one of the aeronauts.

The most widely known flight of Object SS occurred on 1 November 1962, to a height of 25,448 metres. The crew of 'Volga' were to test recent developments in pressure garments and parachute rescue systems. Pilot Pyotr Dolgov was to test high-altitude pressure suit 'SI-3M', which was a prototype of future space suits, and a special parachute system of his own design. Second crewman Yevgeniy Andreyev was to test emergency escape from the gondola and an analogue of a space capsule, and investigate the possibility of free-fall in the thin layers of the upper atmosphere, in the serial production suit 'KKO-3'. Andreyev left the gondola first and landed safely but Dolgov's pressure visor was accidentally cracked on exit causing his death as a result of depressurisation of the helmet and suit. A 'Volga' type capsule is on display at the Monino Air Force Museum near Moscow and displays the number 5 on the gondola, indicating that there were at least four others. Apparently, it was planned that the remaining articles could have been used for further record breaking altitude flights but following the death of Dolgov, a government commission cancelled the programme.

It seems probable that at least part of the 'Volga' programme was associated with the early Soviet man in space programme, Vostok, possibly testing systems and hardware at high-altitude including the cosmonaut ejection and landing system. Whether any cosmonauts were involved in this programme is unclear as full details of the programme are difficult to find. The back-up crew of the November 1962 ascent are known. They were Ivan Kasmyshev and Vasiliy Lazarev, who was a medical doctor and Air Force pilot and was later selected for cosmonaut training. It is clear that parachute training was, and continues to be, an integral element of cosmonaut training and perhaps the Volga series was planned to be part of that training. Lazarev's biography, issued when he flew into space (1973), states that he completed a 28-hour balloon duration mission (not a parachute descent mission), but details of this event are unclear. Volga crew members had to be very highly qualified parachutists and probably Air Force pilots. Several merited test pilots and parachutists of the Soviet Union have been identified in this period, creating records, training cosmonauts or testing Air Force equipment, although direct links to the Volga balloon programme are not so clear. They include:

- Valeriy Golovin who tested the Vostok ejection system and parachute descent hardware wearing a Vostok pressure garment
- Valentin Danikovich, who worked with Andreyev and supported the development of airlock exit and entry techniques for the 1965 Voskhod 2 EVA by Leonov
- Menya Nebovokrug, a famous tester of parachutes
- G. Kondrashev, the first Soviet to test an ejector seat
- Nikolay Gladkov, an Air Force tester

- Nikolay Nikitin, who set a freefall record on 20 August 1957. He later became the parachute instructor for the 1960 First Air Force cosmonaut (Gagarin group) selection and held this post until his death in 1963

Photos of these and many other Air Force testers evaluating space systems or cosmonaut training equipment gave rise to rumours in the West of 'phantom cosmonauts' and cosmonauts lost in space, since they never made spaceflights and were not heard of for years – and in some cases, not at all. Though they were not official cosmonauts, their role in system and procedures evaluations paved the way for developing effective and efficient cosmonaut training systems and procedures that have continued for almost half a century, a testament to their courage and devotion to duty.

One of the stratospheric gondolas produced for the Volga programme (designated gondola 5) on permanent display at the Central Air Force Museum at Monino, Moscow Raion. (Courtesy Bert Vis)

Appendix 4 – Soviet/Russian International Manned Space Flights

Launch Date	Mission Designation	Station Visited	Pos. Flown	Prime Crew	Country of citizenship	Back-Up Crew	Country of citizenship	Landing date	Duration (DDD:HH: MM:SS)	Return Craft
Key: CDR – Commander; CR – Cosmonaut Researcher; FE – Flight Engineer; MS – Mission Specialist (US Shuttle); SFP – Spaceflight Participant; (F) – Female										
1978 Mar 2	Soyuz 28	Salyut 6	CDR	Gubarev	Soviet Union	Rukavishnikov	Soviet Union	1978 Mar 10	007:22:16:00	Soyuz 28
			CR	Remek	Czechoslovakia	Pelchak	Czechoslovakia	1978 Mar 10	007:22:16:00	Soyuz 28
1978 Jun 29	Soyuz 30	Salyut 6	CDR	Klimuk	Soviet Union	Kubasov	Soviet Union	1978 Jul 5	007:22:02:59	Soyuz 30
			CR	Hermaszewski	Poland	Jankowski	Poland	1978 Jul 5	007:22:02:59	Soyuz 30
1978 Aug 26	Soyuz 31	Salyut 6	CDR	Bykovskiy	Soviet Union	Gorbatko	Soviet Union	1978 Sep 3	007:20:49:04	Soyuz 29
			CR	Jähn	GDR	Köllner	GDR	1978 Sep 3	007:20:49:04	Soyuz 29
1979 Apr 10	Soyuz 33	Salyut 6	CDR	Rukavishnikov	Soviet Union	Romanenko Y	Soviet Union	1979 Apr 12	001:23:01:06	Soyuz 33
	Failed to dock with station		CR	Ivanov	Bulgaria	Aleksandrov AP	Bulgaria	1979 Apr 12	001:23:01:06	Soyuz 33
1980 May 26	Soyuz 36	Salyut 6	CDR	Kubasov	Soviet Union	Dzhanibekov	Soviet Union	1980 Jun 3	007:20:45:44	Soyuz 35
			CR	Farkas	Hungary	Magyari	Hungary	1980 Jun 3	007:20:45:44	Soyuz 35
1980 Jul 23	Soyuz 37	Salyut 6	CDR	Gorbatko	Soviet Union	Bykovskiy	Soviet Union	1980 Jul 31	007:20:42:00	Soyuz 36
			CR	Pham Tuan	Vietnam	Liem	Vietnam	1980 Jul 31	007:20:42:00	Soyuz 36
1980 Sep 18	Soyuz 38	Salyut 6	CDR	Romanenko Y	Soviet Union	Khrunov	Soviet Union	1980 Sep 28	007:20:43:24	Soyuz 38
			CR	Tamayo Mendez	Cuba	Lopez Falcon	Cuba	1980 Sep 28	007:20:43:24	Soyuz 38
1981 Mar 22	Soyuz 39	Salyut 6	CDR	Dzhanibekov	Soviet Union	Lyakhov	Soviet Union	1981 Mar 30	007:20:42:03	Soyuz 39
			CR	Guragchaa	Mongolia	Ganzorig	Mongolia	1981 Mar 30	007:20:42:03	Soyuz 39
1981 May 14	Soyuz 40	Salyut 6	CDR	Popov	Soviet Union	Romanenko	Soviet Union	1981 May 22	007:20:41:52	Soyuz 40
			CR	Prunariu	Romania	Dediu	Romania	1981 May 22	007:20:41:52	Soyuz 40

Launch Date	Mission Designation	Station Visited	Pos. Flown	Prime Crew	Country of citizenship	Back-Up Crew	Country of citizenship	Landing date	Duration (DDD:HH: MM:SS)	Return Craft
1982 Jun 24	Soyuz T 6	Salyut 7	CDR	Dzhanibekov	Soviet Union	Kizim	Soviet Union	1982 Jul 2	007:21:50:53	Soyuz T 6
			FE	Ivanchenkov	Soviet Union	Solovyov V	Soviet Union	1982 Jul 2	007:21:50:53	Soyuz T 6
			CR	Chrétien	France	Baudry	France	1982 Jul 2	007:21:50:53	Soyuz T 6
1984 Apr 3	Soyuz T 11	Salyut 7	CDR	Malyshev	Soviet Union	Berezovoy	Soviet Union	1984 Apr 11	007:21:40:06	Soyuz T 11
			FE	Strekalov	Soviet Union	Grechko	Soviet Union	1984 Apr 11	007:21:40:06	Soyuz T 11
			CR	Sharma	India	Malhotra	India	1984 Apr 11	007:21:40:06	Soyuz T 11
1987 Jul 22	Soyuz TM 3	Mir	CDR	Viktorenko	Soviet Union	Solovyov A	Soviet Union	1987 Jul 22	007:23:04:55	Soyuz TM 3
			FE	Aleksandrov A	Soviet Union	Savinykh	Soviet Union	Transferred as a resident crewmember		
			CR	Faris	Syria	Habib	Syria	1987 Jul 22	007:23:04:55	Soyuz TM 3
1988 Jun 7	Soyuz TM 5	Mir	CDR	Solovyov A	Soviet Union	Lyakhov	Soviet Union	1988 Jun 17	009:20:09:19	Soyuz TM 4
			FE	Savinykh	Soviet Union	Serebrov	Soviet Union	1988 Jun 17	009:20:09:19	Soyuz TM 4
	Skipka 88 mission		CR	Aleksandrov AP	Bulgaria	Stoyanov	Bulgaria	1988 Jun 17	009:20:09:19	Soyuz TM 4
1988 Aug 29	Soyuz TM 6	Mir	CDR	Lyakhov	Soviet Union	Beregovoy	Soviet Union	1988 Sep 8	008:20:26:27	Soyuz TM 5
			CR/Dr.	Polyakov	Soviet Union	Arzamazov	Soviet Union	Transferred as a resident crewmember		
			CR	Mohmand	Afghanistan	Dauran	Afghanistan	1988 Sep 8	008:20:26:27	Soyuz TM 5
1988 Nov 26	Soyuz TM 7	Mir	CDR	Volkov A	Soviet Union	Viktorenko	Soviet Union	Transferred as a resident crewmember		
			FE	Krikalev	Soviet Union	Serebrov	Soviet Union	Transferred as a resident crewmember		
	Aragatz mission		CR	Chrétien	France	Tognini	France	1988 Dec 21	024:18:07:25	Soyuz TM 6
1990 Dec 2	Soyuz TM 11	Mir	CDR	Afanasyev	Soviet Union	Artsebarskiy	Soviet Union	Transferred as a resident crewmember		
			FE	Manarov	Soviet Union	Krikalev	Soviet Union	Transferred as a resident crewmember		
			CR	Akiyama	Japan	Kikuchi	Japan	1990 Dec 10	007:21:54:40	Soyuz TM 10
1991 May 18	Soyuz TM 12	Mir	CDR	Artsebarskiy	Soviet Union	Volkov A	Soviet Union	Transferred as a resident crewmember		
			FE	Krikalev	Soviet Union	Kaleri	Soviet Union	Transferred as a resident crewmember		
	Juno mission		CR	Sharman	United Kingdom	Mace	United Kingdom	1991 May 26	007:21:14:20	Soyuz TM 11

1991 Oct 2	Soyuz TM 13	Mir	CDR	Volkov A	Soviet Union	Viktorenko	Soviet Union	Transferred as a resident crewmember		
			CR/FE	Aubakirov	Kazakhstan	Musabayev	Kazakhstan	1991 Oct 10	007:22:12:59	Soyuz TM 12
	AustroMir mission		CR	Viehböck	Austria	Lothaller	Austria	1991 Oct 10	007:22:12:59	Soyuz TM 12
1992 Mar 17	Soyuz TM 14	Mir	CDR	Viktorenko	Russia	Solovyov A	Russia	Transferred as a resident crewmember		
			FE	Avdeyev	Russia	Avdeyev	Russia	Transferred as a resident crewmember		
	Mir92 mission		CR	Flade (DLR)	Germany	Ewald (DLR)	Germany	1992 Mar 25	007:21:56:52	Soyuz TM 13
1992 Jul 27	Soyuz TM 15	Mir	CDR	Solovyov A	Russia	Manakov	Russia	Transferred as a resident crewmember		
			FE	Avdeyev	Russia	Poleshchuk	Russia	Transferred as a resident crewmember		
	Antares mission		CR	Tognini	France	Haigneré J-P	France	1992 Aug 10	013:18:56:14	Soyuz TM 14
1993 Jul 1	Soyuz TM 17	Mir	CDR	Tsibliyev	Russia	Afanasyev	Russia	Transferred as a resident crewmember		
			FE	Serebrov	Russia	Usachev	Russia	Transferred as a resident crewmember		
	Altair mission		CR	Haigneré J-P	France	Andre-Deshays (F)	France	1992 Jul 22	020:16:08:52	Soyuz TM 16
1994 Oct 3	Soyuz TM 20	Mir	CDR	Viktorenko	Russia	Gidzenko	Russia	Transferred as a resident crewmember		
			FE	Kondakova (F)	Russia	Avdeyev	Russia	Transferred as a resident crewmember		
	EuroMir94 ESA mission		CR	Merbold (ESA)	Germany	Duque (ESA)	Spain	1994 Nov 4	031:12:35:56	Soyuz TM 19
1995 Mar 14	Soyuz TM 21	Mir	CDR	Dezhurov	Russia	Solovyov A	Russia	1995 Jul 7	115:08:43:02	STS-71
			FE	Strekalov	Russia	Budarin	Russia	1995 Jul 7	115:08:43:02	STS-71
	NASA-Mir 1		CR	Thagard (NASA)	USA	Dunbar (NASA - F)	USA	1995 Jul 7	115:08:43:02	STS-71
1995 Sep 3	Soyuz TM 22	Mir	CDR	Gidzenko	Russia	Manakov	Russia	1996 Feb 29	179:01:41:46	Soyuz TM 22
			FE	Avdeyev	Russia	Vinogradov	Russia	1996 Feb 29	179:01:41:46	Soyuz TM 22
	EuroMir95 ESA mission		CR	Reiter (ESA)	Germany	Fuglesang (ESA)	Sweden	1996 Feb 29	179:01:41:46	Soyuz TM 22
1996 Aug 17	Soyuz TM 24	Mir	CDR	Korzun	Russia	–	–	Transferred as a resident crewmember		
			FE	Kaleri	Russia	–	–	Transferred as a resident crewmember		
	Cassiopeia mission		CR	Andre-Deshays (F)	France	Eyharts	France	1996 Sep 2	015:18:23:37	Soyuz TM 23
1997 Feb 10	Soyuz TM 25	Mir	CDR	Tsibliyev	Russia	Musabayev	Russia	Transferred as a resident crewmember		
			FE	Lazutkin	Russia	Budarin	Russia	Transferred as a resident crewmember		
	Mir 97 mission		CR	Ewald (DLR)	Germany	Schlegel	Germany	1997 Mar 2	019:16:34:46	Soyuz TM 24

Launch Date	Mission Designation	Station Visited	Pos. Flown	Prime Crew	Country of citizenship	Back-Up Crew	Country of citizenship	Landing date	Duration (DDD:HH: MM:SS)	Return Craft
1998 Jan 29	Soyuz TM 27	Mir	CDR	Musabayev	Russia	Afanasyev	Russia	Transferred as a resident crewmember		
			FE	Budarin	Russia	Treshchev	Russia	Transferred as a resident crewmember		
	Pegasus mission		CR	Eyharts (CNES)	France	Haigneré J-P	France	1998 Feb 19	020:16:36:48	Soyuz TM 26
1999 Feb 20	Soyuz TM 29	Mir	CDR	Afanasyev	Russia	Sharipov	Russia	1999 Aug 28	188:20:16:19	Soyuz TM 29
			CR/FE	Haigneré J-P (CNES)	France	Andre-Deshays (F)	France	1999 Aug 28	188:20:16:19	Soyuz TM 29
			CR	Bella	Slovakia	Fulier	Slovakia	1999 Feb 28	007:21:56:29	Soyuz TM 28
2000 Oct 31	Soyuz TM 31	ISS	CDR	Gidzenko	Russia	Dezhurov	Russia	2001 Mar 21	140:23:38:55	STS-102
			FE-1	Krikalev	Russia	Tyurin	Russia	2001 Mar 21	140:23:38:55	STS-102
	Expedition 1		FE-2	Shepherd (NASA)	USA	Bowersox (NASA)	USA	2001 Mar 21	140:23:38:55	STS-102
2001 Apr 21	Soyuz TM 32	ISS	CDR	Musabayev	Russia	Afanasyev	Russia	2001 Apr 30	007:22:04:08	Soyuz TM 31
			FE	Baturin	Russia	Kozeyev	Russia	2001 Apr 30	007:22:04:08	Soyuz TM 31
			SFP	Tito	USA	–	–	2001 Apr 30	007:22:04:08	Soyuz TM 31
2001 Oct 21	Soyuz TM 33	ISS	CDR	Afanasyev	Russia	Zalyotin	Russia	2001 Oct 31	009:20:00:25	Soyuz TM 32
	Andromede mission		FE-1	Haigneré C (F)	France	Kuzhelnaya (F)	Russia	2001 Oct 31	009:20:00:25	Soyuz TM 32
			FE-2	Kozeyev	Russia	–	–	2001 Oct 31	009:20:00:25	Soyuz TM 32
2002 Apr 25	Soyuz TM 34	ISS	CDR	Gidzenko	Russia	Padalka	Russia	2002 May 5	009:21:25:18	Soyuz TM 33
	Marco Polo mission (ESA)		FE	Vittori (ESA)	Italy	Kononenko	Russia	2002 May 5	009:21:25:18	Soyuz TM 33
			SFP	Shuttleworth	South Africa	–	–	2002 May 5	009:21:25:18	Soyuz TM 33
2002 Oct 30	Soyuz TMA 1	ISS	CDR	Zalotin	Russia	Lonchakov	Russia	2002 Nov 10	010:20:53:09	Soyuz TM 34
	Odessa mission (ESA)		FE-1	De Winne (ESA)		Belgium	Lazutkin	Russia	2002 Nov 10	010:20:53:19
Soyuz TM 34										
			FE-2	Lonchakov	Russia	–	–	2002 Nov 10	010:20:53:19	Soyuz TM 34
	Soyuz TMA 2	ISS	CDR	Malenchenko	Russia	Kaleri	Russia	2003 Oct 28	184:22:46:09	Soyuz TMA 2
	Expedition 7			Lu (NASA)	USA	Foale (NASA)	USA	2003 Oct 28	184:22:46:09	Soyuz TMA 2

2003 Oct 10	Soyuz TMA 3 ISS	CDR	Kaleri	Russia	Tokarev	Russia	2004 Apr 30	194:18:33:43	Soyuz TMA 3
	Expedition 8	FE-1	Foale (NASA)	USA	McArthur (NASA)	USA	2004 Apr 30	194:18:33:43	Soyuz TMA 3
	Cervantes mission (ESA)	FE-2	Duque (ESA)	Spain	Kuipers (ESA)	Netherlands	2003 Oct 28	009:21:02:17	Soyuz TMA 2
2004 Apr 19	Soyuz TMA 4 ISS	CDR	Padalka	Russia	Sharipov	Russia	2004 Oct 24	187:21:16:09	Soyuz TMA 4
	Expedition 9	FE-1	Fincke (NASA)	USA	Chiao (NASA)	USA	2004 Oct 24	187:21:16:09	Soyuz TMA 4
	DELTA mission (ESA)	FE-2	Kuipers (ESA)	Netherlands	Thiele (ESA)	Germany	2004 Apr 30	010:20:52:15	Soyuz TMA 3
2004 Oct 14	Soyuz TMA 5 ISS	CDR	Sharipov	Russia	Tokarev	Russia	2005 Apr 25	192:19:01:59	Soyuz TMA 5
	Expedition 10	FE-1	Chiao (NASA)	USA	McArthur (NASA)	USA	2005 Apr 25	192:19:01:59	Soyuz TMA 5
		FE-2	Shargin	Russia	–	–	2004 Oct 24	009:21:28:41	Soyuz TMA 4
2005 Apr 15	Soyuz TMA 6 ISS	CDR	Krikalev	Russia	Tyurin	Russia	Planned Oct 2005		Soyuz TMA 6
	Expedition 11	FE-1	Phillips (NASA)	USA	Tani (NASA)	USA	Planned Oct 2005		Soyuz TMA 6
	Eneide mission (ESA)	FE-2	Vittori (ESA)	Italy	Thirsk (CSA)	Canada	2005 Apr 25	009:21:21:02	Soyuz TMA 5

Appendix 5: The Full GMVK

УДОСТОВЕРЕНИЕ

№ 97

Настоящее удостоверение выдано Манарову

Мусе Хирамановичу

в том, что он окончил курс общекосмической подготовки

и приказом Министра общего машиностроения

от 8 декабря 1978 года № 439

назначен на должность космонавта-испытателя.

Решением Межведомственной квалификационной

комиссии от „28" ноября 1986 года

Манарову М.Х.

присвоена квалификация космонавта-испытателя.

Председатель Межведомственной

квалификационной комиссии

М. П.

This picture shows the book awarded to all cosmonauts when they complete basic training (OKP). A group of cosmonauts can be seen holding them on page 140

The first part of this table covers all the civilians who passed the internal systems of design bureaus and affiliate agencies, who were then considered for and passed the State Medical Commission. The medicals were mainly conducted by the staff of the Institute of Medical and Biological Problems. Having passed that hurdle, the candidates then had to go before the full State Mandate Commission. This met at irregular intervals and considered additions to the various cosmonaut teams of the USSR. A number of potential candidates never got the call. Careful study of the table shows how the needs of mission planners or for specialist skills influenced who was approved and called up into cosmonaut training. Some candidates waited many years before they were confirmed by the Commission as a trainee cosmonaut.

This table was originally compiled by the staff of Novosti Kosmonavtiki from various Soviet and Russian sources and then reproduced in *Spaceflight* magazine (see Bibliography). It has been updated to take into account recent referrals to the Medical and the State or Mandate Commission.

Cosmonaut	Date of Birth	Affiliation	Medical Commission	State Commission
Fartushniy	1938 Feb 3	IES	1967 Nov 30	1968 May 27
Nikitskiy	1939 Mar 8	TsKBEM	1967 Nov 30	–
Patsayev	1933 Jun 19	TsKBEM	1967 Nov 30	1968 May 27
Berkovich	1935	TsKBM	1968 May 20	–
Smirenniy	1932 Oct 25	IMBP	1968 May 20	1972 Mar 22
Makrushin	1940	TsKBM	1968 Oct 16	1972 Mar 22
Sukhanov	1936	TsKBM	1968 Oct 16	–
Yeremich	1938 Nov 30	TsKBM	1968 Oct 16	–
Demyanenko	1942	NIIAP	1969 Sep 10	–
Velikiy	1939	TsKBEM	1969 Sep 10	–
Yershov	1928 Jun 21	IPM	1969 Sep 10	–
Lebedev	1942 Apr 14	TsKBEM	1969 Dec 26	1972 Mar 22
Lobachek	1939 Oct 3	IMBP	1969 Dec 26	–
Machinskiy	1937 Oct 11	IMBP	1969 Dec 26	1972 Mar 22
Pyotrov	1940	TsNIIMash	1969 Dec 12	–
Polyakov	1942 Apr 27	IMBP	1970 Jul 8	1972 Mar 22
Senkevich	1937 Mar 4	IMBP	1970 Jul 8	–
Abuzyarov	1934 Mar 8	GMTs	1970 Jul 25	–
Grechanik	1939 Mar 25	TsKBM	1971 Feb 25	1978 Dec 1
Ivanyan	1940 Nov 4	LGU	1971 Feb 25	–
Yuyukov	1941 Feb 26	TsKBM	1971 Feb 25	1973 Mar 27
Andreyev	1940 Oct 6	TsKBEM	1971 Nov 1	1972 Mar 22
Ivanchenkov	1940 Sep 28	TsKBEM	1971 Nov 1	1973 Mar 27
Ponomaryov	1932 Mar 24	TsKBEM	1971 Nov 1	1972 Mar 22
Ryumin	1939 Aug 16	TsKBEM	1971 Nov 1	1973 Mar 27
Strekalov	1940 Oct 28	TsKBEM	1971 Nov 1	1973 Mar 27
Aksyonov	1935 Feb 1	TsKBEM	1972 Feb 24	1973 Mar 27
Pyotrenko	1940	TsKBEM	1972 Nov 1	–
Romanov	1946 Aug 18	TsKBEM	1973 Jul 26	1978 Dec 1
Aleksandrov	1943 Feb 20	Energiya	1974 Mar 4	1978 Dec 1
Savinykh	1940 Mar 7	Energiya	1975 Mar 10	1978 Dec 1
Serebrov	1944 Feb 15	NIITP	1975 Mar 10	1978 Dec 1
Bobrov	1948	IMBP	1976 Jun 18	–
Borodin	1953 Mar 3	IMBP	1978 Jun 18	1978 Dec 1
Manarov	1951 Mar 22	Energiya	1976 Jun 18	1978 Dec 1
Potapov	1952 Oct 28	IMBP	1976 Jun 18	1978 Dec 1
Bragin	1945 Oct 19	IMBP	1976 Dec 10	–
Chervyakov	1946	Energiya	1976 Dec 10	–

Morukov	1950 Oct 1	IMBP	1976 Dec 10	1989 Jan 25
Solovyov V	1946 Nov 11	Energiya	1976 Dec 10	1978 Dec 1
Afonin	1949 Apr 28	IMBP	1977 Jan 28	–
Atkov	1949 May 9	VKNTs	1977 Jan 28	1983 Mar 9
Kulik		Energiya	1977 Jan 28	–
Laveykin	1951 Apr 21	Energiya	1977 Aug 3	1978 Dec 1
Gevorkyan	1952 May 28	TsKBMF	1977 Aug 31	1978 Dec 1
Arzamazov	1946 Mar 9	IMBP	1977 Nov 1	1978 Dec 1
Khatulev	1947 Feb 26	TsKBMF	1977 Nov 1	1978 Dec 1
Balandin	1953 Jul 30	Energiya	1978 Aug 3	1978 Dec 1
Bedzyuk	1954	Energiya	1978 Aug 3	–
Kononenko	1938 Aug 16	LII	1978 Aug 3	1980 Jul 30
Levchenko	1941 May 21	LII	1978 Aug 3	1980 Jul 30
Shchukin	1946 Jan 19	LII	1978 Aug 3	1980 Jul 30
Volk	1937 Apr 12	LII	1978 Aug 3	1980 Jul 30
Pyotrov	1945	Energiya	1978 Aug 31	–
Stankyavichus	1944 Jul 26	LII	1979 Jan 19	1980 Jul 30
Kuleshova	1956 Mar 14	Energiya	1979 May 15	1980 Jul 30
Savitskaya	1948 Aug 8	KBYa	1979 May 15	1980 Jul 30
Dobrokvashina	1947 Oct 8	IMBP	1979 Jun 29	1980 Jul 30
Zakharova	1952 Apr 22	VNIIKEKh	1979 Jun 29	1980 Jul 30
Chekh	1954	TsKBM	1979 Jul 9	–
Amelkina	1954 May 22	IMBP	1979 Oct 31	1980 Jul 30
Karlin	1942 Nov 7	Molniya	1979 Oct 31	–
Klyushnikova	1953 Oct 14	IMBP	1979 Oct 31	–
Maksimov		Energiya	1979 Oct 31	–
Morozov	1947 Aug 14	TsKBM	1979 Oct 31	–
Pozharskaya	1947 Mar 15	MONIKI	1979 Oct 31	1980 Jul 30
Sviridova		Energiya	1979 Dec 13	–
Latysheva	1953 Jul 9	IKI	1980 May 13	1980 Jul 30
Pronina	1953 Apr 14	NIITP	1980 May 13	1980 Jul 30
Ivanova	1949 Oct 3	LMI	1980 Nov 3	1983 Mar 9
Chelomey	1952 Aug 2	TsKBM	1981 Apr 14	–
Chuchin	1949 Jun 10	TsKBM	1981 Apr 14	–
Kaleri	1956 May 13	Energiya	1982 Dec 3	1984 Feb 15
Khaustov		Energiya	1982 Dec 3	–
Poleshchuk	1953 Oct 30	Energiya	1982 Dec 3	1989 Jan 25
Yemelyanov	1951 Aug 3	Energiya	1982 Dec 3	1984 Feb 15
Sultanov	1948 Nov 18	LII	1983 Jan 25	1983 Mar 9
Tolboyev	1951 Jan 20	LII	1983 Jan 25	1983 Mar 9

Zabolotskiy	1946 Apr 19	LII	1983 Apr 4	1984 Feb 15
Krikalev	1958 Aug 27	Energiya	1983 Jun 7	1985 Sep 2
Zaytsev	1957 Aug 5	Energiya	1983 Jun 7	1985 Sep 2
Stepanov	1936 Sep 27	IMBP	1983 Mov 11	1985 Sep 2
Sheffer	1947 Jun 30	LII	1984 Jul 8	1985 Sep 2
Melua	1950 Feb 7	IIET	1985 Apr 17	–
Tresvyatskiy	1954 May 6	LII	1985 Apr 17	1985 Sep 2
Budarin	1953 Apr 29	Energiya	1986 Feb 26	1989 Jan 25
Avdeyev	1956 Jan 1	Energiya	1987 Feb 2	1987 Mar 26
Fursov	1959	AMN	1988 Jul 20	–
Karashtin	1962 Nov 18	IMBP	1988 Jul 20	1989 Jan 25
Kondakova	1957 Mar 30	Energiya	1988 Jul 20	1989 Jan 25
Lukyanyuk	1958 Sep 22	IMBP	1988 Jul 20	1989 Jan 25
Prikhodko	1953 Nov 15	LII	1988 Oct 21	1989 Jan 25
Usachev	1957 Oct 9	Energiya	1988 Oct 21	1989 Jan 25
Vinogradov	1953 Aug 31	Energiya	1988 Oct 21	1992 Mar 3
Musabayev	1951 Jan 7	MGA	1989 Feb 27	1990 May 11
Khokhlachova	1954	Energiya	1989 Mar 4	–
Murashov	1959	AMN	1989 Mar 4	–
Treshchev	1958 Aug 18	Energiya	1989 Mar 4	1992 Mar 3
Karatyev	1954	VNTsKh	1989 Jul 17	–
Severin	1956 Nov 20	Zvezda	1989 Jul 17	1990 May 11
Lazutkin	1957 Oct 30	Energiya	1989 Sep 14	1992 Mar 3
Krikun	1963 Jun 3	Journalist	1990 Mar 27	1990 May 11
Mukhortov	1966 Mar 11	Journalist	1990 Mar 27	1990 May 11
Omelchenko	1951 Aug 20	Journalist	1990 Mar 27	1990 May 11
Sharov	1953 Dec 26	Journalist	1990 Mar 27	1990 May 11
Andryushkov	1947 Oct 6	Journalist	–	1990 May 11
Baberdin	1948 Oct 28	Journalist	–	1990 May 11
Aubakirov	1946 Jul 17	MZM	1991 Feb 19	1991 Jan 21
Moshchenko	1954 Jan 12	KB Salyut	1992 Jun 16	1997
Tyurin	1960 Mar 2	Energiya	1992 Jun 16	1994 Apr 1
Kuzhelnaya	1962 Nov 6	Energiya	1993	1994 Apr 1
Yurchikhin	1959 Jan 3	Energiya	1993	1997 Jul 28
Kozeyev	1967 Dec 1	Energiya	–	1996 Feb 9
Revin	1966 Jan 12	Energiya	–	1996 Feb 9
Kononenko	1964 Jun 21	Progress	–	1996 Feb 9
Skripochka	1969 Dec 24	Energiya	–	1997 Jul 28
Korniyenko	1960 Apr 15	Energiya	–	1998 Feb 24

Artemyev	1970 Dec 28	Energiya	2000 Spring	2003 May 29
Zavyalova	1975	Energiya	2000 Spring	–
Loktionov	1950 Dec 12	Private	2000 Oct 19	–
Zhukov	1956 Sep 8	Private	2002 May 31	2003 May 29
Borisenko	1964 Apr 17	Energiya	2002 Sep 12	2003 May 29
Polonskiy	1972	Private	2002 Sep 12	–
Serov	1974 May 13	Energiya	2002 Dec 17	2003 May 29
Aimbetov	1972 Jul 27	Kazakhstan	2002 Dec 17	–
Aymakhanov	1967 Jan 1	Kazakhstan	2002 Dec 17	–
Ryazanskiy	1974 Nov 13	IMBP	2003 Mar 26	2003 May 29

Air Force candidates

The Air Force used a different system for selecting their cosmonauts. They did hold intensive medical screening (which is detailed is the chapter The Cosmonaut Group of the RGNII TsPK). They were then subject to the State Commission but were transferred into the team by order of the Air Force and the Ministry of Defence. This changed in 1985 when they were subject to the full and public scrutiny of the Mandate Commission. We do not know the identity of failed short-listed candidates, except for a few cases. Since 1985 we are only aware of one candidate who passed the Medical Commission but was not passed by the Mandate Commission.

Group	Admission Order	Admitted Candidates
1962 Group	1962 Mar 12. Order No. 67	Solovyova, Kuznetsova and Tereshkova
	1962 Apr 3. Order No. 92	Yerkina and Ponomaryova
1963 Group	1963 Jan 10. Order No. 14	Full group
1965 Group	1965 Oct 28. Order No. 942	Full group
1966 Group	1966 Jan 1. Order No. 37	Lazarev
1967 Group	1967 Apr 12. Order No. 282	Alekseyev and Burdayev
	1967 May 7. Order No. 369	Remainder of the group
1970 Group	1970 Apr 27. Order No. 505	Full Group
1976 Group	1976 Aug 23. Order No. 686	Full Group
1978 Group	1978 May 28. Order No. 374	Grekov

Cosmonaut	Date of Birth	Affiliation	Medical Commission	State Commission
Artsebarskiy	1956 Sep 9	Air Force	–	1985 Sep 2
Afanasyev	1948 Dec 31	Air Force	–	1985 Sep 2
Manakov	1950 Jun 1	Air Force	–	1985 Sep 2
Gidzenko	1962 Mar 26	Air Force	–	1987 Mar 26
Dezhurov	1962 Jul 30	Air Force	–	1987 Mar 26
Korzun	1953 Mar 5	Air Force	–	1987 Mar 26
Malenchenko	1961 Dec 22	Air Force	–	1987 Mar 26
Tsibliyev	1954 Feb 20	Air Force	–	1987 Mar 26
Krichevskiy	1955 Jul 9	Air Force	–	1989 Jan 25
Onufriyenko	1961 Feb 6	Air Force	–	1989 Jan 25
Padalka	1958 Jun 21	Air Force	–	1989 Jan 25
Vozovikov	1958 Apr 17	Air Force	–	1990 May 11
Zalyotin	1962 Apr 21	Air Force	–	1990 May 11
Sharipov	1964 Aug 24	Air Force	–	1990 May 11
Shargin	1960 Mar 20	Air Force	–	1996 Feb 9
Kotov	1965 Oct 27	Air Force	–	1996 Feb 9
Valkov	1971 Nov 11	Air Force	–	1997 Jul 28
Volkov	1973 Apr 1	Air Force	–	1997 Jul 28
Kondratyev	1969 Apr 25	Air Force	–	1997 Jul 28
Lonchakov	1966 Mar 4	Air Force	–	1997 Jul 28
Moshkin	1964 Apr 23	Air Force	–	1997 Jul 28
Romanenko	1971 Aug 9	Air Force	–	1997 Jul 28
Skvortsov	1966 May 6	Air Force	–	1997 Jul 28
Surayev	1972 May 24	Air Force	–	1997 Jul 28
Tokarev	1952 Oct 29	Air Force	–	1997 Jul 28
Baturin	1949 Jun 12	Air Force	1997 Sep	1997 Sep 15
Kotik		Air Force	2001 Aug 29	–
Tarelkin	1974 Dec 29	Air Force	2002 Mar 1	2003 May 29
Shkaplerov	1972 Feb 20	Air Force	2002 Sep 12	2003 May 29
Samokutyayev	1970 Mar 13	Air Force	2002 Sep 12	2003 May 29
Ivanishin	1969 Jan 15	Air Force	–	2003 May 29

Appendix 6 – The Cosmonaut Team 1960–2005

Air Force Pilots and Engineers

Name		D.O.B	First Flight Year	Cosmonaut No. (USSR /Russia)	No. of Russian Flights	Status/Retirement*	Died/Killed*
1960 Group First Air Force Selection (Mar–Jun 1960)							
Anikeyev IN	Pilot	1933 Feb 12	–	–	0	1963 Apr 17	1992 Aug 20
Belyayev PI	Pilot	1925 Jun 26	1965	10	1	Deceased	1970 Jan 10
Bondarenko VV	Pilot	1937 Feb 16	–	–	0	Deceased	1961 Mar 23
Bykovskiy VF	Pilot	1934 Aug 2	1963	5	3	1982 Jan 26	
Filatyev VI	Pilot	1930 Jan 21	–	–	0	1963 Apr 17	1990 Sep 15
Gagarin YA	Pilot	1934 Mar 9	1961	1	1	Deceased	1968 Mar 27
Gorbatko VV	Pilot	1934 Dec 3	1969	21	3	1982 Aug 28	
Kartashov AY	Pilot	1932 Aug 25	–	–	0	1961 Apr 7	
Khrunov YV	Pilot	1933 Sep 10	1969	16	1	1980 Dec 25	2000 May 19
Komarov VM	Pilot	1927 Mar 16	1964	7	2	Deceased	1967 Apr 24
Leonov AA	Pilot	1934 May 30	1965	11	2	1982 Jan 26	
Nelyubov GG	Pilot	1934 Mar 31	–	–	0	1963 May 4	1966 Feb 18
Nikolayev AG	Pilot	1929 Sep 5	1962	3	2	1982 Jan 26	2004 Jul 3
Popovich PR	Pilot	1930 Oct 5	1962	4	2	1982 Jan 26	
Rafikov MZ	Pilot	1933 Sep 29	–	–	0	1962 Mar 24	2000 Jul 23
Shonin GS	Pilot	1935 Aug 3	1969	17	1	1979 Apr 28	1997 Apr 6
Titov GS	Pilot	1935 Sep 11	1961	2	1	1970 Jun 17	2000 Sep 20
Varlamov VS	Pilot	1934 Aug 15	–	–	0	1961 Mar 6	1980 Oct 2
Volynov BV	Pilot	1934 Dec 18	1969	14	2	1990 Mar 17	

Name		D.O.B	First Flight Year	Cosmonaut No. (USSR /Russia)	No. of Russian Flights	Status/Retirement*	Died/Killed*
Zaikin DA	Pilot	1932 Apr 29	–	–	0	1969 Oct 25	
1962 Group Female Cosmonauts (Mar–Apr 1962)							
Kuznetsova TD	Female	1941 Jul 14	–	–	0	1969 Oct 17	
Ponomaryova VL	Female	1933 Sep 18	–	–	0	1969 Oct 1	
Solovyova IB	Female	1937 Sep 6	–	–	0	1969 Oct 1	
Tereshkova VV	Female	1937 Mar 6	1963	6	1	1997 Apr 30	
Yerkina ZD	Female	1939 May 6	–	–	0	1969 Oct 1	
1963 Group Second Air Force Selection (10 Jan 1963)							
Dobrovolskiy GT	Pilot	1928 Jun 1	1971	24	1	Deceased	1971 Jun 30
Filipchenko AV	Pilot	1928 Feb 26	1969	19	2	1982 Jan 26	
Gubarev AA	Pilot	1931 Mar 29	1975	33	2	1981 Sep 1	
Kuklin AP	Pilot	1932 Jan 3	–	–	0	1975 Sep 15	
Shatalov VA	Pilot	1927 Dec 8	1969	13	3	1971 Jun 25	
Vorobyov LV	Pilot	1931 Feb 24	–	–	0	1974 Jun 28	
Artyukhin YP	Engineer	1930 Jul 22	1974	30	1	1982 Jan 26	1998 Aug 4
Buinovskiy EI	Engineer	1936 Feb 26	–	–	0	1964 Dec 11	
Demin LV	Engineer	1926 Jan 11	1974	32	1	1982 Jan 26	1998 Dec 18
Gulyayev VI	Engineer	1937 May 31	–	–	0	1968 Mar 6	1990 Apr 19
Kolodin PI	Engineer	1930 Sep 23	–	–	0	1983 Apr 20	
Kugno EP	Engineer	1935 Jun 27	–	–	0	1964 Apr 16	1994 Feb 24
Matinchenko AN	Engineer	1927 Sep 4	–	–	0	1972 Jan 19	1999 Jun 18
Voronov AF	Navigator	1930 Jun 11	–	–	0	1979 May 25	1993 Oct 31
Zholobov VM	Engineer	1937 Jun 18	1976	35	1	1981 Jan 7	

1963 Group Second Air Force Selection – Supplemental (25 Jan 1964)							
Beregovoy GT	Pilot	1921 Apr 15	1968	12	1	1982 Feb 25	1995 Jun 30
1965 Group Third Air Force Selection (25 Jan 1965)							
Fedorov AP	Pilot	1941 Apr 14	–	–	0	1974 May 28	2002 Mar 21
Kizim LD	Pilot	1941 Aug 5	1980	48	3	1987 Jun 13	
Klimuk PI	Pilot	1942 Jul 10	1973	28	3	1982 Mar 3	
Kramarenko AY	Pilot	1942 Nov 8	–	–	0	1969 Apr 30	2002 Apr 13
Pyotrushenko AY	Pilot	1942 Jan 1	–	–	0	1973 Jun 15	1992 Nov 11
Sarafanov GV	Pilot	1942 Jan 1	1974	31	1	1986 Jul 7	
Sharafutdinov AI	Pilot	1939 Jun 26	–	–	0	1968 Jan 5	
Shcheglov VD	Pilot	1940 Apr 9	–	–	0	1972 Oct 18	1973 Jul 16
Skvortsov AA	Pilot	1942 Jun 8	–	–	0	1968 Jan 5	
Voloshin VA	Pilot	1942 Apr 24	–	–	0	1969 Apr 9	
Yakovlev OA	Pilot	1940 Dec 31	–	–	0	1973 May 22	1990 May 2
Zudov VD	Pilot	1942 Jan 8	1976	37	1	1987 May 14	
Belousov BN	Engineer	1930 Jul 24	–	–	0	1968 Jan 5	1998 Jun 27
Degtyaryov VA	Doctor	1932 Apr 4	–	–	0	1966 Jan 17	
Glazkov YN	Engineer	1939 Oct 2	1977	39	1	1982 Jan 26	
Grishchenko VA	Navigator	1942 Apr 26	–	–	0	1968 Feb 5	1992 May 4
Khludeyev YN	Engineer	1940 Sep 10	–	–	0	1988 Oct 11	1995 Sep 19
Kolesnikov GM	Engineer	1936 Oct 7	–	–	0	1967 Dec 16	
Lisun MI	Engineer	1935 Sep 5	–	–	0	1989 Sep 19	
Preobrazhenskiy VY	Engineer	1939 Feb 3	–	–	0	1980 Nov 18	1993 Oct 25
Rozhdestvenskiy VI	Engineer	1939 Feb 13	1976	38	1	1986 Jun 24	
Stepanov EN	Engineer	1937 Apr 17	–	–	0	1992 Oct 31	
1966 Group Third Air Force Selection – Supplemental (17 Jan 1966)							
Lazarev VG	Pilot/Doctor	1928 Feb 23	1973	26	2	1985 Nov 27	1990 Dec 31

Name		D.O.B	First Flight Year	Cosmonaut No. (USSR /Russia)	No. of Russian Flights	Status/Retirement*	Died/Killed*
1967 Group Fourth Air Force Selection (Apr-May 1967)							
Beloborodov VM	Pilot	1939 Oct 26	–	–	0	1969 Aug 29	2004 Sep 20
Kovalyonok VV	Pilot	1942 Mar 3	1977	40	3	1984 Jun 23	
Kozelskiy VS	Pilot	1942 Jan 12	–	–	0	1983 Apr 20	
Lyakhov VA	Pilot	1941 Jul 20	1979	45	3	1994 Aug 19	
Malyshev YV	Pilot	1941 Aug 27	1980	47	2	1988 Jul 2	1999 Nov 8
Pisarev VM	Pilot	1941 Aug 15	–	–	0	1968 May 21	
Alekseyev VB	Engineer	1933 Aug 19	–	–	0	1983 Apr 20	
Burdayev MN	Engineer	1932 Aug 27	–	–	0	1983 Apr 20	
Gaydukov SN	Navigator	1936 Oct 31	–	–	0	1978 Dec 4	
Isakov VT	Navigator	1940 Apr 4	–	–	0	1983 Apr 20	
Porvatkin NS	Engineer	1932 Apr 15	–	–	0	1983 Apr 20	
Sologub MV	Navigator	1936 Nov 6	–	–	0	1968 Sep 20	1996 Aug 4
1970 Group Fifth Air Force Selection (27 Apr 1970)							
Berezovoy AN	Pilot	1942 Apr 11	1982	51	1	1992 Oct 31	
Dedkov AI	Pilot	1944 Jul 27	–	–	0	1983 Apr 20	
Dzhanibekov VA	Pilot	1942 May 13	1978	43	5	1986 Jun 24	
Isaulov YF	Pilot	1943 Aug 31	–	–	0	1982 Jan 29	
Kozlov VI	Pilot	1945 Oct 2	–	–	0	1973 May 28	
Popov LI	Pilot	1945 Aug 31	1980	46	3	1987 Jun 13	
Romanenko YV	Pilot	1944 Aug 1	1977	42	3	1988 Oct 11	
Fefelov NN	Engineer	1945 May 20	–	–	0	1995 Nov 9	
Illarionov VV	Engineer	1939 Jun 2	–	–	0	1992 Oct 30	1999 Mar 10

1976 Group Sixth Air Force Selection (23 Aug 1976)

Ivanov LG	Pilot	1950 Jun 25	–	–		0	Deceased	1980 Oct 24
Kadenyuk LK	Pilot	1950 Jan 28	–	–		0	1983 Mar 22	
Moskalenko NT	Pilot	1949 Jan 1	–	–		0	1986 Jun 30	
Protchenko SF	Pilot	1947 Jan 3	–	–		0	1979 Apr 28	
Saley YV	Pilot	1950 Jan 1	–	–		0	1987 Oct 1	
Solovyov AV	Pilot	1948 Jan 16	1988	65		5	1999 May 3	
Titov VG	Pilot	1947 Jan 1	1983	54		4	1998 Aug 20	
Vasyutin VV	Pilot	1952 Mar 8	1985	59		1	1986 Feb 25	2002 Jul 20
Volkov AA	Pilot	1948 May 27	1985	60		3	1998 Aug 20	

1978 Group Seventh Air Force Selection (23 May 1978)

Grekov NS	Pilot	1950 Feb 15	–	–		0	1986 Dec 30	
Viktorenko AS	Pilot	1947 Mar 29	1987	62		4	1997 Jul 21	

1987 Group Eighth Air Force Selection (26 Mar 1987)

Dezhurov VN	Pilot	1962 Jul 30	1995	81	9	2	2004 Jul 12	
Gidzenko YP	Pilot	1962 Mar 26	1995	83	11	3	2002	
Korzun VG	Pilot	1953 Mar 5	1996	85	13	2	2003 Sep 9	
Malenchenko YI	Pilot	1961 Dec22	1994	78	6	3	Active	
Tsibliyev VV	Pilot	1954 Feb 20	1993	76	4	2	1998 Jun 19	

1987 Group Ex-Buran Selection, Transferred to TsPK (7 Aug 1987)

Bachurin II	Pilot Buran	1942 Jan 29	–	–		0	1992 Nov 28	
Boroday AS	Pilot Buran	1947 Jul 28	–	–		0	1993 Dec 29	

1988 Group Ninth Air Force Selection, transferred to TsPK (8 Jan 1988)

Afanasyev VM	Test Pilot	1948 Dec 31	1990	70		4	Active	
Artsebarskiy AP	Test Pilot	1956 Sep 9	1991	71		1	1993 Sep 7	
Manakov GM	Test Pilot	1950 Jun 1	1990	69		2	1996 Dec 20	

Name		D.O.B	First Flight Year	Cosmonaut No. (USSR /Russia)		No. of Russian Flights	Status/Retirement*	Died/Killed*
1988 Group Buran re-selection (25 Oct 1989)								
Kadenyuk LK	Pilot Buran	1950 Jan 28	1997 (as a Ukrainian Cosmonaut)			1	1996 Feb 14	
1989 Group Tenth Air Force Selection (22 Apr 1989)								
Krichevskiy SV	Pilot	1955 Jul 9	–	–		0	1998 Apr 30	
Onufriyenko YI	Pilot	1961 Feb 6	1996	84	12	2	2004 Mar 17	
Padalka GI	Pilot	1958 Jun 21	1998	89	18	2	Active	
1989 Group Special Test Pilot Selection (completed training in Apr 1991 and reported back to Test Pilot School). Four transferred to TsPk in 1992 (8 Apr), 1993 (30 Jan) and 1995 (6 Feb)								
Maksimenko VY	Test Pilot	1950 Jul 16	–	–		0	1993	
Polonskiy AB	Test Pilot	1956 Jan 1	–	–		0	1993	
Puchkov AS	Test Pilot	1948 Oct 15	–	–		0	1996 Nov 28	
Pushenko NA	Test Pilot	1952 Aug 10	–	–		0	1996 Nov 28	
Tokarev VT	Test Pilot	1952 Oct 29	–	–		0	1996 Nov 28	
Yablontsev AN	Test Pilot	1955 Apr 3	–	–		0	1996 Nov 28	
1990 Group Eleventh Air Force Selection (8 Aug 1990)								
Sharipov SS	Pilot	1964 Aug 24	1998	88	16	2	Active	
Vozovikov SY	Pilot	1958 Apr 17	–	–		0	Deceased	1993 Jul 11
Zalyotin SV	Pilot	1962 Apr 21	2000	92	21	2	2004 Sep 20	
1991 Group Attached to the Air Force Team (6 Mar 1991)								
Musabayev TA	Pilot	1951 Jan 7	1994	79	7	3	2003	

1996 Group Attached to the Air Force Team (7 Jul 1996)

Kotov OV	Doctor	1965 Oct 27	–	–		0	Active

1997 Group Twelfth Air Force Selection (28 Jul 1997)

Kondratyev DY	Pilot	1969 May 25	–	–		0	Active
Lonchakov YV	Pilot	1965 Mar 4	2001	94	25	2	Active
Moshkin OY	Pilot	1964 Apr 23	–	–		0	2002 Mar
Romanenko RY	Pilot	1971 Aug 9	–	–		0	Active
Skvortsov AA	Pilot	1966 May 6	–	–		0	Active
Surayev MV	Pilot	1972 May 24	–	–		0	Active
Valkov KA	Pilot	1971 Nov 11	–	–		0	Active
Volkov SA	Pilot	1973 Apr 1	–	–		0	Active

1997 Re-selection – Supplement to the Twelfth Air Force Selection (28 Jul 1997)

Tokarev VT	Pilot	1952 Oct 29	1999	91	20	1	Active

1998 Selection Civilian Attached to the Air Force Team (30 Apr 1998)

Baturin YM	Engineer	1949 Jun 12	1998	90	17	2	Active

1998 Selection Transferred from the Rocket Forces to the Air Force Team (2 Sep 1998)
Transferred in May 2002 to the team of the Defence Ministry Space Force

Shargin YG	Engineer	1960 Mar 20	–	–		0	Transferred – Active

2003 Group Thirteenth Air Force Selection (29 May 2003)

Ivanishin AA	Pilot	1969 Jan 15	–	–		0	Candidate
Samokutyayev AM	Pilot	1970 Mar 13	–	–		0	Candidate
Shkaplerov AN	Pilot	1972 Feb 20	–	–		0	Candidate
Tarelkin YI	Engineer	1974 Dec 29	–	–		0	Candidate

Energiya Engineers 1966-2005

Name		D.O.B	First Flight Year	Cosmonaut No. (USSR /Russia)	No. of Russian Flights	Status/Retirement*	Died/Killed*
1966 Group First Selection (TsKBEM 23 May 1966)							
Anokhin SN	TsKBEM	1910 Apr 1	–	–	0	1968 May 27	1986 Apr 15
Bugrov VY	TsKBEM	1933 Jan 18	–	–	0	1968 Jul 12	
Dologpolov GA	TsKBEM	1935 Nov 14	–	–	0	1967 May 3	
Grechko GM	TsKBEM	1931 May 25	1975	34	3	1986 Jun 5	
Kubasov VN	TsKBEM	1935 Jan 7	1969	18	3	1989 Dec 29	
Makarov OG	TsKBEM	1933 Jan 6	1973	27	4	1986 Apr 4	2003 May 28
Volkov VN	TsKBEM	1935 Nov 23	1969	20	2	Deceased	1971 Jun 30
Yeliseyev AS	TsKBEM	1934 Jul 13	1969	15	3	1986 Jan 10	
1967 Group Supplemental Selection (TsKBEM Jan-Feb 1967)							
Rukavishnikov NN	TsKBEM	1939 Sep 18	1971	23	3	1987 Jul 7	2002 Oct 19
Sevastyanov VI	TsKBEM	1935 Jul 8	1970	22	2	1993 Dec 30	
1967 Group Second Selection (TsKBEM 18 Aug 1967)							
Nikitskiy VP	TsKBEM	1939 Mar 8	–	–	0	1968 May 27	
Patsayev VI	TsKBEM	1933 Jun 19	1971	25	1	Deceased	1971 Jun 30
Yazdovskiy VA	TsKBEM	1930 Jun 8	–	–	0	1982 Jul 1	
1972 Group Third Selection (TsKBEM 27 Mar 1972)							
Andreyev BD	TsKBEM	1940 Oct 6	–	–	0	1983 Sep 5	
Lebedev VV	TsKBEM	1942 Apr 14	1973	29	2	1989 Nov 4	
Ponomaryov YA	TsKBEM	1932 Mar 24	–	–	0	1983 Apr 11	2005 Apr 16

1973 Group Fourth Selection (TsKBEM 27 Mar 1973)								
Aksyonov VV	TsKBEM	1935 Feb 1	1976	36		2	1988 Oct 17	
Ivanchenkov AS	TsKBEM	1940 Sep 28	1978	44		2	1993 Nov 3	
Ryumin VV	TsKBEM	1939 Aug 16	1977	41		4	1987 Oct 28	
Strekalov GM	TsKBEM	1940 Oct 28	1980	49		5	1995 Jan 17	2004 Dec 25
1977 Supplemental Selection (NPO Energiya 1 Dec 1977)								
Feoktistov KP	Energiya	1926 Feb 7	1964	8		1	1987 Oct 28	
1978 Group Fifth Selection (NPO Energiya 1 Dec 1978)								
Aleksandrov AP	Energiya	1943 Feb 20	1983	55		2	1993 Oct 12	
Balandin AN	Energiya	1953 Jul 30	1990	68		1	1994 Oct 17	
Laveikin AI	Energiya	1951 Apr 21	1987	61		1	1994 Mar 28	
Manarov MK	Energiya	1951 Mar 22	1987	63		2	1992 Jul 23	
Savinykh VP	Energiya	1940 Mar 7	1981	50		3	1989 Feb 9	
Serebrov AA	Energiya	1944 Feb 15	1982	52		3	1995 Apr 27	
Solovyov VA	Energiya	1946 Nov 11	1984	56		2	1994 Feb 18	
1980 Group Female (Sixth) Selection (NPO Energiya 30 Jul 1980)								
Kuleshova ND	Energiya	1956 Mar 14	–	–		0	1992 Jul 23	
Pronina IR	Energiya	1953 Apr 14	–	–		0	1992 Jul 23	
1983 Transferred from MMZ Skorost to NPO Energiya (16 May 1983)								
Savitskaya SY	Energiya	1948 Aug 8	1982	53		2	1993 Oct 27	
1984 Group Seventh Selection (NPO Energiya 15 Feb 1984)								
Kaleri AY	Energiya	1956 May 13	1992	73	1	4	Active	
Yemelyanov SA	Energiya	1951 Aug 3	–	–		0	1992 Jul 9	1992 Dec 5

Name		D.O.B	First Flight Year	Cosmonaut No. (USSR /Russia)		No. of Russian Flights	Status/Retirement*	Died/Killed*
1985 Group Eighth Selection (NPO Energiya 2 Sep 1985)								
Krikalev SK	Energiya	1958 Aug 27	1988	67		6	Active	
Zaitsev AY	Energiya	1957 Aug 5	–	–		0	1996 Mar 14	
1987 Group Ninth Selection (NPO Energiya 6 Mar 1987)								
Avdeyev SV	Energiya	1956 Jan 1	1992	74	2	3	2003 Feb 14	
1989 Group Tenth Selection (NPO Energiya 25 Jan 1989)								
Budarin NM	Energiya	1953 Apr 29	1995	82	10	3	2004 Sep 7	
Kondakova YV	Energiya	1957 Mar 30	1994	80	8	2	1999 Dec 30	
Poleshchuk AF	Energiya	1953 Oct 30	1993	75	3	1	2004 Mar 25	
Usachev YV	Energiya	1953 Aug 31	1994	77	5	4	2004 Apr 5	
1992 Group Eleventh Selection (NPO Energiya 3 Mar 1992)								
Lazutkin AI	Energiya	1957 Oct 30	1997	86	14	1	Active	
Treshchev SY	Energiya	1958 Aug 18	2002	97	27	1	Active	
Vinogradov PV	Energiya	1953 Aug 31	1997	87	15	1	Active	
1994 Group Twelfth Selection (NPO Energiya 1 Apr 1994)								
Kuzhelnaya NV	Energiya	1962 Nov 6	–	–		0	2004 May 27	
Tyurin MV	Energiya	1960 Mar 2	2001	95	24	1	Active	
1996 Group Thirteenth Selection (RKK Energiya 9 Feb 1996)								
Kononenko OD	TsSKB	1964 Jun 21	–	–		0	Transferred	
Kozeyev KM	Energiya	1967 Dec 1	2001	96	23	1	Active	
Revin SN	Energiya	1966 Jan 12	–	–		0	Active	

1997 Group Fourteenth Selection (RKK Energiya 28 Jul 1997)								
Skripochka OI	Energiya	1969 Dec 24	–	–		0	Active	
Yurchikhin FN	Energiya	1959 Jan 3	2002	98	26	1	Active	
1998 Supplemental to Fourteenth Selection (RKK Energiya 24 Feb 1998)								
Korniyenko MB	Energiya	1960 Apr 15	–	–		0	Active	
1999 Transferred from TsSKB Samara to RKK Energiya (5 Jan 1999)								
Kononenko OD	Energiya	1964 Jun 21	–	–		0	Active	
2003 Group Fifteenth Selection (RKK Energiya 29 May 2003)								
Artemyev OG	Energiya	1970 Dec 28	–	–		0	Candidate	
Borisenko AI	Energiya	1964 Apr 17	–	–		0	Candidate	
Serov MK	Energiya	1974 May 23	–	–		0	Candidate	

Cosmonaut Team of the Institute of Medical Biological Problems (Formed in 1972 to give medical support in orbit)

1972 Group First Selection (22 Mar 1972)								
Machinskiy GV	Doctor	1937 Oct 11	–	–		0	1974 Jun 4	
Polyakov VV	Doctor	1942 Apr 27	1988	66		2	1995 Jun 1	
Smirenniy LN	Engineer	1932 Oct 25	–	–		0	1986 Oct 23	
1978 Group Second Selection (1 Dec 1978)								
Arzamazov GS	Doctor	1946 Mar 9	–	–		0	1995 Dec 1	
Borodin AV	Doctor	1953 Mar 3	–	–		0	1993 Mar 10	
Potapov MG	Doctor	1952 Oct 28	–	–		0	1985 May 27	
1980 Group Third Selection (30 Jul 1980)								
Amelkina GV	Doctor	1954 May 22	–	–		0	1983 May	
Dobrokvashina YI	Doctor	1947 Oct 8	–	–		0	1993 Mar 10	
Pozharskaya LG	Doctor	1947 Mar 15	–	–		0	1993 Mar 10	2002 Feb 18
Zakharova TS	Doctor	1952 Apr 22	–	–		0	1995 Sep 1	

Name		D.O.B	First Flight Year	Cosmonaut No. (USSR	/Russia)	No. of Russian Flights	Status/Retirement*	Died/Killed*
1985 Group Fourth Selection (2 Sep 1985)								
Stepanov YN	Engineer	1936 Sep 27	–	–		0	1995 Mar 20	
1989 Group Fifth Selection (25 Jan 1989)								
Karashtin VV	Doctor	1962 Nov 18	–	–		0	2002 Jan 17	
Lukyanyuk VY	Doctor	1958 Sep 22	–	–		0	2003 Feb 18	
Morukov BV	Doctor	1950 Oct 1	2000	93	22	1	Active	
2003 Group Sixth Selection (29 May 2003)								
Ryazanskiy SN	Biochemist	1974 Nov 13	–	–		0	Candidate	

Cosmonauts of the Flight Research Institute of the Ministry of the Aviation Industry (LII MAP)

(All civilian test pilots. The original group was selected by the GMVK on 1 December 1978. Later candidates were approved by the GMVK then officially included in the group of LII Buran Test Pilots by MAP. Despite a recommendation from the GMVK in 1995 to disband the group, it still exists on paper)

Name		D.O.B	First Flight Year	Cosmonaut No. (USSR	/Russia)	No. of Russian Flights	Status/Retirement*	Died/Killed*
1978 Group First Selection (GMVK 1 Dec 1978; MAP 10 Aug 1981)								
Kononenko OG	Test Pilot	1938 Aug 16	–	–		0	Deceased	1980 Sep 8
Levchenko AS	Test Pilot	1941 May 21	1987	64		1	Deceased	1988 Aug 6
Shchukin AV	Test Pilot	1946 Jan 19	–	–		0	Deceased	1988 Aug 18
Stankyavichus RA	Test Pilot	1944 Jul 26	–	–		0	Deceased	1990 Sep 9
Volk IP	Test Pilot	1937 Apr 12	1984	58		1		
1983 Group Second Selection (GMVK 9 Mar 1983; MAP 25 Apr 1983)								
Sultanov UN	Test Pilot	1948 Nov 18	–	–		0		
Tolboyev MO	Test Pilot	1951 Jan 20	–	–		0	1994 Jan 12	

1984 Group Third Selection (GMVK 2 Sep 1984; MAP 12 Apr 1984)

Zabolotskiy VV	Test Pilot	1946 Apr 19	–	–	0		

1985 Group Fourth Selection (GMVK 2 Sep 1985; MAP 21 Nov 1995)

Sheffer YP	Test Pilot	1947 Jun 30	–	–	0	Deceased	2001 Jun 5
Tresvyatskiy SN	Test Pilot	1954 May 6	–	–	0		

1989 Group Fifth Selection (GMVK 25 Jan 1989; MAP 22 Mar 1989)

Prikhodko YV	Test Pilot	1953 Nov 15	–	–	0	1994 Apr 27	2001 Jul 26

Cosmonauts of the State Red Banner Scientific Research Institute of the Air Force (GKNII VVS)

(All military test pilots. The first group was selected by the GMVK in 1978 but it was not until 1987 that a group of GKNII Buran test pilots was officially created by the Ministry of Defence (MO). Not all the pilots selected by the GMVK were officially transferred and the group was disbanded by the Air Force on 30 September 1996)

1978 Group (GMVK 1 Dec1978; MO 7 Aug 1987)

Bachurin II	Test Pilot	1942 Jan 29	–	–	0	1987 Aug
Boroday AS	Test Pilot	1947 Jul 28	–	–	0	1987 Aug
Chirkin VM	Test Pilot	1944 Jul 13	–	–	0	1981
Mosolov VY	Test Pilot	1944 Feb 21	–	–	0	1987 Aug
Sattarov NS	Test Pilot	1942 Dec 23	–	–	0	1980 May
Solovykh AM	Test Pilot	1944 Jan 12	–	–	0	1986

1985 Group (GMVK Aug 1985. All transferred to the TsPK Air Force Team on 8 Jan 1988)

Afanasyev VM	Test Pilot	1948 Dec 31	1990	70	4	Active
Artsebarskiy AP	Test Pilot	1956 Sep 9	1991	71	1	1993 Sep 7
Manakov GM	Test Pilot	1950 Jun 1	1990	69	2	1996 Dec 20

Name		D.O.B	First Flight Year	Cosmonaut No. (USSR /Russia)		No. of Russian Flights	Status/Retirement*	Died/Killed*
1988 Group (MO 25 Oct 1988. Originally members of the Air Force Group between 1976 and 1983)								
Kadenyuk LK	Test Pilot	1950 Jan 28	Flew as Ukrainian Cosmonaut aboard the US Shuttle in 1997					
1989 Group (GMVK 25 Jan 1989; MO 8 Apr 1992-30 Jan 1993)								
Polonskiy AB	Test Pilot	1956 Jan 1	–	–		0	1991 Apr	
Tokarev VI	Test Pilot	1952 Oct 29	1999	91	20	1	Transferred	
Yablontsev AN	Test Pilot	1955 Apr 3	–	–		0	1991 Apr	
1990 Group (GMVK 11 May 1990; Air Force 6 Feb 1995)								
Maksimenko VY	Test Pilot	1950 Jul 16	–	–		0	1991 Apr	
Puchkov AS	Test Pilot	1948 Oct 15	–	–		0	1991 Apr	
Pushenko NA	Test Pilot	1952 Aug 10	–	–		0	1991 Apr	
Cosmonaut Testers GKNII VVS								
First Group (7 Aug 1987 MO of the Soviet Union CCCP)								
Bachurin II	Test Pilot	1942 Jan 29	–	–		0	1992 Nov 28	
Boroday AS	Test Pilot	1947 Jul 28	–	–		0	1993 Dec 29	
Second Group (25 Oct 1988 MO CCCP)								
Kadenyuk LK	Test Pilot	1950 Jan 28	–	–		0	1996 Feb 14	
Third Group (Aug 1992 MO of the Russian Federation RF)*								
Puchkov AS	Test Pilot	1948 Oct 15	–	–		0	1996 Nov 28	
Yablontsev AN	Test Pilot	1955 Apr 3	–	–		0	1996 Nov 28	

Fourth Group (30 Jan 1993 MO RF)							
Tokarev VI	Test Pilot	1952 Oct 29	–	–	0	1996 Nov 28	
Fifth Group (6 Feb 1995 VVS)							
Pushenko NA	Test Pilot	1952 Aug 10	–	–	0	1996 Nov 28	

Cosmonaut Team of the Academy of Sciences

Group 1 (22 May 1967)							
Fatkullin MN	Scientist	1939 May 14	–	–	0	1970	2003 Apr 16
Gulyayev RA	Scientist	1934 Nov 14	–	–	0	1968	
Kolomitsev OP	Scientist	1939 Jan 29	–	–	0	1968	
Yershov VG	Scientist	1928 Jun 21	–	–	0	1974	1998 Feb 15
Supplementary Selection (May 1968)							
Katys GP	Scientist	1926 Aug 31	–	–	0	1972	
Group 2 (30 Jul 1980)							
Latysheva ID	Scientist	1953 Jul 9	–	–	0	1993 Feb 25	
Supplementary Selection (6 Jul 1986, transferred from NPO Energiya)							
Grechko GM	Engineer	1931 May 25	–	–	0	1992 Mar 1	
Supplementary Selection (4 Nov 1989, transferred from NPO Energiya)							
Lebedev VV	Engineer	1942 Apr 14	–	–	0	1993 Feb 25	
Supplementary Selection (7 Sep 1993, transferred from the TsPK team)							
Artsebarskiy AP	Pilot	1956 Sep 9	–	–	0	1994 Jul 28	

Name		D.O.B	First Flight Year	Cosmonaut No. (USSR /Russia)	No. of Russian Flights	Status/Retirement*	Died/Killed*
Supplementary Selection (20 Mar 1995, transferred from the IMBP group)							
Stepanov YN	Engineer	1936 Sep 27	–	–	0	Active	
Other Cosmonaut Selections							
Cosmonaut Team of NPO Mashinostroyenyia							
(Civilian engineers from OKB-52 were selected to fly on the Almaz military space stations. The group was disbanded in 1986)							
Group 1 (22 Mar 1972)							
Makrushin VG	Engineer	1940 Jan 14	–	–	0	1987 Apr 8	
Group 2 (27 Mar 1973)							
Yuyukov DA	Engineer	1941 Feb 26	–	–	0	1987 Apr 8	
Group 3 (1 Dec 1978)							
Gevorkyan VM	Engineer	1952 May 28	–	–	0	1987 Apr 8	
Grechanik AA	Engineer	1952 Mar 25	–	–	0	1987 Apr 8	
Khatulev VA	Engineer	1947 Feb 26	–	–	0	1980	
Romanov VA	Engineer	1946 Aug 18	–	–	0	1987 Apr 8	
Cosmonaut Selections for Special Missions or Political Reasons							
Voskhod 1 (May-Jun 1964)							
Benderov was a member of the Tupolev OKB; Katys was a member of the Academy of Science; Lazarev was a member of NII-7; Feoktistov was a senior design engineer from OKB-1; two of the doctors came from IMBP and the other from TsPK							
Benderov VN	Test Pilot	1924 Aug 4	–	–	0	1964 Jul	1973 Jun 3

Feoktistov KP	Engineer	1926 Feb 7	1964	8	1	1964 Oct	
Katys GP	Scientist	1926 Aug 31	–	–	0	1964 Oct	
Lazarev VG	Air Force	1928 Feb 23	–	–	0	1964 Oct	1990 Dec 31
Polyakov BI	Dr. IMBP	1938 May 10	–	–	0	1964 Jul 2	
Sorokin AV	Dr. TsPK	1931 Mar 30	–	–	0	1964 Oct	
Yegorov BB	Dr. IMBP	1937 Nov 26	1964	9	1	1964 Oct	1994 Sep 12
Voskhod-3 (Apr 1965)							
Katys GP	Scientist	1926 Aug 31	–	–	0	1965 Dec	
Voskhod Biomedical Mission (1965) *All doctors were working at IMBP*							
Ilyin YA	Doctor	1937 Aug 17	–	–	0	1966	
Kisilev AA	Doctor	1935 Jun 13	–	–	0	1966	
Senkevich YA	Doctor	1937 Mar 4	–	–	0	1966	2003 Sep 25
Paton Institute of Electric Welding ('Vulkan' welding experiment on Soyuz, 27 May 1968)							
Fartushniy VG	Engineer	1938 Feb 3	–	–	0	1970	
Various Institutes All-Female Group *Visiting mission to Salyut 7, selected Dec 1979; GMVK 30 Jul 1980*							
Amelkina GV	Doctor	1954 May 22	–	–	0	Transferred IMBP 1980 Aug 5	
Dobrokvashina EI	Doctor	1947 Oct 8	–	–	0	Transferred IMBP 1980 Aug 5	
Kuleshova ND	Engineer	1956 Mar 14	–	–	0	Transferred Energiya 1980 Sep 26	
Latysheva ID	Engineer	1953 Jul 9	–	–	0	Transferred Academy of Sciences	
Pozharskaya LG	Doctor	1947 Mar 15	–	–	0	Transferred IMBP 1980 Aug 5	
Pronina IR	Engineer	1953 Apr 14	–	–	0	Transferred Energiya 1981 Mar 16	
Savitskaya SY	Engineer	1948 Aug 8	1982	53	1	Transferred Energiya 1983 May 16	
Zakharova TS	Doctor	1952 Apr 22	–	–	0	Transferred IMBP 1980 Aug 5	

Name		D.O.B	First Flight Year	Cosmonaut No. (USSR /Russia)	No. of Russian Flights	Status/Retirement*	Died/Killed*
Supplement to All-Female Group (9 Mar 1983)							
Ivanova YA	Engineer	1949 Oct 3	–	–	0	1987 Apr	
All-Union Cardiology Scientific Centre Academy of Medical Sciences							
(long-duration mission to Salyut 7, 9 Mar 1983)							
Atkov OY	Doctor	1949 May 9	1984	57	1	1984 Oct	
Soviet Journalists							
(Mir visiting mission 11 May 1990)							
Andryushkov AS	Journalist	1947 Oct 6	–	–	0	1992 Feb	
Baberdin VV	Journalist	1948 Oct 28	–	–	0	1992 Feb	2003 Oct 2
Krikun YY	Journalist	1963 Jun 3	–	–	0	1992 Feb	
Mukhortov PP	Journalist	1966 Mar 11	–	–	0	1992 Feb	
Omelchenko SO	Journalist	1951 Aug 20	–	–	0	1992 Feb	
Sharov VY	Journalist	1953 Dec 26	–	–	0	1992 Feb	
MMZ Zvezda							
11 May 1990 (reported in Oct 1990)							
Severin VG	Engineer	1956 Nov 20	–	–	0	1995	
Ministry of Civil Air Fleet of Kazakhstan							
(Kazakh visiting mission to Mir 11 May 1990)							
Musabayev TA	Pilot	1951 Jan 7	1994	79 7	3	Transferred TsPK 1991 Mar	

MMz Mikoyan *(Kazakh visiting mission to Mir 21 Jan 1991)*							
Aubakirov TO	Test Pilot	1946 Jul 17	1991	72		1	1991 Sep
Military Space Forces of the Russian Federation *(9 Feb 1996)*							
Shargin YG	Engineer	1960 Mar 20	See below				Transferred TsPK 1998 Sep 2
TsSKB/Progress – Samara *(29 Mar 1996)*							
Kononenko OG	Engineer	1964 Jun 21					Transferred to Energiya 1999 Jan 5
GKNITs Khrunichev *(Feb 1997, reported Jan 1998)*							
Moshchenko SI	Engineer	1954 Jan 12	–	–		0	Active
Political Representative *(Visiting mission to Mir, 15 Sep 1997)*							
Baturin YM	Engineer	1949 Jun 12	1998	90	17	2	Transferred TsPK 1998 Apr 30
Mir Film Project *(Dec 1997)*							
Steklov VA	Actor	1948 Jan 3	–	–		0	2000 Mar
Defence Ministry Space Force Cosmonaut Team *(May 2002)*							
Shargin YG	Engineer	1960 Mar 20	2004	99	28	1	Active

Name		D.O.B	First Flight Year	Cosmonaut No. (USSR /Russia)	No. of Russian Flights	Status/Retirement*	Died/Killed*
ZAO Centre for Technology Transfer Rosaviakosmos (TsPT)							
(2003 selection, reported 29 May 2004)							
Zukhov SA	Engineer	1956 Sep 8	–	–		0	Candidate
Kazakhstan Government Selections							
(4 Jan 2001)							
Mukhamedrakhimov RR			1972 Apr 11	–	–		0
(9 Nov 2002, reported 29 May 2003)							
Aimbetov AA	Pilot	1972 Jul 27	–	–		0	Candidate
Aymakhanov MR	Pilot	1967 Jan 1	–	–		0	Candidate
Shaydullin YB	Pilot	1980 Aug 9					
2004 Tourist Training OAO 'Stroymontazh'							
Polonskiy S	Tourist	1972 Dec 1	–	–		0	

* Dates shown in the Status/Retirement column are the dates of retirement, as far as is known. A separate date under the Died/Killed column indicates the date of death of the former cosmonaut after standing down from the team. Where the Status/Retirement column shows 'Deceased', the date in the Died/Killed column indicates the date of death while the cosmonaut was either active or undergoing cosmonaut training

Bibliography

The authors used the following books, articles and reports as leading sources of information, in conjunction with reference material from their own personal archives collected from Interkosmos, ESA, NASA, CSA, JAXA, and Russian and Soviet sources.

Research Visits to TsPK

Rex Hall has visited Star City on nine occasions since 1996, eight times in the company of Bert Vis and once in 2003 with Dave Shayler and Bert Vis. Bert Vis has been on fourteen occasions; the first visit occurred in 1991

Reports

1976 *Soviet Space Programs 1971–1975*, Vol 1, Manned Space Programs and Space Life Sciences, US congress

1981 *The Soviet Year in Space*, Teledyne Brown Engineering, Nicholas L. Johnson. Annually published 1981–90

1984 *Soviet Space Programs 1976–1980*, Vol 2, Overview, Facilities and Hardware, Manned and Unmanned space programmes, US congress

1988: *Soviet Space Programs 1981–1987*, Vol 1, Piloted Space Activities, Launch Vehicles, Launch Sites and Tracking Support, US Congress

1992 *Europe and Asia in Space*, USAF Phillips Lab, Kaman Sciences Corp, Nicholas L. Johnson, published 1991–2 and 1993–4

1992 *The Shuttlenauts*, Vol 2, Flight Crew Assignments, David J. Shayler, Astro Info Service Publications (with unpublished updates through 2005)

1995 *US-Russian Cooperation in Space*, US Congress , Office of Technology Assessment, OTA-ISS-618

1999 *Phase 1 Program Joint Report*, Editors George C. Nield and Pavel M. Vorobiev, NASA SP-1999–6108 (in English)

2000 *Towards the Future* , ESA BR-192

Periodicals

Spaceflight, British Interplanetary Society, London

Journal of the British Interplanetary Society, London

Flight International

Aviation Week and Space Technology

ESA Bulletin
On Station: The ESA newsletter of the directorate of Human Spaceflight, Zero-gravity and Exploration

English Language Books
1971 *Russians in Space*, Yevgeniy Riabchikov, Novosti Press Agency, Moscow (English edition)
1976 *Apollo Soyuz*, Walter Froehlich, NASA SP-109
1977 *The Sun's Wind*, Aleksey Leonov, Progress Publishers, Moscow
1978 *The Partnership, A History of the Apollo Soyuz Test Project*, Edward Ezell and Linda Ezell, NASA SP-4209
1979 *Rendezvous in Space: Soyuz Apollo*, Lev Lebedev and Aleksander Romanov, Progress Publishers, Moscow
1988 *Orbits of Peace and Progress*, Edited by P. Popovich, Mir Publishers, Moscow
1988 *The Soviet Manned Space Programme*, Phillip Clark, Salamander Books
1990 *The Soviet Cosmonaut Team*, Gordon R. Hooper, GRH Publications; Vol 1, Background Sections, Vol 2, Cosmonaut Biographies
1993 *Seize the Moment*, Helen Sharman with Christopher Priest, Victor Gollancz
1994 *Deke: US Manned Space from Mercury to the Shuttle*, Donald K 'Deke' Slayton with Michael Cassutt, Tom Doherty Associates Books, New York, 1995
1997 *Walking to Olympus: An EVA Chronology*. David Portree and Robert C Trevino, NASA History Office, October 1997
1998 *Dragonfly*, Bryan Burrough, Harper Collins
1999 *Who's Who in Space: The International Space Station Edition*, Michael Cassutt, Macmillan.
2000 *Off the Planet*, Jerry M. Linenger, McGraw Hill
2000 *The History of Mir 1986–2000*, Editor Rex Hall, British Interplanetary Society, also *Mir: The Final Year* supplement, Editor Rex Hall, BIS 2001
2000 *Challenge to Apollo, The Soviet Union and the Space Race, 1945–1974*, Asif Siddiqi NASA SP-2000–4408
2002 *The International Space Station: From Imagination to Reality*, Editor Rex Hall, British Interplanetary Society
2002 *Star Crossed Orbits: Inside the US-Russian Space Alliance*, James Oberg, McGraw Hill.
2002 *We Have Capture*, Thomas P. Stafford with Michael Cassutt, Smithsonian

Russian Language Magazines
Novosti Kosmonavtiki, a monthly magazine

Russian Language Books:
1977 *Zvyozdnyy Gorodok – Star City*, edited by V.A Shatalov and G.T. Beregovoy, Mashinostroyenyia
1981 *Starry*, Andreev and others, Mashinostroyenyia

1993 *Space Academy*, G.T. Beregovoy and others, Mashinostroyenyia
2000 *Soviet & Russian Cosmonauts 1960–2000*, Novosti Kosmonavtiki, Editors I.A. Marinin, S.X. Shamsutdinov and A.V.Glushko
2001 *Testers LII*, biographys of test pilots V.P.Vacin and A.A.Simonev, Aviation Pechatnick Dvor
2004 *A Dream Come True*, Eduard Buinovskiy, Molodaya Gvardiya

Publications of the Cosmonaut Training Centre named for Yuri Gagarin

1987 *Booklet of the museum of TsPK*
Pamphlets of the training centre, 1993, 1997, 2001 and 2004
2000 *Centre of the Training of Cosmonauts named for Yuri Gagarin 40 years*
2002 *Centre of the Training of Cosmonauts named for Yuri Gagarin, 2nd edition*

Springer-Praxis Space Sciences Series

2000 *Disasters and Accidents in Manned Spaceflight*, David J. Shayler
2000 *The Challenges of Human Space Exploration*, Marsha Freeman
2001 *The Rocket Men, Vostok & Voskhod, The First Soviet Manned Spaceflights*, Rex Hall and David J. Shayler
2002 *The Continuing Story of the International Space Station*, Peter Bond
2002 *Creating the International Space Station*, David M. Harland and John E. Catchpole
2003 *Soyuz: A Universal Spacecraft*, Rex Hall and David J. Shayler
2003 *Russia's Spacesuits* Isaak P. Abramov and Å. Ingemaar Skoog
2004 *China's Space Program: From Concept to Manned Spaceflight*, Brian Harvey
2004 *Walking In Space*, David J. Shayler
2004 *The Story of the Space Shuttle* David M. Harland
2005 *The Story of Space Station Mir* David M. Harland
2005 *Women in Space: Following Valentina*, David J. Shayler and Ian Moule

Index

Printing: Mercedes-Druck, Berlin
Binding: Stein+Lehmann, Berlin